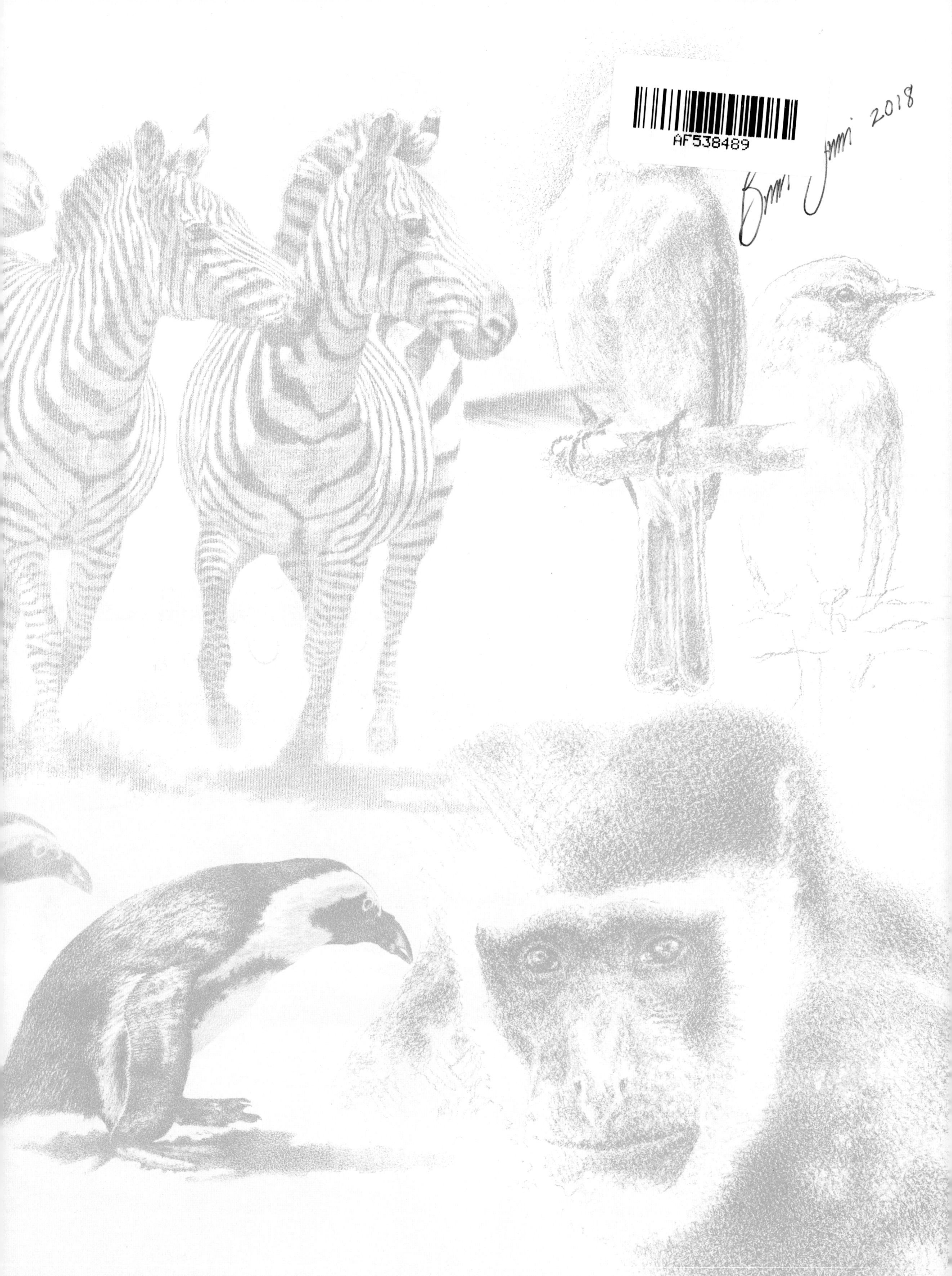

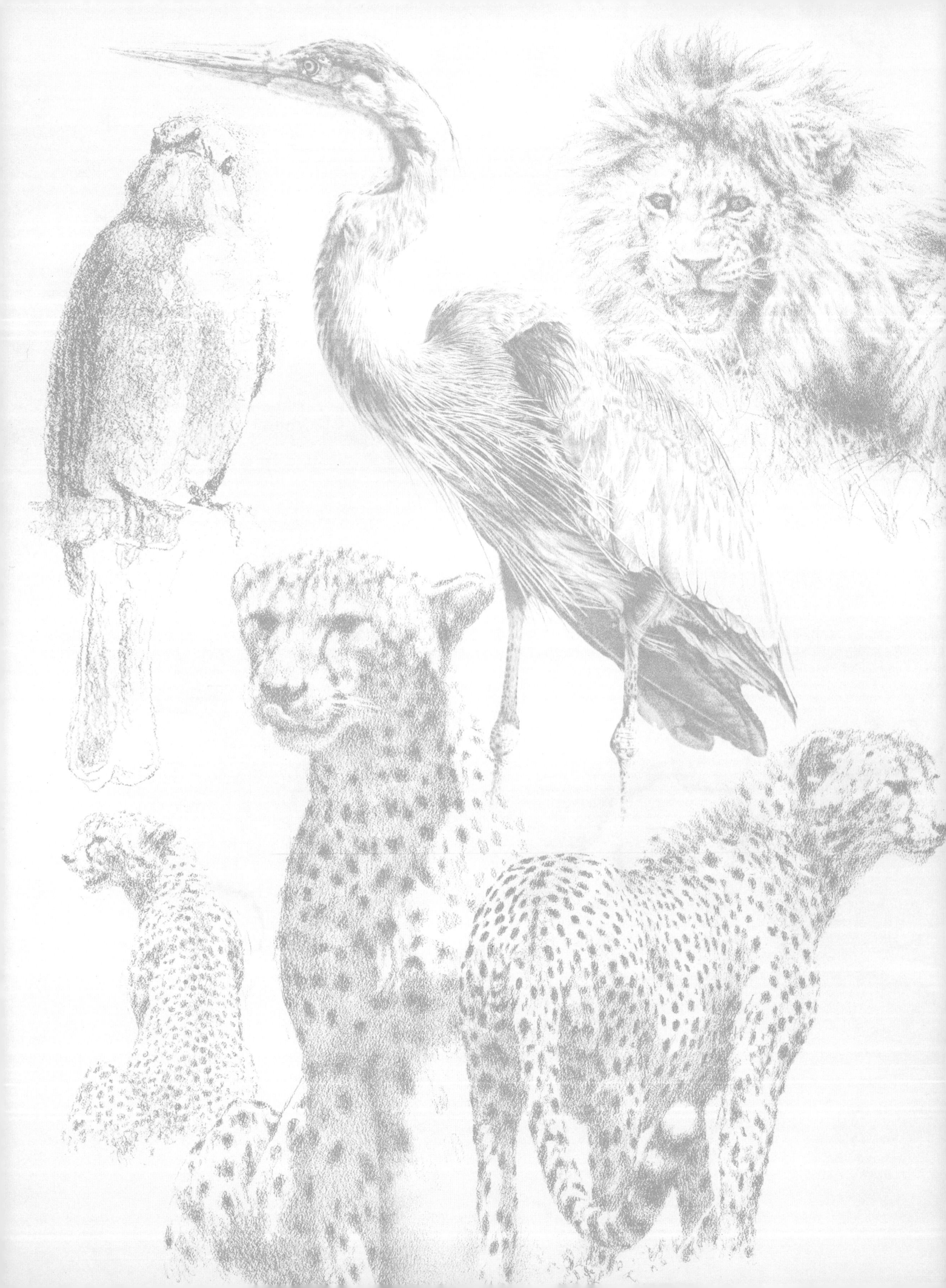

AFRICAN MENAGERIE

A Celebration of Nature

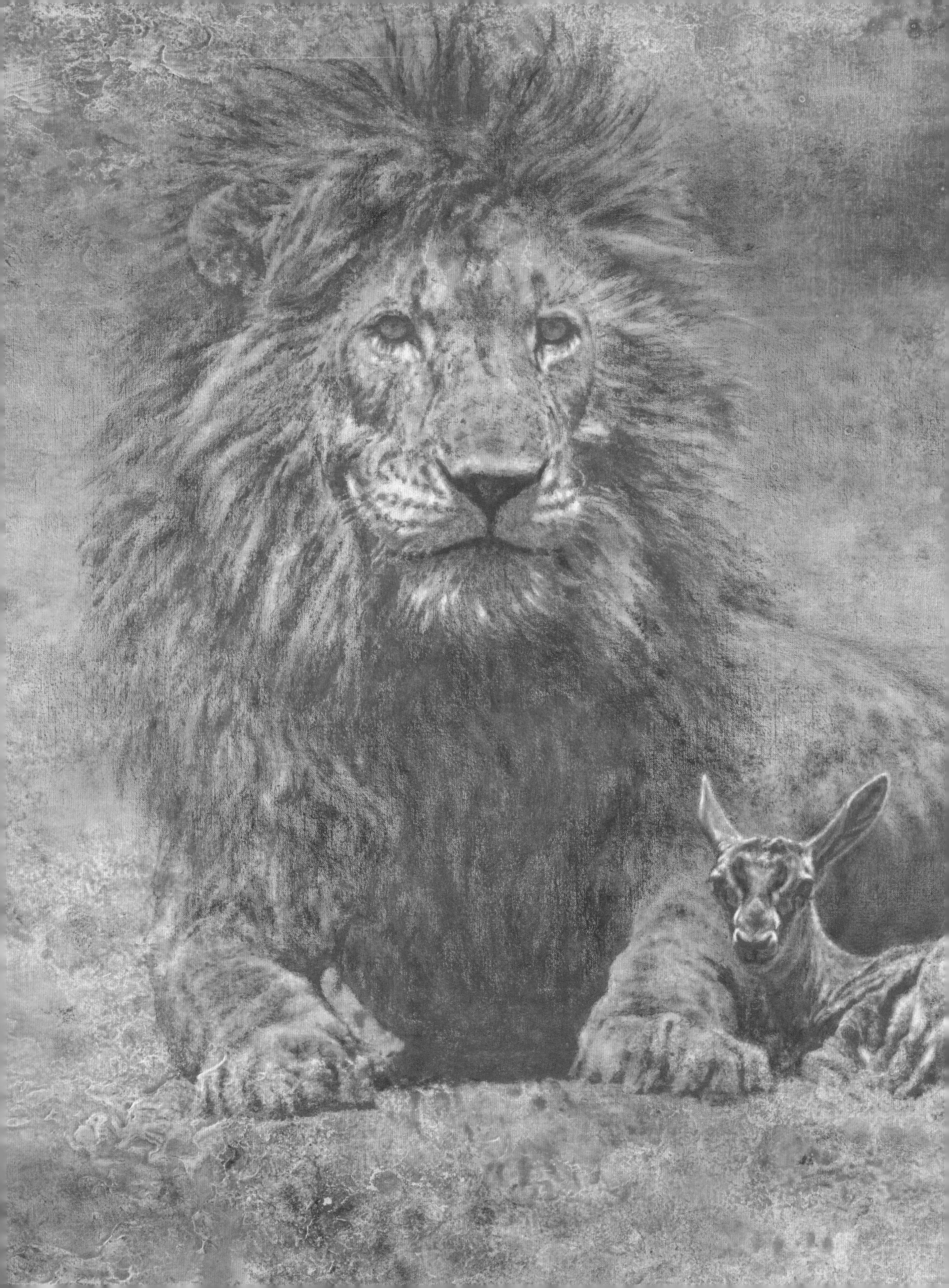

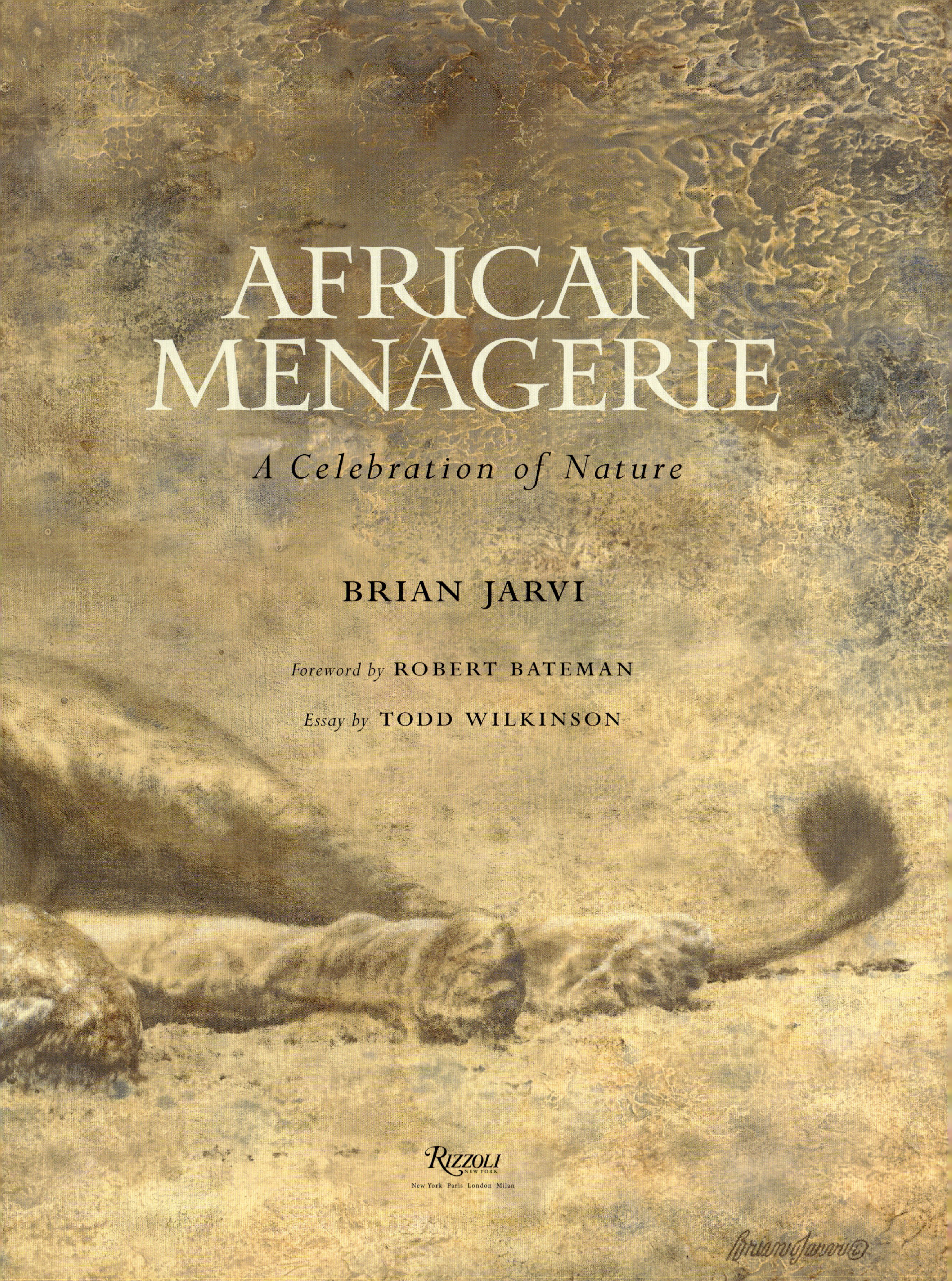

AFRICAN MENAGERIE

A Celebration of Nature

BRIAN JARVI

Foreword by ROBERT BATEMAN

Essay by TODD WILKINSON

RIZZOLI
NEW YORK
New York Paris London Milan

CONTENTS

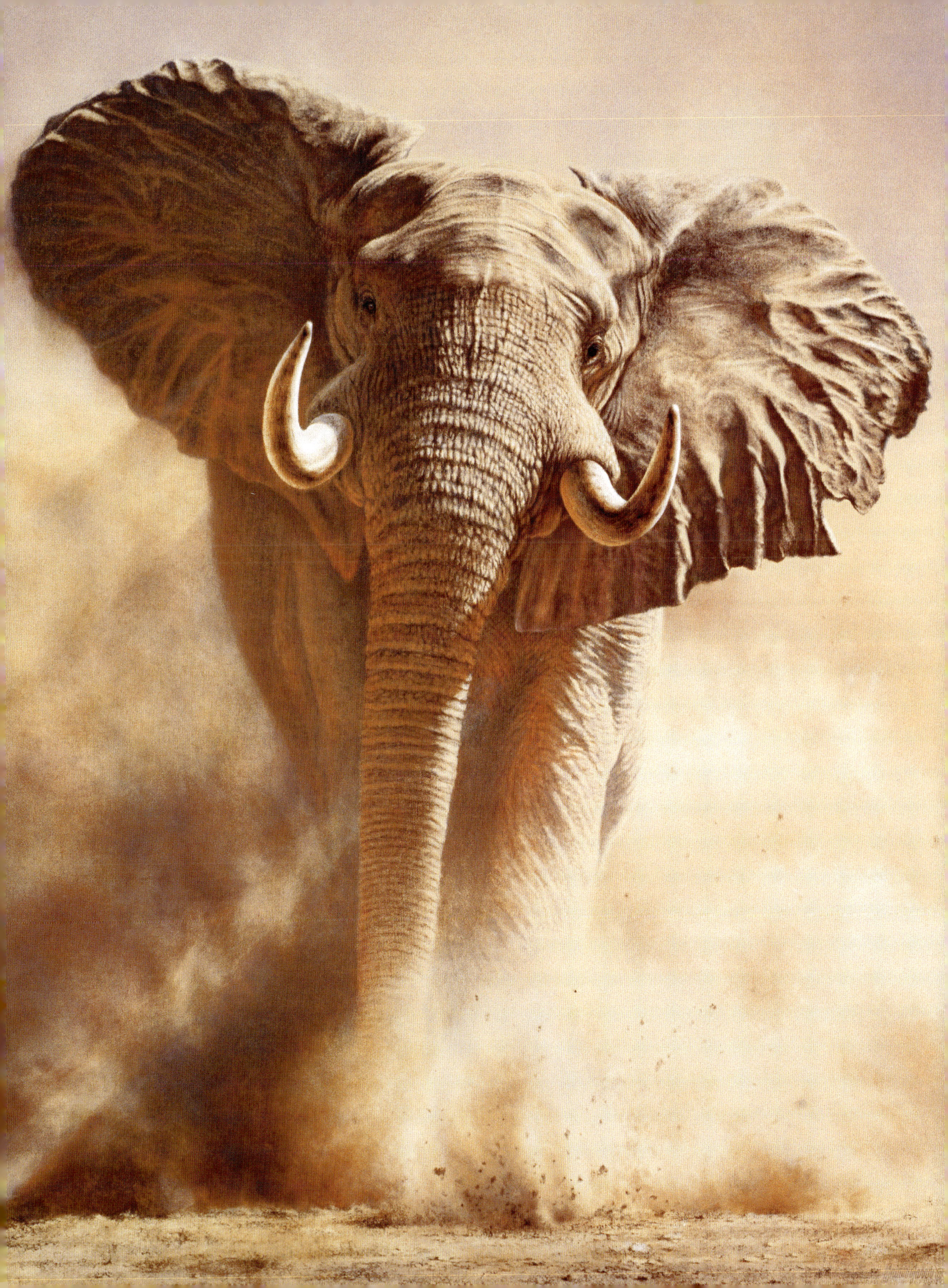

FOREWORD

Robert Bateman

I have always believed that wildlife art needs to be more than merely pretty pictures documenting animals in their habitats. It should spur viewers into action, into taking personal responsibility, and considering ways they can support the natural world that gives so much to them. What Brian Jarvi has set out to do with *African Menagerie* is deliver that message on an impressive and monumental scale.

Over the course of my own career, I have never approached my paintings as "wildlife art," but rather as artworks that speak to my own time on earth yet will hold up over the years, the same as natural history, with its evolutions and revelations. With *African Menagerie*, Jarvi not only showcases his talent and skill to render his subjects realistically, but also inspires us to stop and take notice.

Together, we have reached a critical juncture in human history when all of us hold the future of the natural world in our hands. Without conscientious thinking about how we can steward iconic species, starting with preserving their habitats, they will be lost. It is remarkably similar to the challenges that great American and Canadian painters confronted at the end of the 19th century when we came very close to losing certain species on this continent.

Art can serve as a powerful rallying cry to save the natural world, and it's important that we heed the message Jarvi is delivering.

Silent Song, The Birds, 2015
African Birds
Oil on Belgian Linen
48 x 36"

In 2015, this piece was juried into the Leigh Yawkey Woodson Art Museum's *Birds in Art* exhibition.

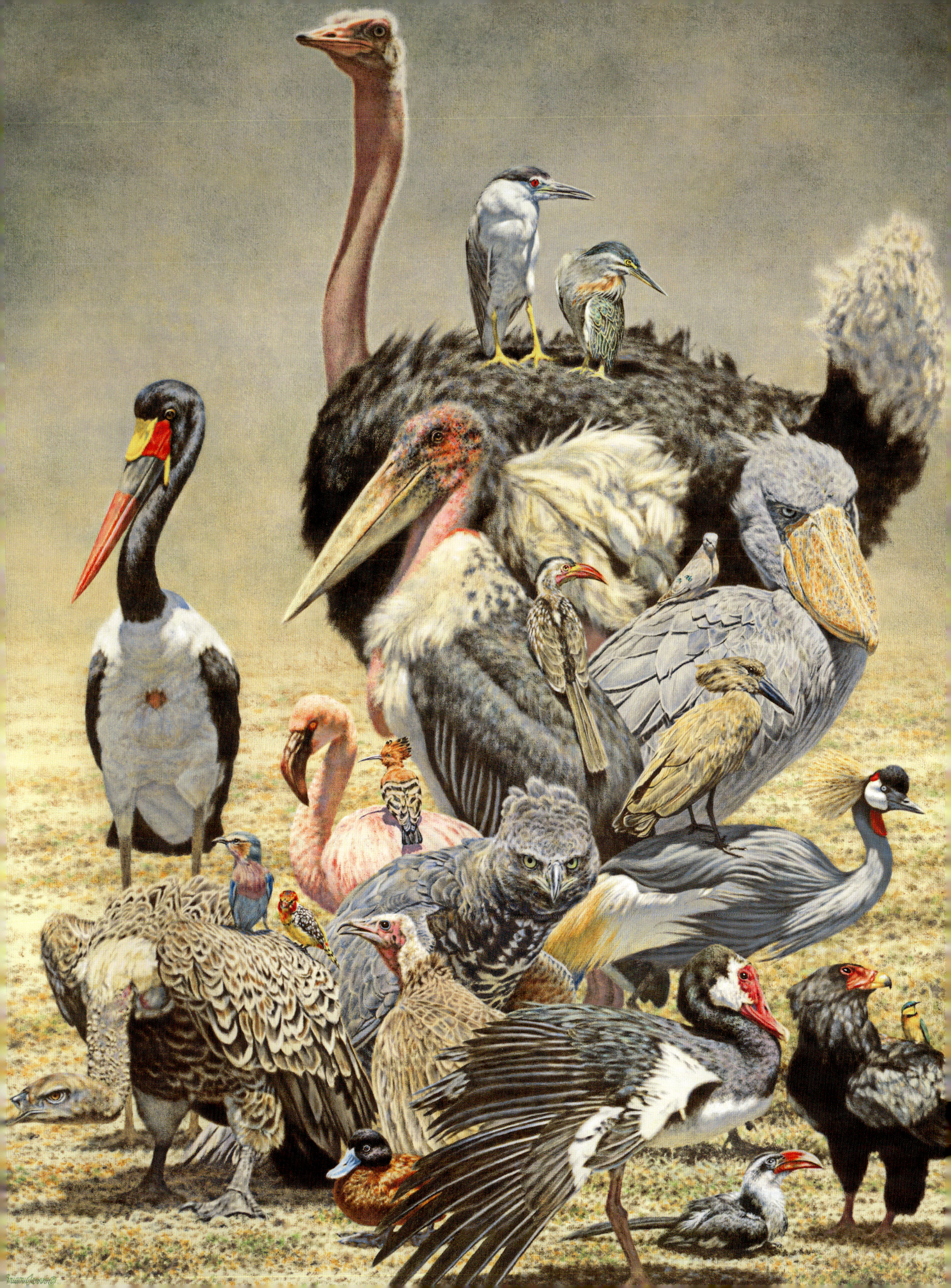

INTRODUCTION

Todd Wilkinson

Hieronymus Bosch
The Garden of Earthly Delights, 1505
Oil on Wood Panel
87 x 156"

Imagine it is the year 1509. Raffaello Sanzio da Urbino has just pressed a crude stick of graphite down on paper. He is about to sketch lines that will ripple through history.

Standing in front of a blank surface, his "canvas" is roughly the size of a modern billboard; metaphorically, it looms larger than life. Such epic dimension is essential because the artist knows monumental impact cannot be achieved with anything smaller.

Less is definitely not more. Size is more. Scale is more. So is the undertaking. Indeed, Sanzio da Urbino's assignment is audacious. It involves exploring the convergence of Christian spirituality and an adjoining blend of classical philosophy and emerging scientific logic. As one might presume, it is laden with controversy, with ideas that could make some viewers uncomfortable. The icons he selects represent political notions that challenge convention, and concepts that some people, dead set in their ways, might find heretical. Such is the risky pathway, however, of courageous art destined to become great. Every art movement in history, after all, is ignited by fearless innovators who challenge the status quo.

Sanzio da Urbino, a master draftsman, spends months using mathematical measurements to ensure the figures in his composition are precisely scaled. He creates shading to serve as reference points of value, tonality, and harmony in his sketches. During the next two years, he continually applies layers of paint to a wall, perfecting a magnetism of sublime aesthetics. Slowly, his fresco emerges, its story line speaking to poles of opinion that not only define his time, but also surface daily in ours. Also embedded in the scene is a complicated code of imagery and cues, visual messages that he leaves open-ended for viewers to decipher, symbolic allusions intended to spark deeper contemplation of our own mortal purpose.

If the artist's name doesn't immediately ring with familiarity, perhaps one may recognize Sanzio da Urbino by another. He is more famously known as Italian Renaissance painter Raphael. His magnum opus, *The School of Athens*, rises from floor to ceiling in an old papal residence—Stanza della Segnatura—inside the Vatican. Raphael's pictorial features 21 legendary thinkers of antiquity—great Western philosophers such as Aristotle, Socrates, Heraclitus, and Plato.

Around the same time, another European painter, Dutchman Hieronymus Bosch, conceives of his own masterwork, a

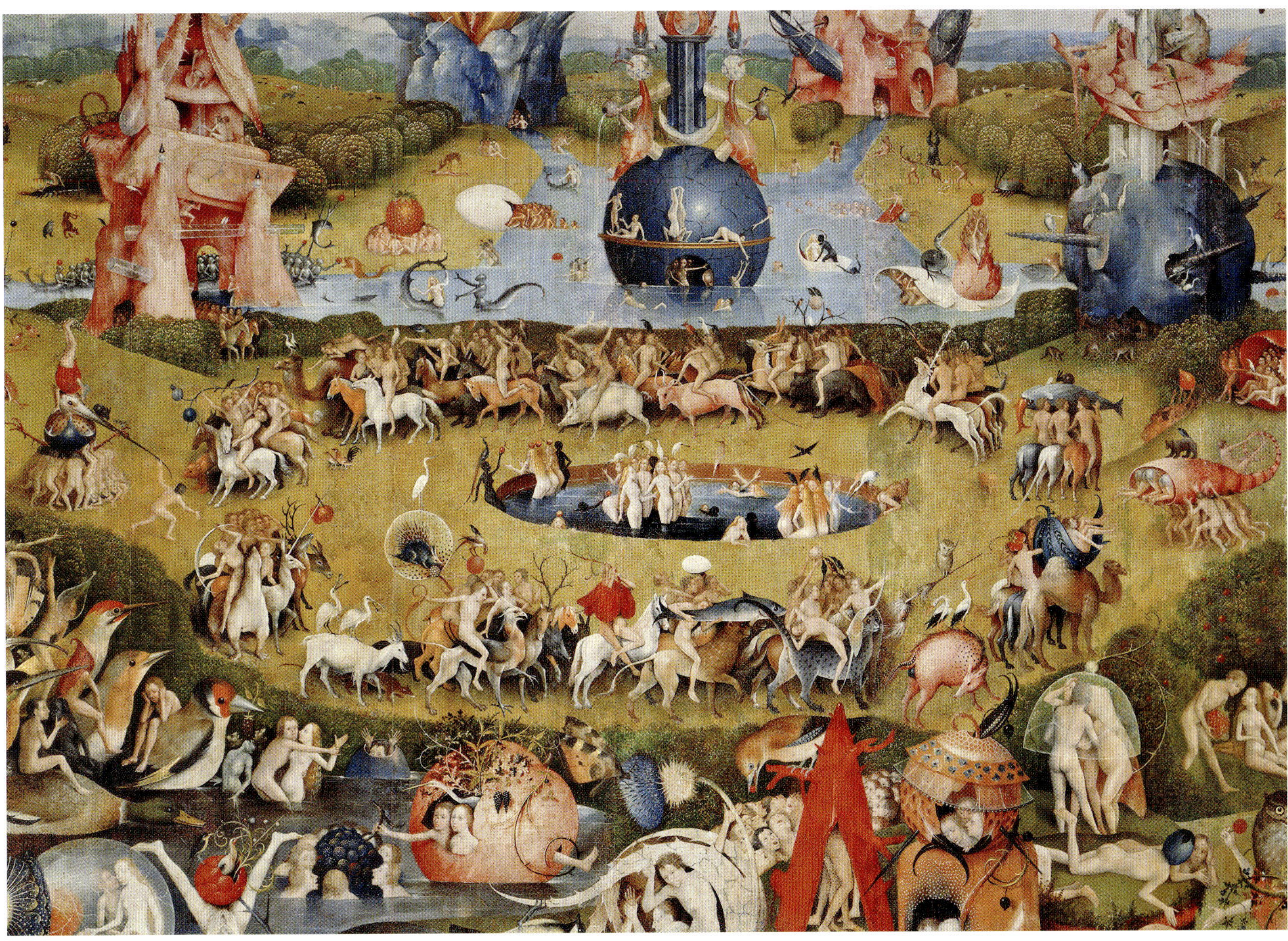

scene expressed in oil on three giant interconnected oak panels. Titled *The Garden of Earthly Delights*, this seminal triptych is also freighted with visual symbols meant to entice speculation about the artist's intended meaning. Bosch's tour de force is widely interpreted to be a celebration of beauty in God's original canvas of creation. It is also a cautionary tale about the corrupting forces of human self-indulgence—the destruction of nature wrought by not paying attention to the miracle of the divine, also known as the miracle of life on earth.

When executed skillfully, allegorical paintings are powerful vehicles for transporting thought. They abound throughout the arc of civilization as timeless touchstones. In ways more resonant than words, they help us understand big archetypal concepts dealing with survival that transcend generations. Both Raphael and Bosch dwelled in a past and time zone far removed from the Northwoods of Minnesota, yet they have direct bearing here, inspiring another artist to push new boundaries. More than five hundred years after Raphael finished his epic painting, contemporary artist Brian Jarvi completed his magnum opus, *African Menagerie*.

With *The School of Athens* and *The Garden of Earthly Delights*, Raphael and Bosch knew that what they were doing was bold, but they were not interested in merely making pretty pictures to serve as wall ornaments. They had no desire to produce anodyne two-dimensional illusions that put viewers to sleep. As geniuses, they possessed the rarefied talent—call it a gift—of conjuring imagery that soothes the eyes, tugs at the heart, reaches into the soul, and leaves viewers speechless. Foremost, they wanted to boldly *provoke*, to wake up humanity, to compel us to reflect on where we fit into the bigger picture. They wanted our vicarious encounter with their visions to be heightened by having a communion with one of the most potent human emotions of all—*awe*. Perhaps those same elusive emotions will be rekindled through *African Menagerie*.

BRIAN JARVI: ANATOMY OF AN ARTIST AND HIS WILDLIFE MASTERPIECE

Todd Wilkinson

Jarvi works on the "underpainting" of *African Menagerie*, Panel 3. Think of the underpainting as the underlying superstructure of the design.

On a September night in 2017, seven hundred people gathered at the Reif Performing Arts Center in Grand Rapids, Minnesota. No one could mistake this timber town, hard along the Mississippi River, for pretentious Rome, Madrid, or Paris. This is a community where the legend of Paul Bunyan looms large. Grand Rapids was the hometown of actress Judy Garland, who played Dorothy in *The Wizard of Oz*. Up the road in the neighboring Iron Range burg of Hibbing, Robert Zimmerman started his metamorphosis into Bob Dylan.

The mood in the packed auditorium was punctuated by expectation and brimmed with the din of conversation. People were talking about the changing autumn hues of the hardwoods, their last pursuit of walleyes on local tarns before the ice arrived, and the status of the ruffed grouse population. As the lights faded, all fell quiet.

A stage curtain was pulled open, marking the culmination of thousands of hours and innumerable sleepless nights. Unveiled in the next moment was a wall of colors, forms, and patterns. It gleamed like the side of a mountain flashed with brightening alpenglow at dawn. There was a long pause, a collective expression of astonishment as seven hundred souls attempted to absorb what they were seeing. The curving panorama was a surface filled with animals. A second later came a massive communal gasp, followed by everyone in attendance spontaneously rising from their seats in a standing ovation.

What happened at the Reif Performing Arts Center was no mere art opening as one might experience at a gallery. Something extraordinary had taken place. While the unveiling was certainly worthy of occurring in a city—debuting in a cultural capital, perhaps, with guests dressed in formal gowns and black ties—Brian Jarvi, by his own stubborn insistence, had wanted to first share his composition with a roomful of people he regarded as kindreds. "I hoped to unveil it here first, to give my friends a glimpse before it went away," he said. An upcoming traveling museum tour would center around this, his greatest work—*African Menagerie: An Inquisition*. It would bring his creation in front of a national audience and untold thousands.

Though inspired by the large-scale paintings of Renaissance artists such as Raphael and Bosch, Jarvi would never infer that his talent compares to them. As a 21st-century painter, he only aspires to make his own declarative

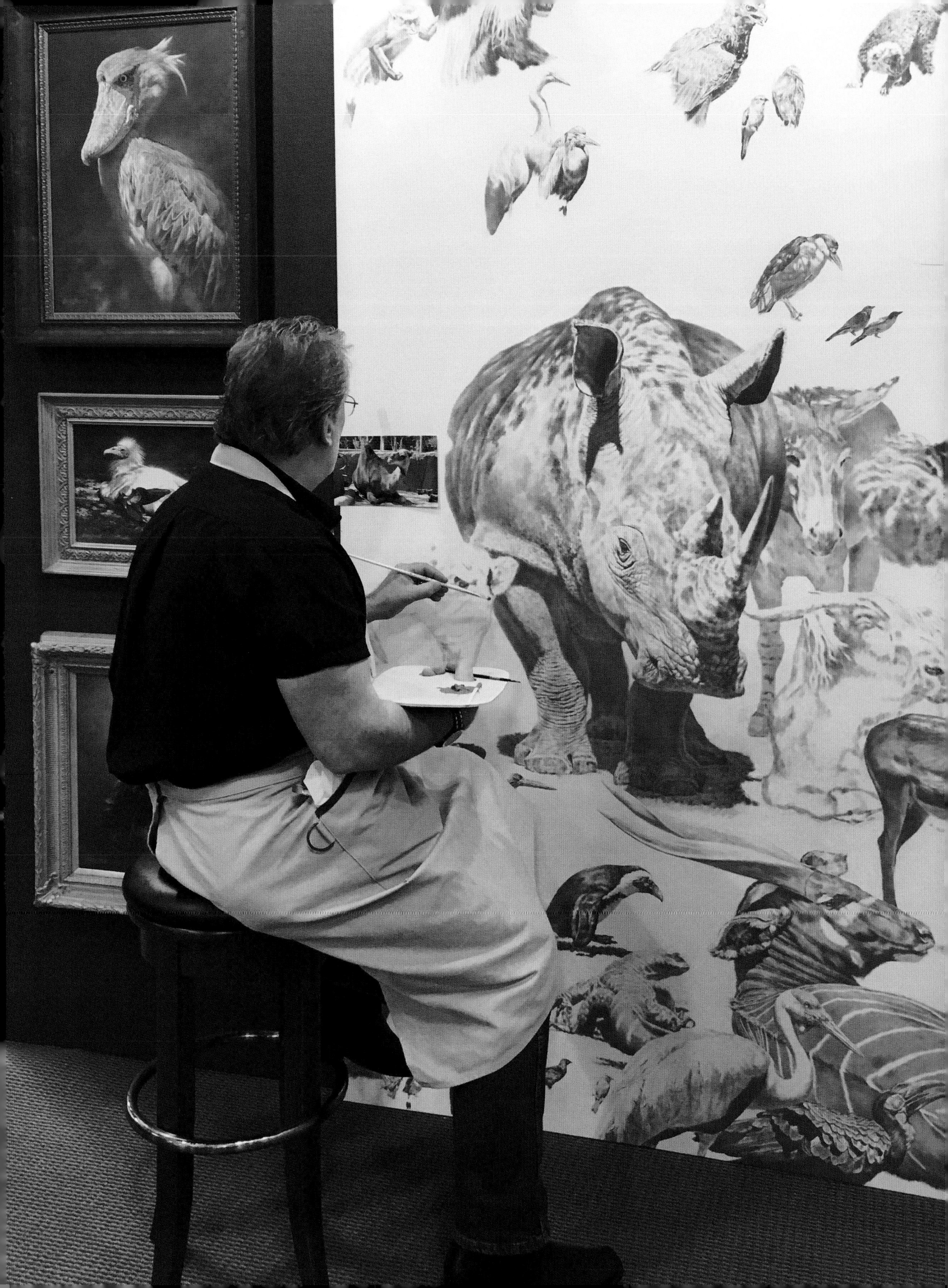

statement—challenging us in the here and now to ponder how people 50 years from today will judge our actions. Jarvi is laying down a marker with *African Menagerie*—a breathtaking one that at first inspires and then inserts a question into our conscience: How many of the species depicted in this painting, given current trend lines, will still be alive in two generations?

African Menagerie has no rival in the rich tradition of wildlife art. It measures 28 feet wide and nearly a story and a half tall. It sprawls across seven interconnected panels. It is full of figures, fit to scale and proportionality, stretching from foreground to background—a feat that is far more difficult than it appears. Jarvi presents more than 200 African animals and invites us to consider several questions: How many do we recognize? How does one relate to the other? Where does each animal live? What is its natural history?

One could literally spend hours, days, even years moving through this work and discovering something new. It doesn't matter if a viewer is young or old, rich or poor, from this nation or that one—the questions above appeal as ones that transcend human differences. "I believe in the principle that we care about things we love, but we will never learn how to love them unless they are made visible to us," Jarvi said. "For me, that's what my calling as an artist is—to illuminate the unseen."

African Menagerie is a statement that can no longer be ignored. To make the challenge palpable, Jarvi delivers us to the continent with arguably the most at stake—Africa. Its four-legged and feathered denizens represent the greatest reservoir of both large and charismatic wildlife on the planet, hundreds of millions of years in the making.

By meticulous design, Jarvi presents *African Menagerie* as a puzzle exploring the interrelationship between humankind and the other beings on earth. He knows most people will never make it to the Serengeti plain, trek up Mount Kilimanjaro, visit the Okavango Delta, disappear into the rain forests of the Congo, or explore the great wildlife reserves of Zimbabwe, Mozambique, South Africa, or Namibia. He knows that out-of-sight can translate in our busy modern lives to out-of-mind and therefore yield a lack of conscious awareness. That's why he has aspired to bring a fountainhead of earthly wildlife to us.

African Menagerie's subtitle, *An Inquisition*, is not intended to be an overt religious reference; it is instead about an artist and viewer enjoying an interlude of self-examination, spurring introspection on why nature matters so much in our lives and how it bonds us to one another. Jarvi is deeply concerned about the plight of the natural world on a planet where the human population will grow from 7.5 billion to more than 10 billion in half a century.

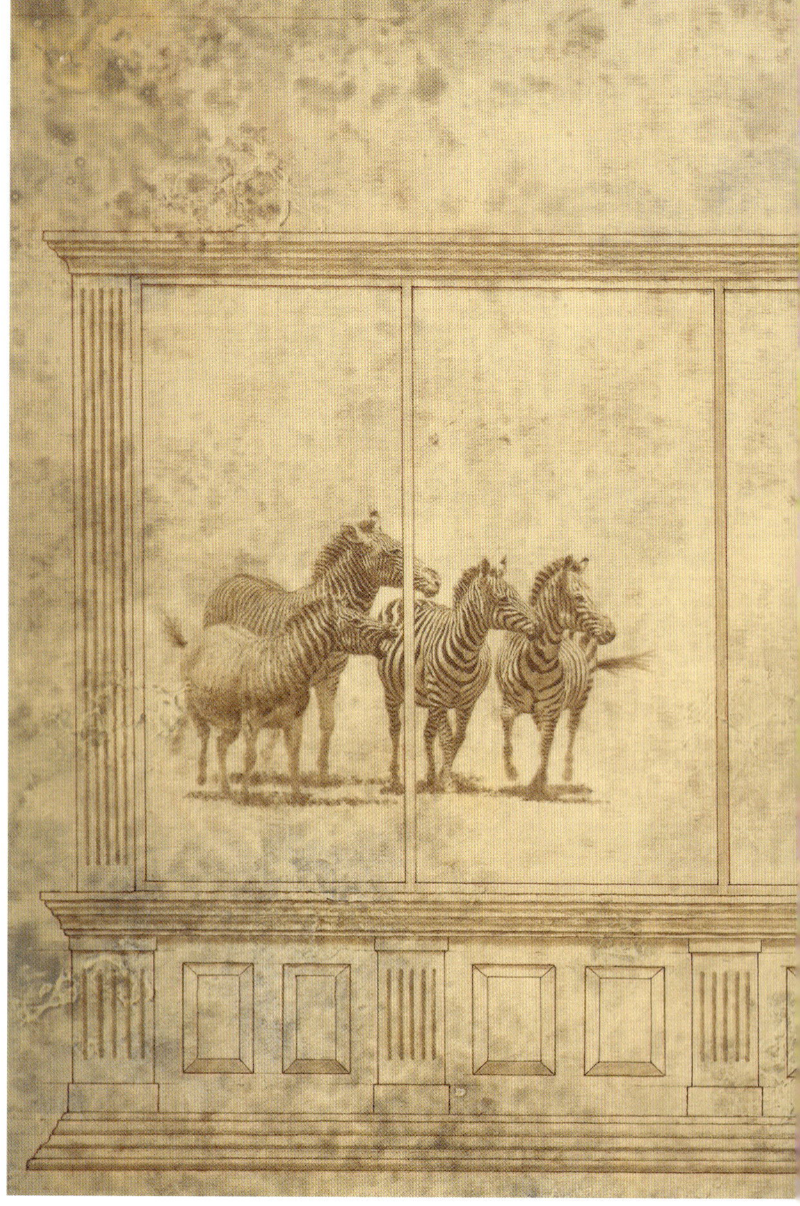

Extinction can be a heavy, depressing subject to consider. We are at the front end of the sixth major extinction episode and the only one caused by a single species—*Homo sapiens*. Yet it is impossible to ponder a world without African elephants, mountain gorillas, rhinos, chimpanzees, lions, leopards, giraffes, cheetahs, or many others. We are not

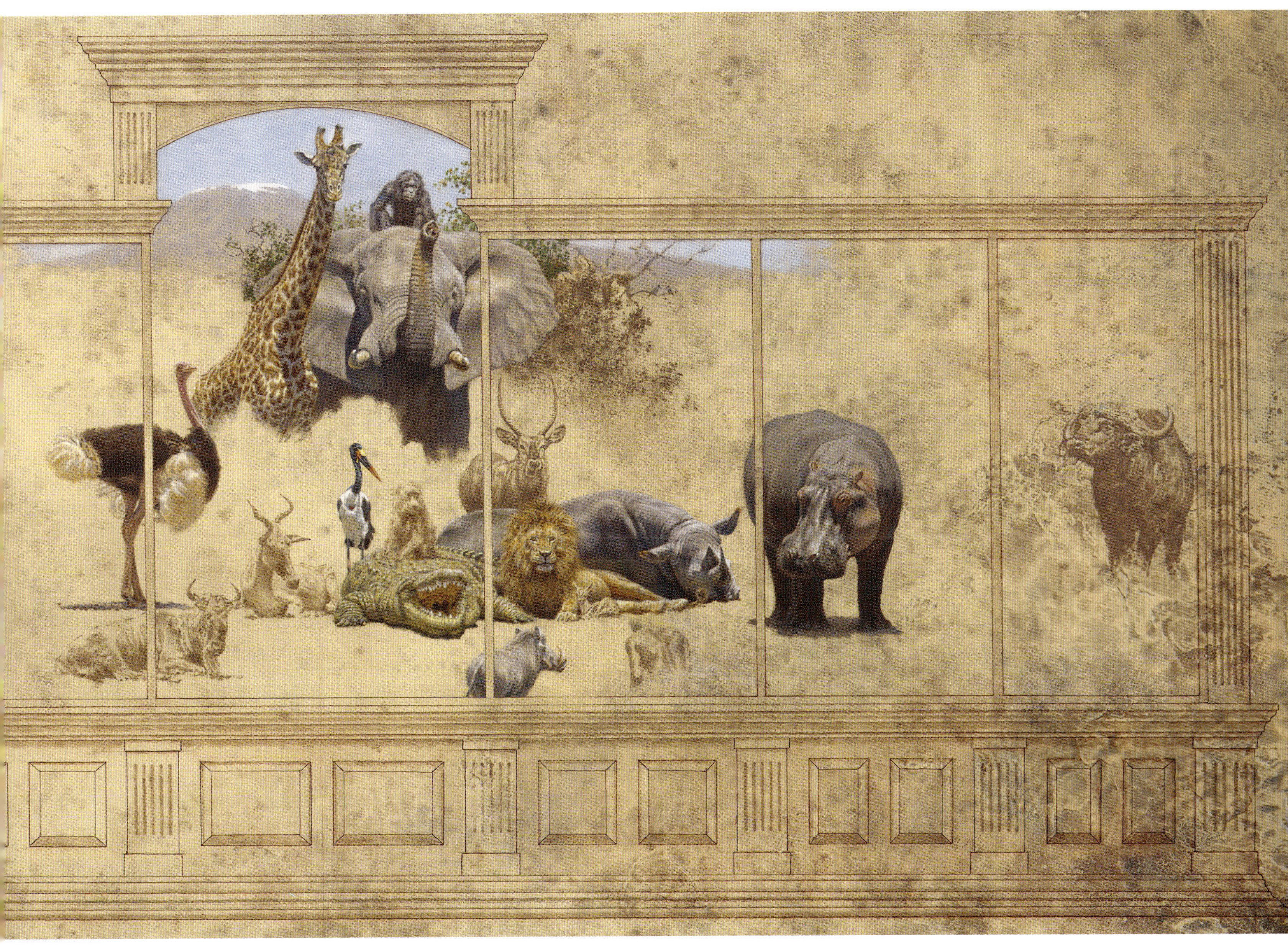

the first to confront finality. From the dodo to the Carolina parakeet, from the passenger pigeon to the near-complete annihilation of 35 million North American bison, humankind has stared into the abyss of vanishing species before.

ARTISTS AS CONSERVATIONISTS

On this continent, artists metamorphosed from being mere documentarians of nature—as evidenced by the frontier paintings of George Catlin, the aquatints of Karl Bodmer, and the masterful avian lithographs of John James Audubon—to helping ignite the movement of wildlife conservation. On March 1, 1872, Yellowstone became the first national

The Enhanced Menagerie, 2016
African Species
Oil on Canvas
27 x 54"
Collection of Jay and Karen Autio

OVERLEAF
Risky Crossing, 1989
Timber Wolf
Oil on Panel
24 x 36"
Private Collection

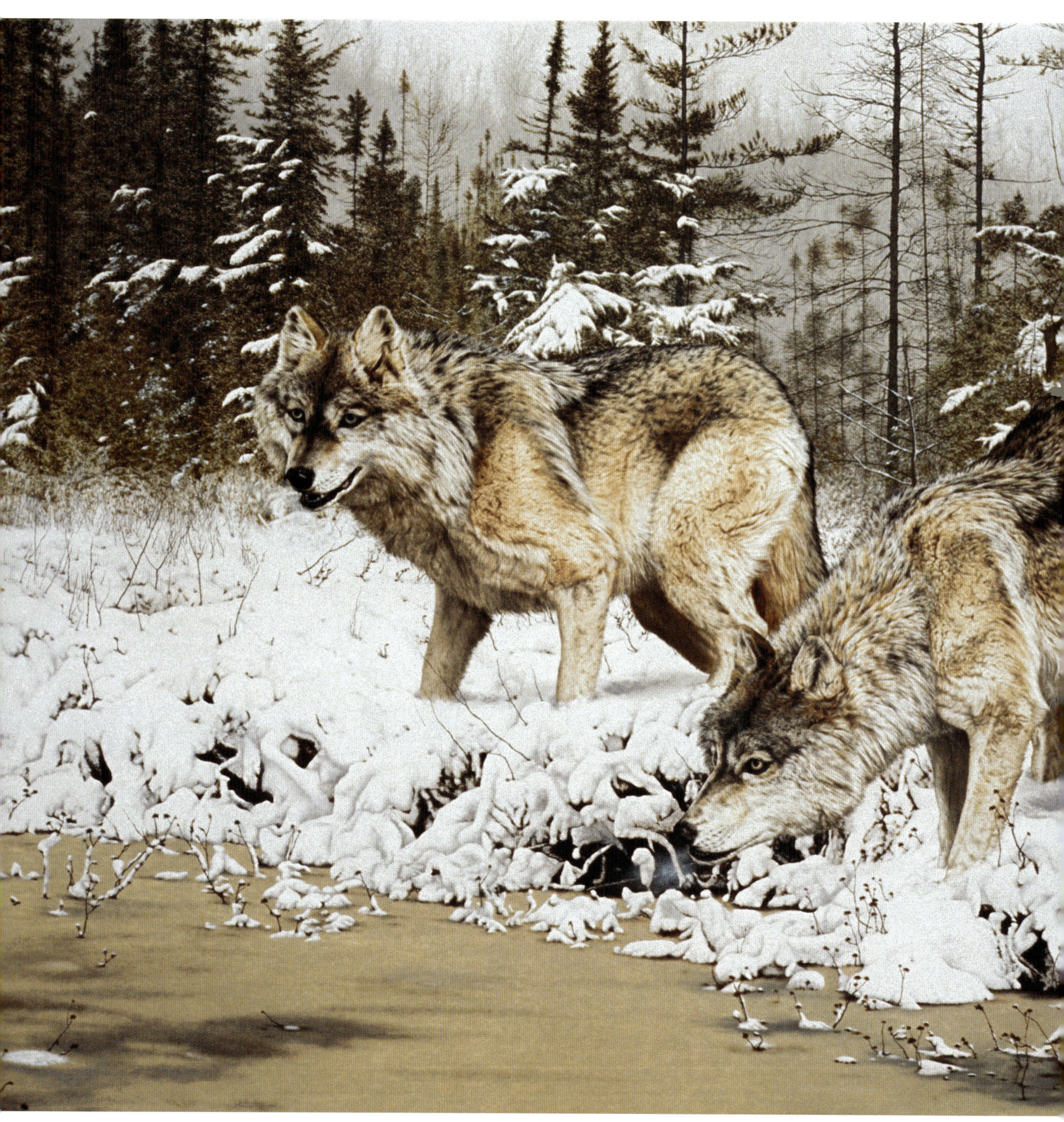

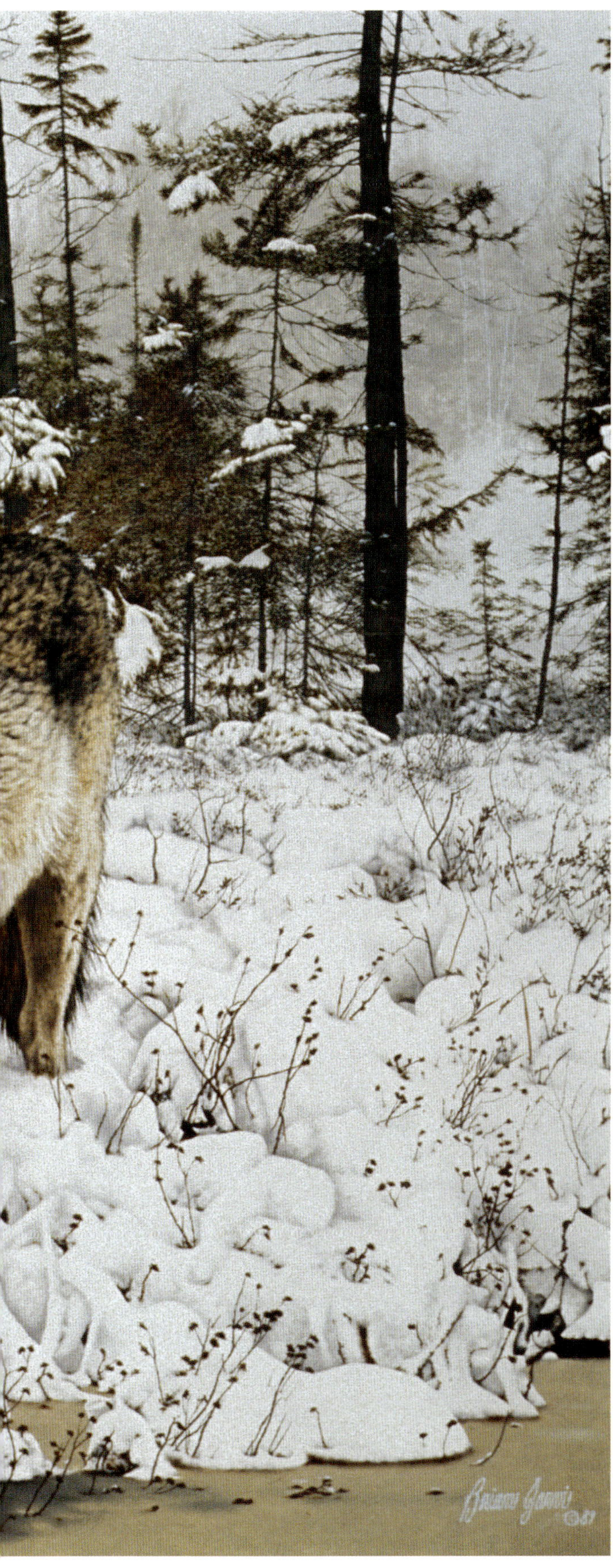

park in the world based largely on the strength of paintings by Thomas Moran, who presented them to Congress and President Ulysses S. Grant. Artists have made the case for the protection of species and extraordinary landscapes.

David J. Wagner received his PhD in wildlife art, and his thesis was transformed into the definitive book *American Wildlife Art*. Wagner, curator of the special museum exhibition of *African Menagerie*, has represented many of the best contemporary wildlife artists and has seen a large number of private collections. According to Wagner, while Jarvi's masterpiece holds together as a single scene, it is actually a labyrinth of paintings set inside of paintings.

"I've never seen any other contemporary piece like it. *African Menagerie* makes any room it inhabits live larger because it has an almost interactive quality. You don't want to take a photograph of it. You want to take a photograph and put yourself in it," Wagner explained. "It is so novel and its range of species so visually accessible and intriguing in a variety of ways, that I see it not only as Jarvi's artistic pièce de résistance, but also the kind of stand-alone piece that would be a capstone of any collection."

M. Stephen Doherty, former editor in chief of *American Artist* and *Plein Air* magazines, expressed his admiration for *African Menagerie* and stated that Jarvi is carrying on a virtuous tradition. "The challenge of painting massive canvases obviously did not originate with American artists. Long before, European masters such as Rubens, Tintoretto, Delacroix, and Monet proved their talents and physical capabilities by filling cathedrals, palaces, and castles with series of large-scale paintings," he said. "Their murals, frescoes, mosaics, and canvases celebrated successful battles, conquests by a monarch, miracles performed by a saint, or wonders of the natural world. And while they brought attention to the events depicted in the works of art, they also elicited wide respect for the talented artists who accomplished such amazing artistic feats."

In 1857, the great Hudson River School painter Frederic E. Church unveiled his 40-by-90-inch painting of Niagara Falls. The fascination with and publicity surrounding the painting brought out more than one hundred thousand people, who each paid 25 cents to see what was then considered a colossal masterwork. "That enthusiastic public response was not accidental or unexpected," said Doherty. "From the very beginning of his celebrated career, Church attracted attention to his work by creating large, blockbuster-sized landscape paintings of exotic, unknown, and captivatingly dramatic locations."

Others who followed suit were Moran, Albert Bierstadt, Frederic Remington, Charles M. Russell, and dozens of others who included wildlife in their scenes. Sometimes the work was done to point out what was being lost. Artistic portrayals of the plunging buffalo population, for example, helped inspire naturalists-sportsmen such as William T. Hornaday and President Theodore Roosevelt to found the American Bison Society in 1905.

Jarvi is venerating nature in the here and now. He's calling attention to other sentient beings. Doherty noted that *African Menagerie*, when one understands its intention, possesses a sense of modern urgency, informed by enlightened science and knowledge of evolution that did not exist in the middle of the 19th century. A century from now, this work will be coveted just like the other megaworks that precede it.

WHO IS BRIAN JARVI?

Prior to completing *African Menagerie*, Jarvi was already lauded. His depictions of predators and prey were hailed for being dynamic and innovative, his handling of avian subjects praised for their elegance, and his draftsmanship considered first-rate. Even the late Bob Kuhn, considered during his life one of the premier painters of North American and African subjects, took notice of Jarvi. Flipping through the catalog assembled by the Society of Animal Artists for its nationally touring *Art and the Animal* exhibition, Kuhn paused when he arrived at Jarvi's page, which featured a portrayal of an African lion. "Now that guy can paint," he said, "but more importantly you can tell he knows how to draw." Kuhn's skills as a draftsman were considered nonpareil; for him to take notice of Jarvi was the highest of praise.

But really, who *is* Brian Jarvi? What are the crucial ingredients necessary for a person to come out of nowhere and deliver a piece of important art that will weather the test of time? What was the *je ne sais quoi* quality in Grand Rapids girl Judy Garland that enabled her to find fame as a movie actress and singer? How did Hibbing's Robert Zimmerman become Bob Dylan? In the most unlikely of ways, Jarvi emerged from the firmament of the same riddle.

The first thing you need to know about Jarvi is this: he is an incorrigibly stoic Scandinavian, a self-admitted hardheaded Finlander and grandson of immigrant stock. Built

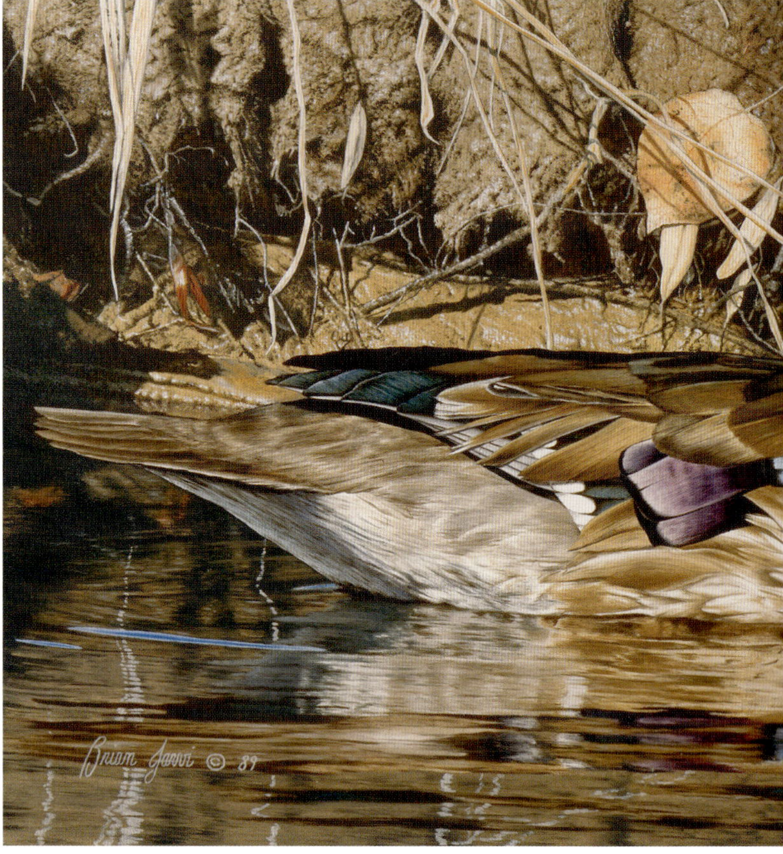

like an aging NFL linebacker with massive biceps and a chest shaped by decades of weight lifting, he is, at the same time, incredibly shy, a fine orator when he has to be, but one who is exceedingly uncomfortable about being the center of attention. He would prefer that his artwork do the talking.

Nature's rhythms course through Jarvi, and he is a product of place. For those fortunate enough to have wild water in their lives—the curve of the ocean and sound of crashing surf on the horizon, or maybe the mellifluous hum of a passing river current or a lakefront view—there are things one sees that are simply unexplainable. The word *ineffable* applies as much to the view that Jarvi and his wife, Raelene, behold from their home on the hill above Sugar Lake as to the elements present in *African Menagerie*.

Of all the hours in a day, arguably the most hallowed for dwellers of northern Minnesota are those of eventide. From the Jarvi dock—installed in May after ice-out and kept up until late October—the seasons are constantly calling. Nothing is more melodious and primordial than the fluted trilling of a loon song, or the late-night howl of a wolf, or the V formations of waterfowl quacking or honking in the

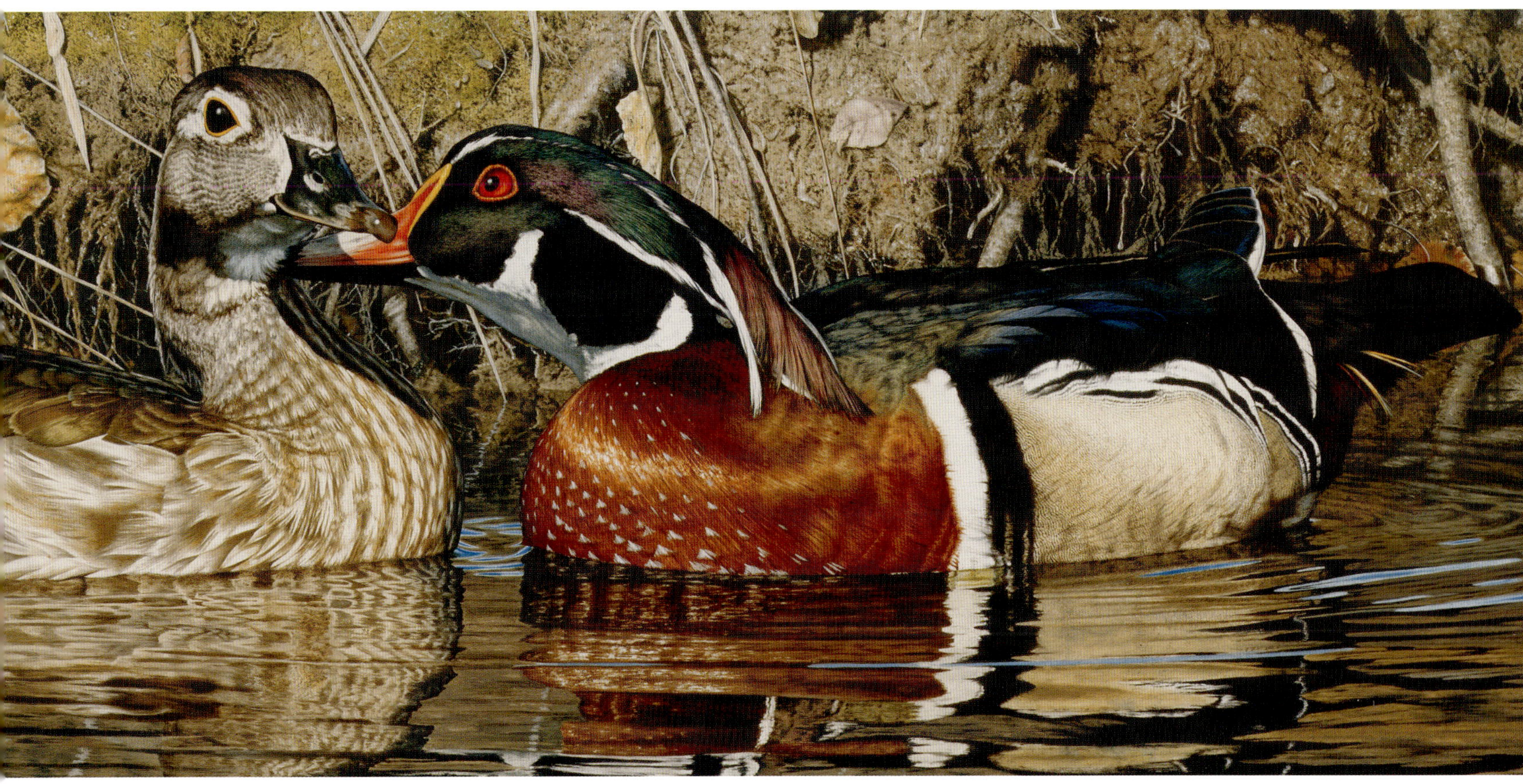

sky above. These kinds of natural acoustics and visual theater stimulate Jarvi's response to the environment.

Not in his wildest dreams did Jarvi ever imagine that one day he would have a painting studio on Sugar Lake. He had no formal art education. He never attended a prestigious academy or spent years taking classical studio classes painting and drawing the nude human figure. In nature art, Jarvi is comparable to an unexplained comet suddenly passing across the sky. His inexplicable rise is the stuff of Horatio Alger—that is, if Alger had grown up in Floodwood, Minnesota, and ended up squeezing sweat equity out of tubes of oil paint.

Jarvi's hometown, current population five hundred, is distinct for a couple of things. First, it's a community held together by tough working-class people who strive to achieve the American dream. Generations of Floodwoodians toiled to make ends meet by working on the railroad or cutting trees in the forest. People in Floodwood do not project smug airs. They judge people by their sincerity, work ethic, and ability to follow through on promises. Second, Floodwood is a hamlet that has the St. Louis River—a classic Northwoods river that empties into Lake Superior—running through it. The Jarvi

Spring Courtship, 1988
Wood Duck
Oil on Panel
10 x 30"
Private Collection

OVERLEAF
Confrontation, 1990
Moose
Oil on Panel
20 x 30"
Private Collection

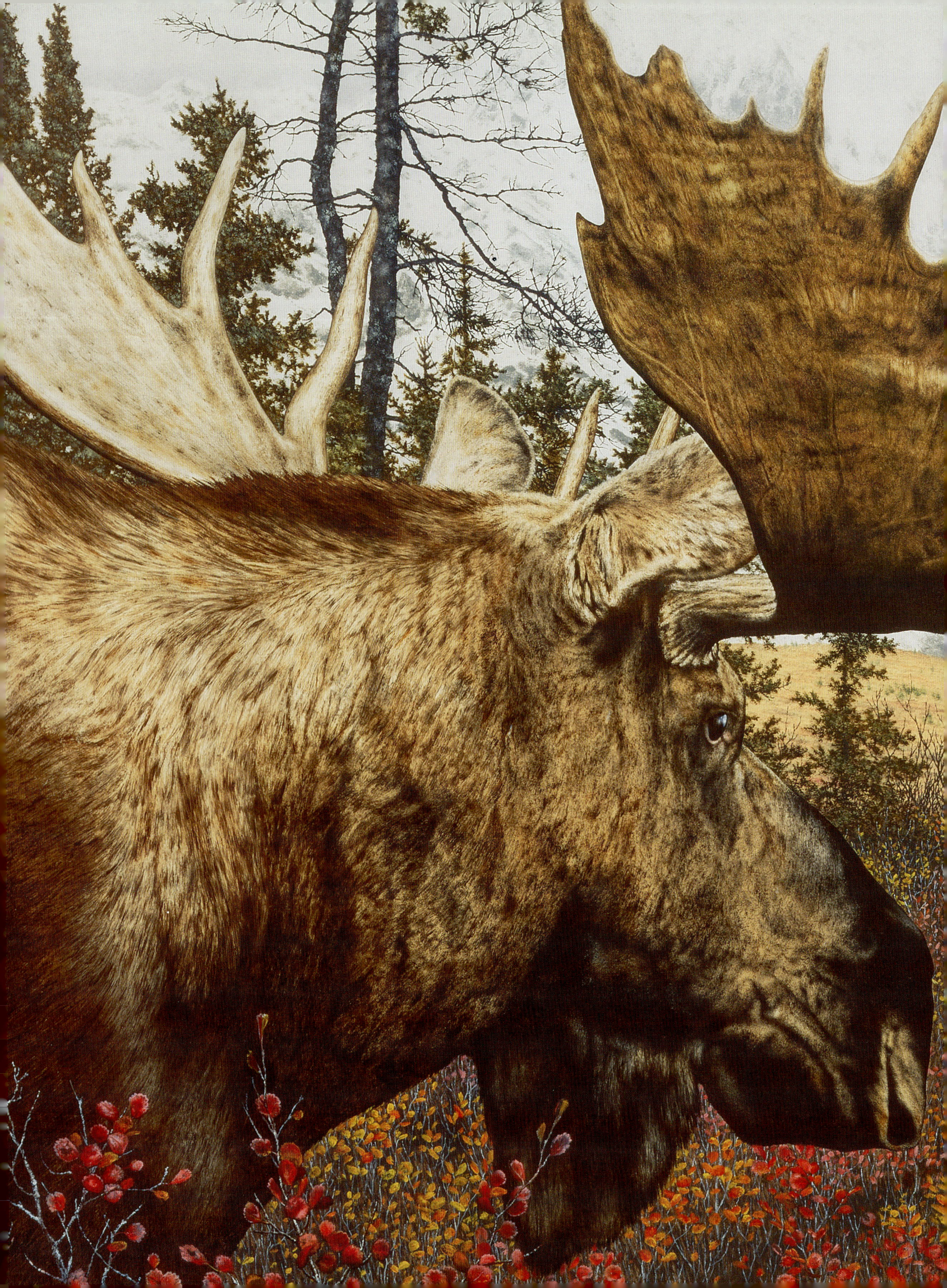

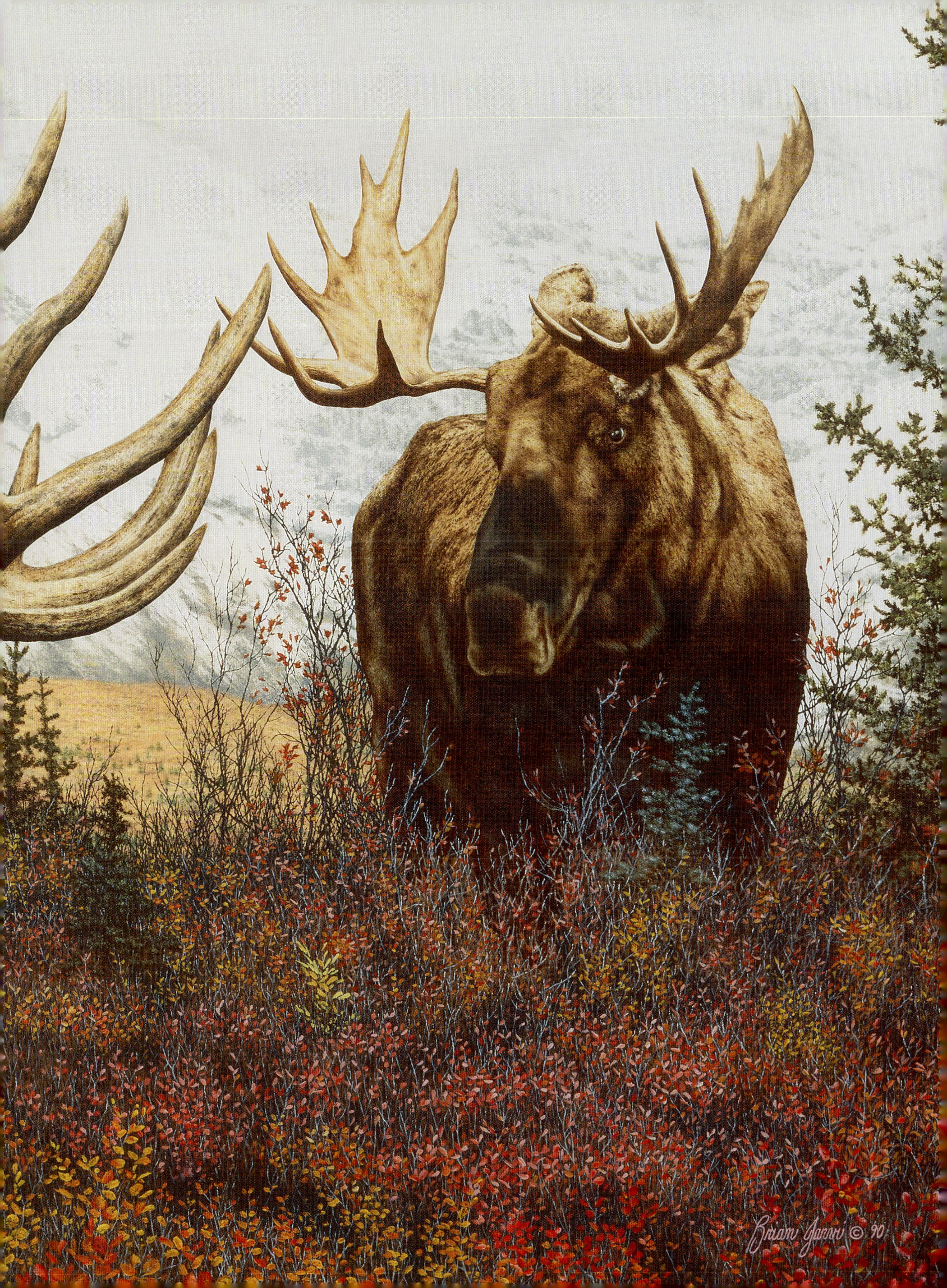
Brian Jarvi © 90

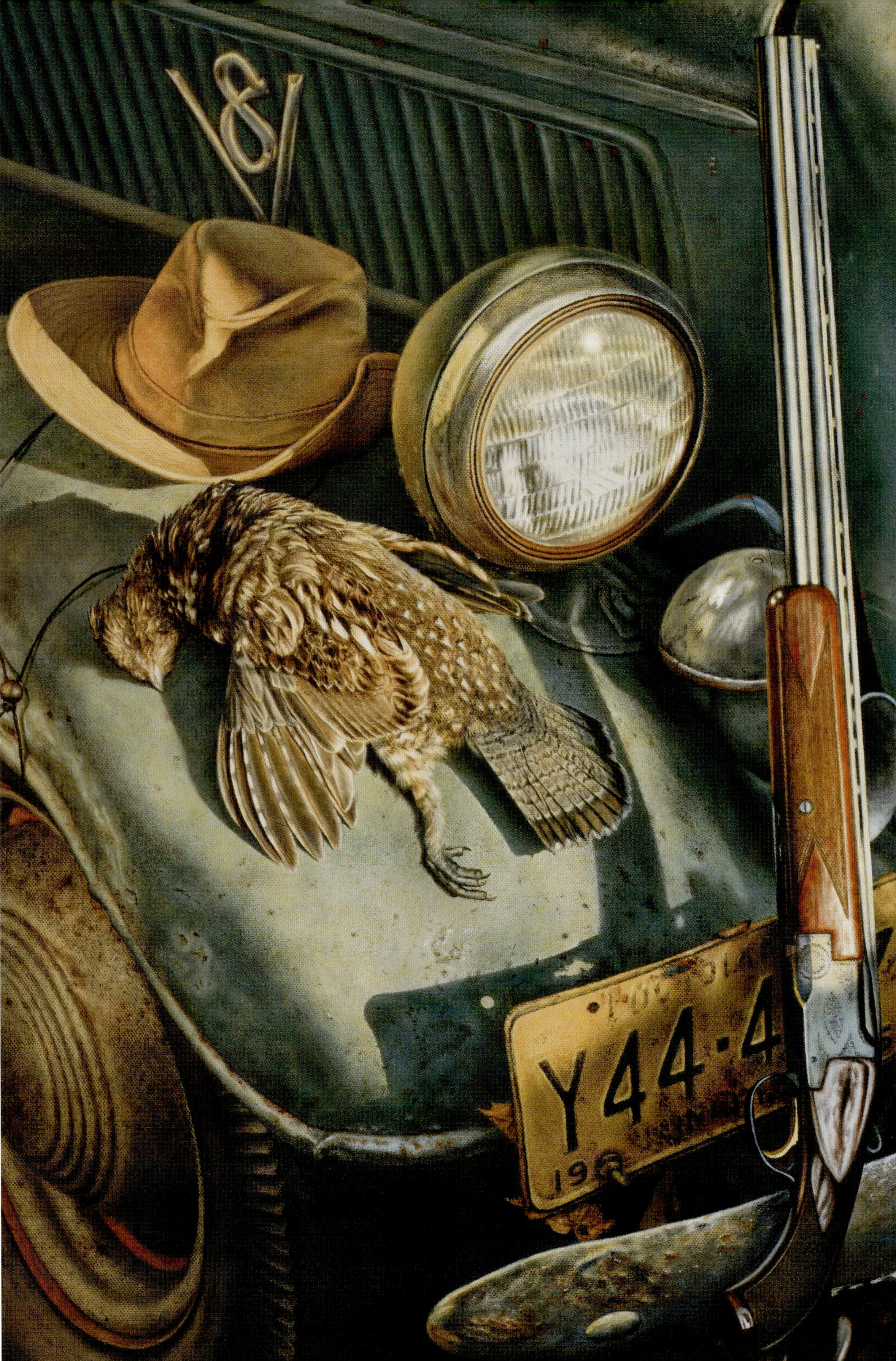
V8
Y44-4

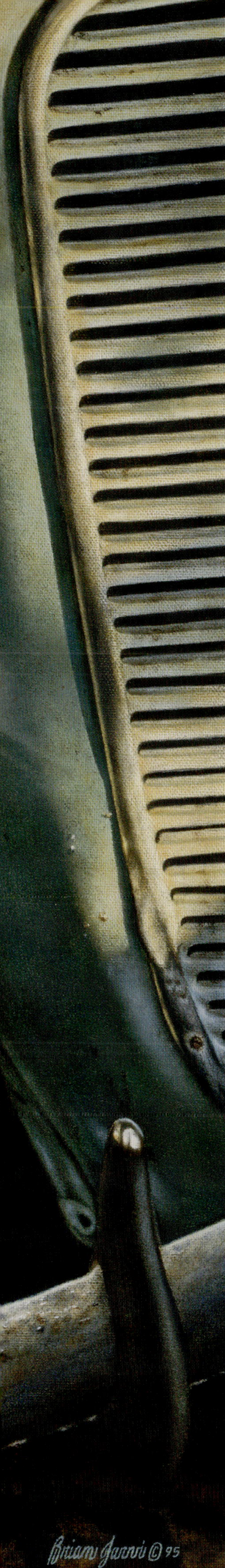

Seasons Past, 1995
Ruffed Grouse
Oil on Panel
20 x 16"
Private Collection

OVERLEAF
Fishing the Shallows, 1991
Bald Eagle
Oil on Panel
18 x 36"
Private Collection

clan was shaped mightily by the presence of these two converging forces; they galvanized the way Jarvi sees the world.

In Ralph Axel Jarvi and Virginia Elizabeth Jarvi's big Lutheran family, Brian Jarvi, born in 1956, was the third youngest of nine children grouped relatively close in age. The family lived in a small clapboard house with two bedrooms—two and a half bedrooms if you count a makeshift overflow area. With 11 souls and a couple of pets sharing the same small space, there wasn't an abundance of elbowroom. Jarvi and his brothers shared beds, the same for his sisters. The setup could be likened to a scene lifted out of *The Waltons*, a popular 1970s television show that dramatized life in Appalachia during the Great Depression. They didn't have much materially, but the Jarvis had each other and remain tight-knit to this day.

"Our mom came from a poor family. She was very shy as a young girl, and there was not a lot of food to eat in the house," Jarvi's sister Kim Mrosla shared. In the 1940s, during Virginia's junior year in high school, her parents moved to California to work in war factories. "Being in a big urban school broadened her outlook and exposed her to a lot of different cultural influences. That's where her lifelong interest in art, literature, and music really started."

"Our mother," Jarvi reflected, "was the closest thing we had in our lives to a living saint." It's an assessment shared by each of his siblings. With her short-tempered husband away for long stretches on railroad jobs, Virginia had her hands full trying to make ends meet, keeping her brood fed and clothed on a limited budget, and making sure the roughhousing boys were distracted to keep them out of trouble. "You can guess what it was like," Jarvi's brother Guy said. "There were many moments of bedlam." Disputes would sometimes arise between Jarvi and his brothers. They would jostle into melees. If it was cold outside, Virginia would send each child to a different corner of the small house, hand them a couple of crayons or lead pencils with paper, and tell them to go settle down by drawing.

Yes, amid the chaos, Virginia grounded her children in art as a form of self-expression and meditation. Like a schoolmarm, she would make her rounds among all the Jarvi kids, encouraging them to copy pictures they cut out of magazines and the newspaper. As they drew, she would read them stories out of books and, at night, have them read to her. She would talk about her favorite artists. She would

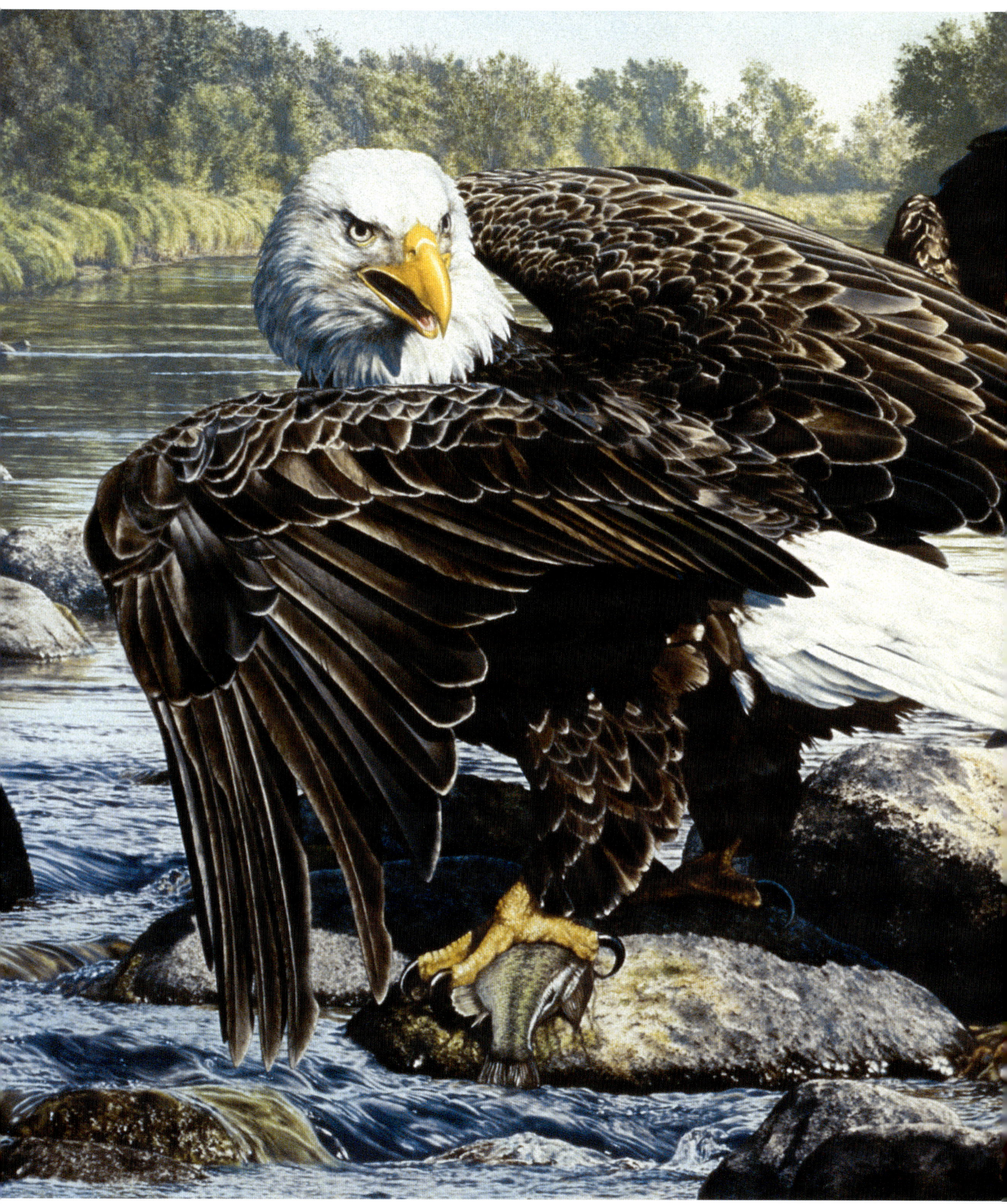

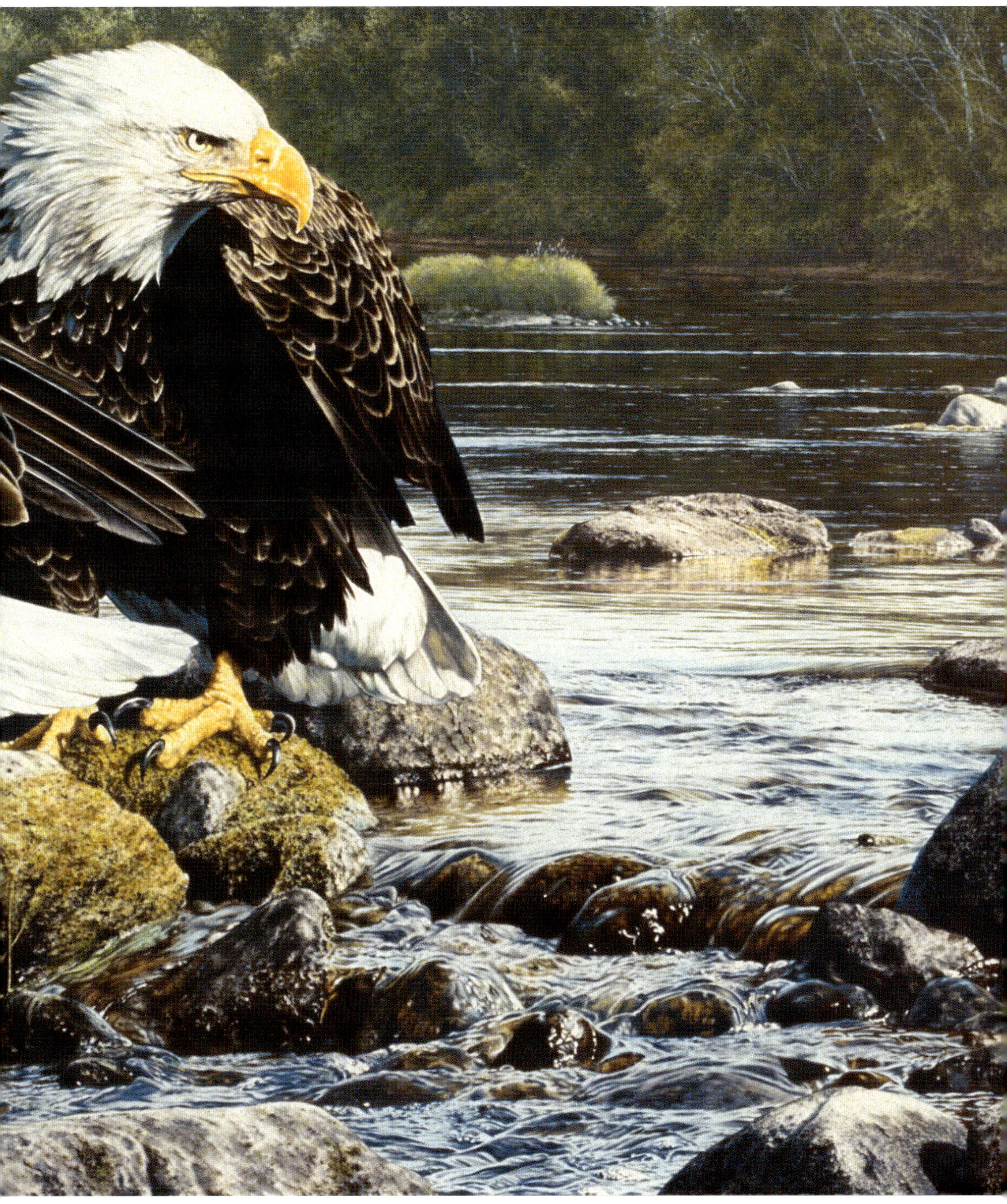

play classical music records on the phonograph. And, as the youngsters disappeared into the challenge of bringing subjects to life in two dimensions, she would sew, keeping their clothes mended. When Ralph was home, he pulled out a harmonica, an instrument he taught himself to play.

Brian Jarvi was always quiet. Undistracted, he would disappear mentally into his own world, conjuring up figures and scenes in drawings. It's how he made sense of things. Long winters were a creative time. When the snow started to melt, Virginia pushed the kids out the door and encouraged them to explore. She was neither inclined nor did she have time to be a helicopter parent. "We knew we were loved. That's all that mattered, and we were turned loose to run wild," Jarvi said. "The biggest gift she gave me was art. She would show me pictures of what great works were."

Something else Virginia nurtured was the idea of nature leaving a profound imprint. The Jarvi kids spent years roaming the riverside. Every month brought changes in water levels, color schemes along the tree-lined corridor, the buzzing of mosquitoes, and the din of croaking frogs. The river was the kids' playground. They explored a series of islands; bathed and swam in the pools, whether warm or cold; fished for northern pike, walleye, and smallmouth bass; built forts out of lumber they scavenged; and walked the ice.

"My first inkling of what a menagerie was emerged during those years," Jarvi said. Dating to his earliest recollection, he was always fascinated by the concept of animal collections—zoos, circuses, groups of species kept on private estates, or even his own. He put frogs and turtles he caught along the river into aquariums and temporary cages. "Over the years, I've become more intrigued by the contrast between wildness and captivity and the differences you see with animals that continue to freely roam in their habitat versus those that are tamed. We're really at a point in human history where we are exerting our presence and leaving little terrain that hasn't been conquered. What are we gaining? What are we giving up?"

What attracted Jarvi's young eyes most of all were birds—passerines (the colorful boisterous songbirds that arrived in their Northwoods breeding grounds in the spring after spending winters in the southern United States or even Central America); crows and jays; raptors, such as hawks, owls, and eagles; and, of course, the multitude of waterfowl that poured through that corner of the Central Flyway.

Solitary Whitetail, 1988
Whitetail Deer
Oil on Panel
36 x 24"
Private Collection

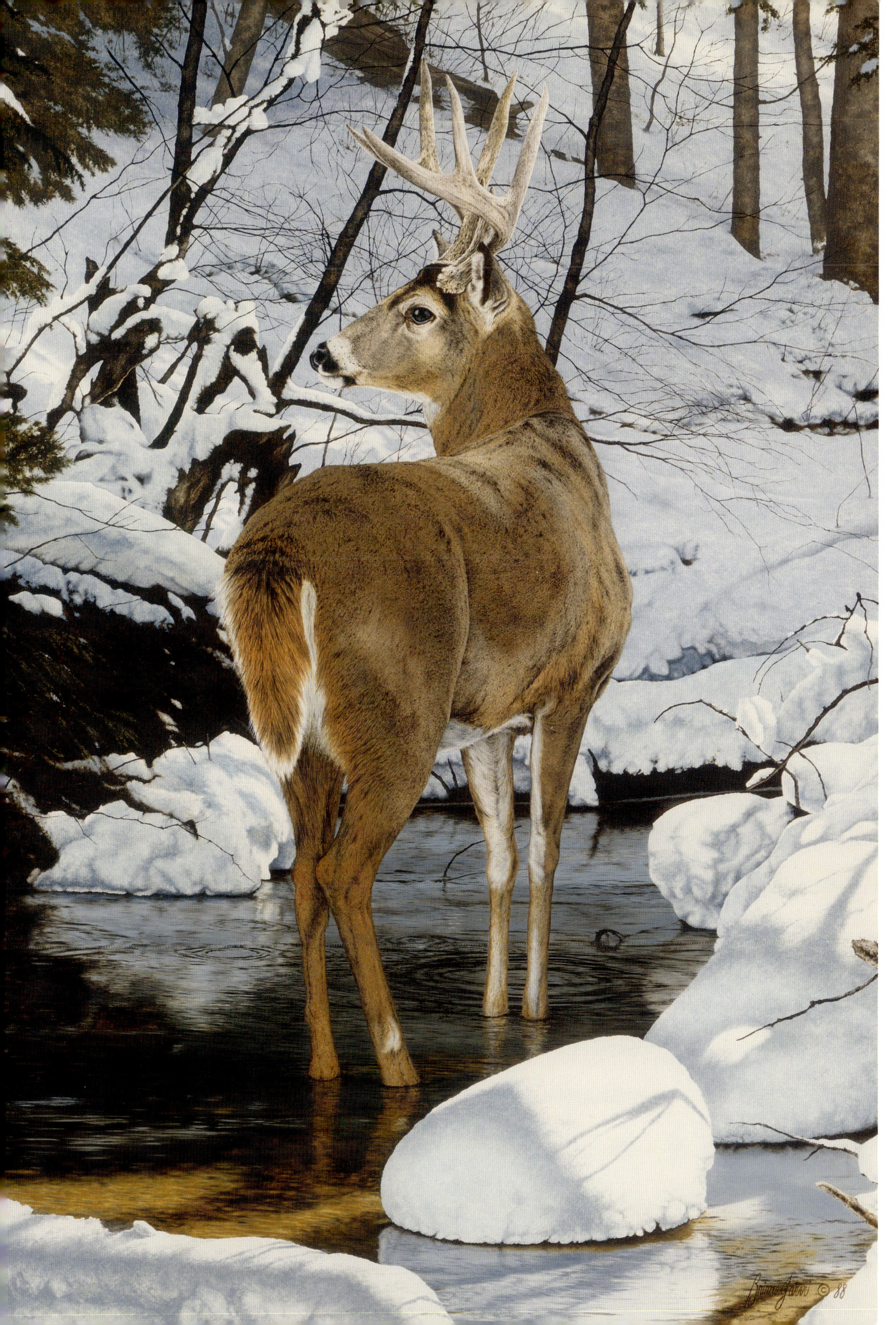

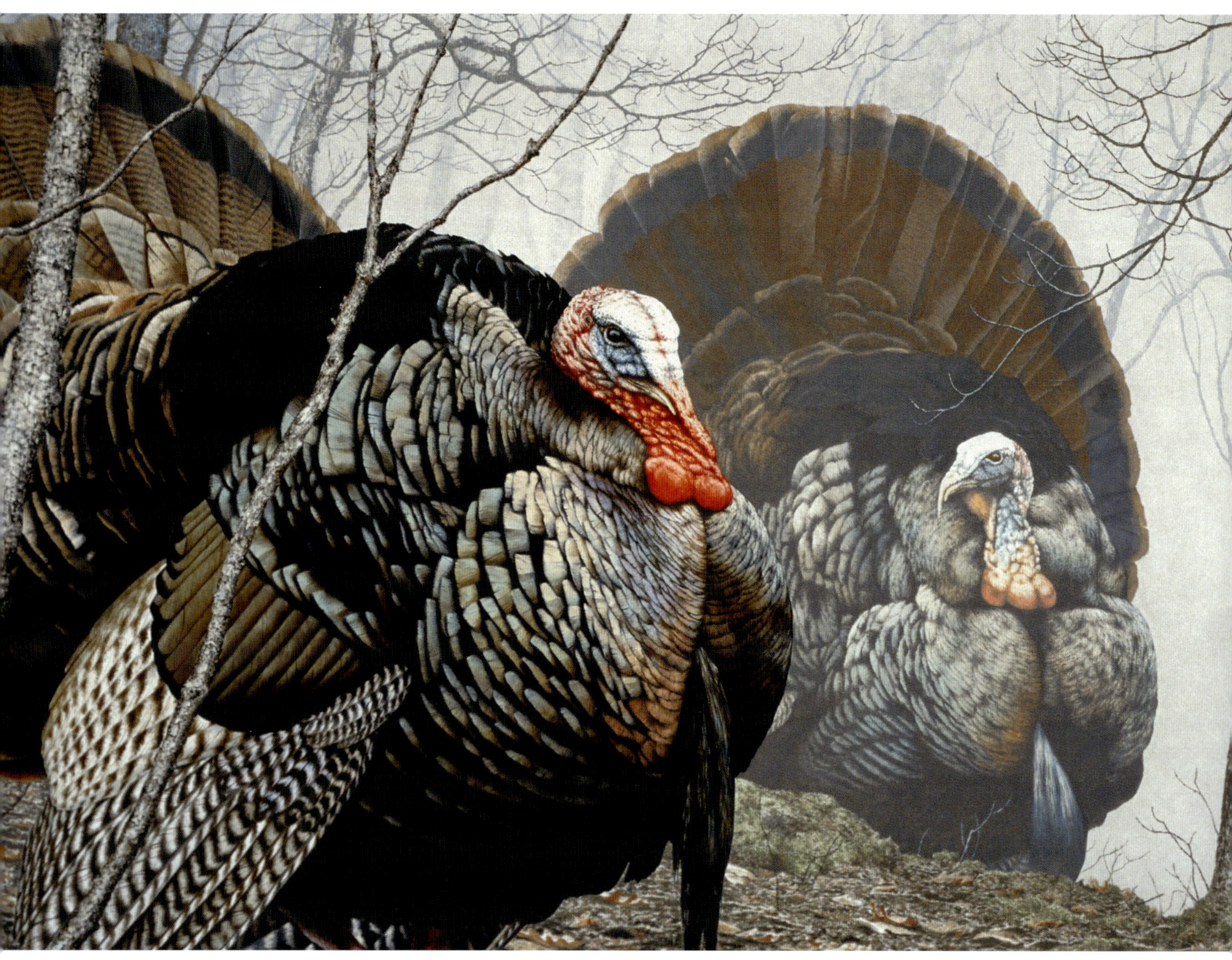

Mystic Warriors, 1990
Wild Turkey
Oil on Panel
18 x 24"
Private Collection

OVERLEAF
Quiet Water Canadas, 1986
Canada Goose
Oil on Panel
20 x 30"
Private Collection

To this day, Jarvi remains endlessly fascinated by the avian form, the personalities of individual birds, the diversity of plumages, and how their arrival and departure brings a rich pageant of seasonality to the forest and riparian areas.

"It was a totally different playground than what normal people in cities had and a simpler, nonpretentious life," Jarvi's brother Rick said. "Floodwood had no bowling alley or theater or recreation center. All the action was down at the river. The more you walked through the woods, the more you witnessed."

Rick, who vied with Jarvi and their brother Bruce to earn the title of top artist, is a passionate lifelong deer hunter and angler. "Brian would go out on the water and accompany us to the deer stand, but he wasn't interested in taking the animal. He was noticing everything happening around the hunt. He appreciates hunting and, in a way, his art makes us more aware of everything about the experience we might be missing or taking for granted."

The brothers displayed solid talent as portrait artists (visible today in Jarvi's numerous sketches of Masai tribal members in Africa). "In those early years, we were pretty competitive because that's how you really won Mom's attention," Bruce said. "I was always trying to match my drawings to the images I saw in print publications, and I was pretty decent. Brian started with drawing people and then expanded into animals. He was darned good, even when he was in grade school."

Jarvi would pour himself into drawing animals, amazing his brothers and sisters, but they all mentioned one quirk: he never completed any of them. He would be three-quarters of the way into making an impressive drawing and stop because he wasn't satisfied. "The piece would get tossed in the trash or set aside in a stack of paper and we would plead with him, 'Aren't you going to finish it? You've done the hard part. You're so close!' And he would just shake his head as if to say, 'No, not good enough,' and then move on to another," said Jarvi's sister Jean Melloch.

"When I think of Brian, the earliest indication that he had talent was when he was still practically a toddler," his sister Kim said. "He would fill pieces of paper with tiny circles that created these pictures and he would work on them obsessively. Looking back, I think it was an indication of his fine motor skills. When he set his mind to creating something only he could see, he went for it." Another sister, Candy MacRostie, was amazed at her brother's ability to render a scene at small scale with precise detail. "We knew Brian was the real thing when it came to art," she said. "I think he could paint anything if he put his mind to it."

Jarvi admired the cover art and illustrations that adorned outdoor magazines in the Floodwood barber shop and drugstore. He remembered savoring the predicament scene paintings of Bob Kuhn and Lynn Bogue Hunt in magazines such as *Sports Afield*, *Outdoor Life*, and *Field & Stream*. Unfathomable in his own mind was any notion of becoming a professional artist. The thought of going away to college to get a degree in fine art? And then doing what? Those were reveries that did not inhabit his realm of possibility. And yet his mother constantly encouraged him to keep drawing.

Following in the footsteps of his dad and brothers, Jarvi landed a grunt job with the railroad during his teenage years. The Jarvi boys were tough and tall, known for their work ethic. It was easy for them to get put on a crew maintaining tracks, both mainlines and the smaller stems that circuited the Iron Range and stretched westward toward the Dakotas and southward into the constellation of farming communities throughout the Upper Midwest. "Not working for the railroad? I never gave it much of a second thought," said Jarvi. "It was what teenagers from Floodwood did. You could earn fairly decent spending money on a railroad crew."

There weren't any profound Hamlet-like "to be or not to be" moments of contemplation. It's a script of necessity that generations of American families know well: Work for the railroad, mining company, factory, construction outfit, or marine shipping company. Show up for work, don't bellyache, put in the hours, receive the paycheck, and repeat until retirement.

Jarvi was a manual laborer who heaved a lot of backbreaking weight. He pounded railroad spikes and replaced broken ties. He and his crew members re-contoured track lines that were washed out by rain and dropped steel in place. In summer, on infernally hot, humid days, they would be quartered in rooms at night. They headed to the bars, where air conditioners would cool them off. All along, he drew pictures as a way to unwind and relax.

On trips home to Floodwood, Jarvi took notice of a woman he had known in school, a sweetheart named Raelene who had blossomed into a beauty. They had been in the same circle of friends. According to Raelene, Jarvi did not possess profound social skills in asking a woman out on a date or to dance, but she did have a crush on him. They started spending time together, fell in love, and got married. "She is the best thing that ever happened to me," Jarvi said. "I'm lucky she has put up with me all these years."

Raelene played a critical role in encouraging him to return to his art. She saw how it stilled the unrest that welled up with a job that gave him little satisfaction. The couple lived in a tiny house in Floodwood, and Jarvi carved out a small nook in their home. There he drew and painted when he could steal away the time, producing figurative pieces of

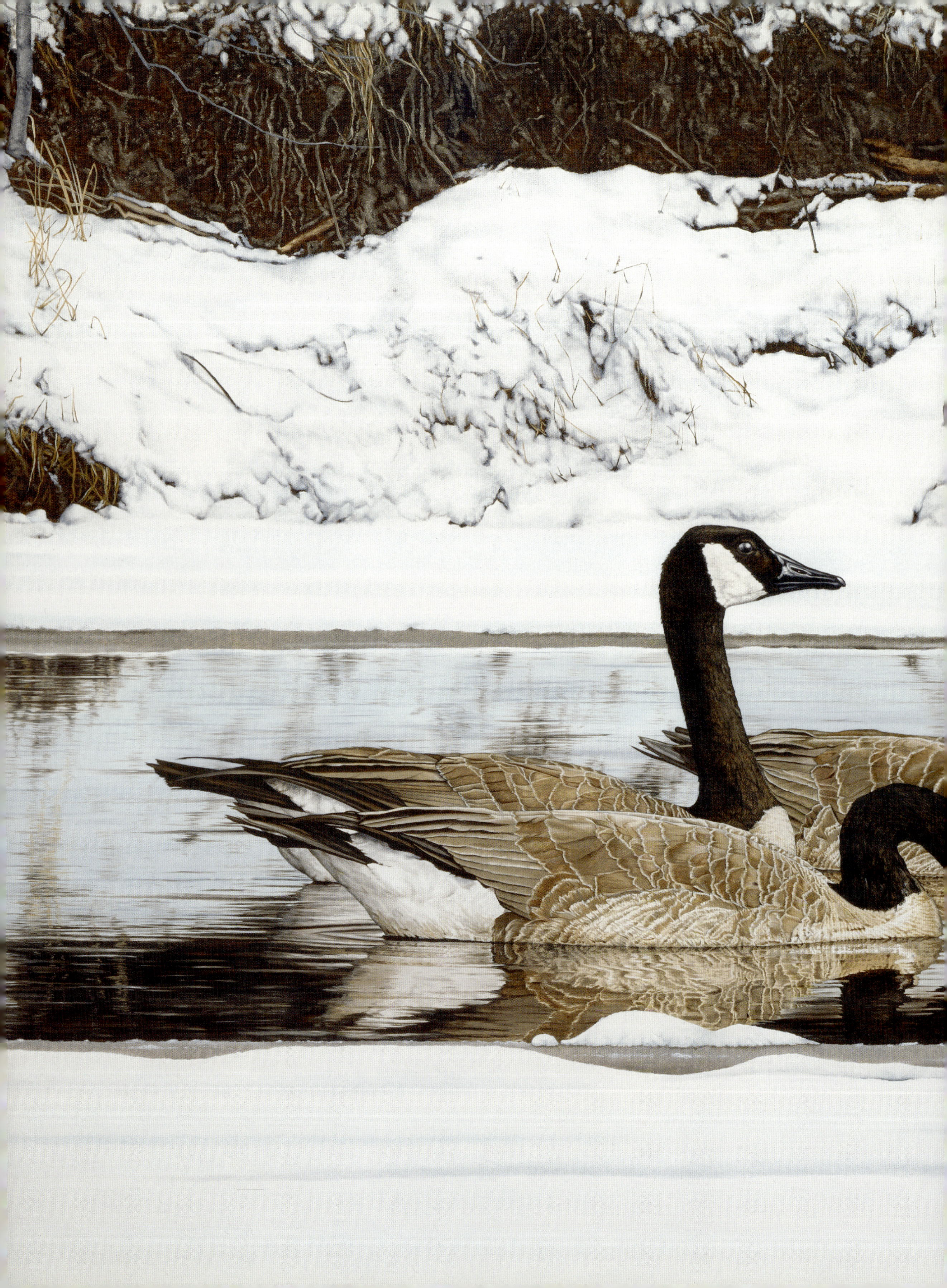

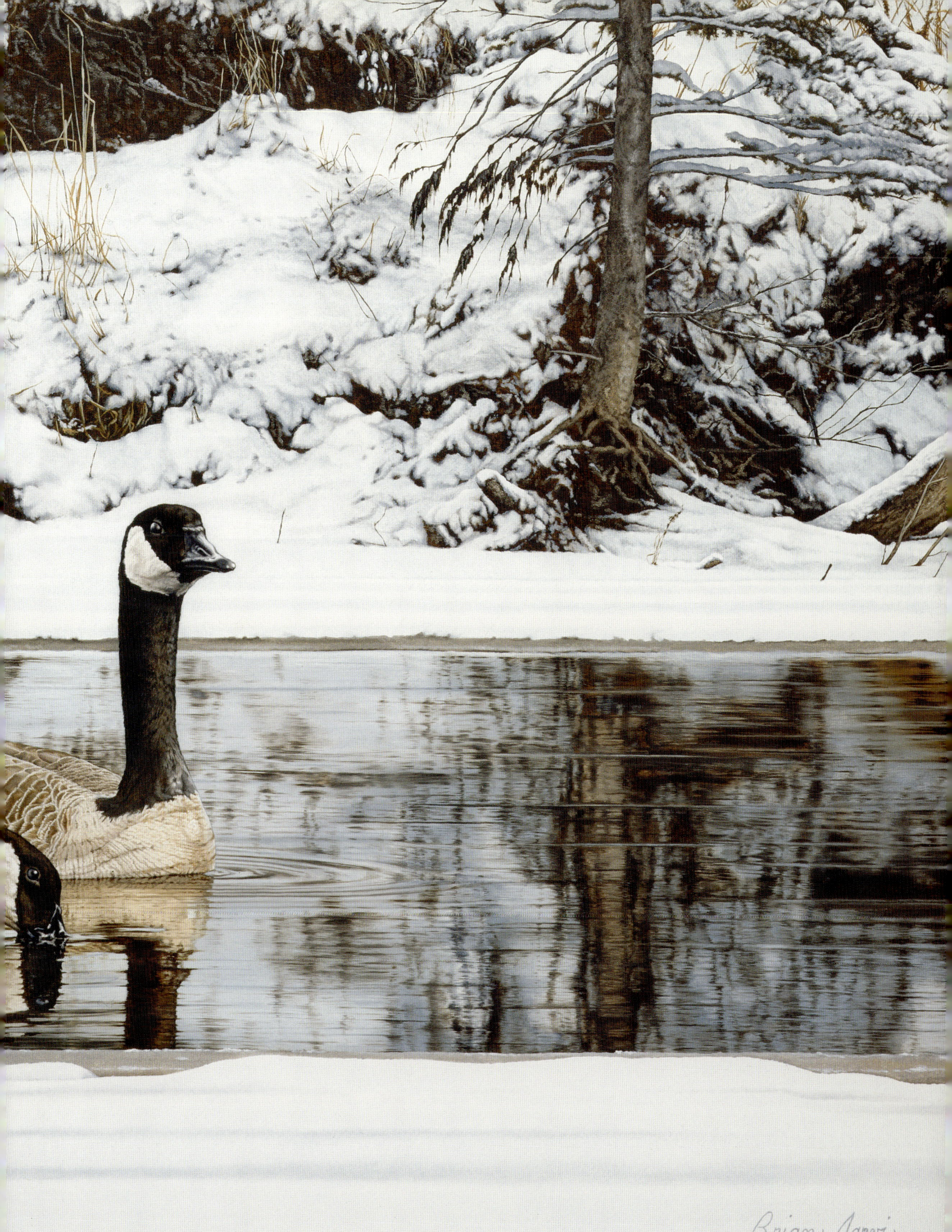
Brian Jarvi ©
86

both people and animals. A few collectors took notice, sometimes aided by Raelene bringing the pieces to their attention. Jarvi was shocked that buyers wanted to hang the pieces in their homes and pay money for them. But the income wasn't much—too little to even ponder the radical action of quitting his day job.

BECOMING A DUCK STAMP ARTIST

The 1970s and 1980s—with its emergence of limited-edition art prints and the presence of a waterfowl protection group called Ducks Unlimited—put "wildlife art" in vogue among the masses that couldn't afford to buy original paintings. Framed portrayals of flying ducks, fish, and white-tailed deer were hung decoratively in the dens of homes and weekend cabins. It was art made for working-class folk and a way for such collectors to express their love of nature.

Wildlife art was also used as a tool to advance the cause of conservation. Dollars generated through the sale of hunting and fishing licenses were used to protect vast sweeps of habitat and pay for professional wildlife management. At Ducks Unlimited fund-raising banquets, originals and lithographs were auctioned off to raise money for wetlands protection. At the same time, Minnesota arose as a mecca for sporting art, not only being home to collectors but also producing a group of perennial favorites in the Federal Duck Stamp Contest. Such artists were folk heroes and rock stars.

Raelene remembered her husband getting this outlandish notion in his head that he would submit an entry. So lucrative was winning the Federal Duck Stamp Contest—given that artists could sell lithographic reproductions of their work—that generally it was thought of as a million-dollar payday. "Brian was in his mid-20s," Jay Autio, his best friend, remembered. "He felt the clock was ticking. He said, 'If I don't give it my all with art now, it won't happen.'"

Next to the federal competition, the Minnesota Migratory Waterfowl Stamp Contest ranked up there with national prestige. In fact, there were a number of artists who were hailed for having their works on both of these stamps. Competition was fierce. The sense was that no matter what else a person did, being a duck stamp contest winner would allow one to die with satisfaction. Jarvi believed it was true. In 1982, two years after beginning his quest, he nervously entered his third Minnesota duck stamp contest. He came in fourth with a piece so realistic that one of the judges reportedly claimed it was somehow copied from a photograph. It was so exquisite, so finely rendered, that the judge believed it wasn't possible Jarvi had produced it.

Devastated when he learned of the judge's unfounded claims, he thought of quitting. Jarvi was just beginning to attend art shows to not only assess the competition but also sell pieces. "He had no name recognition," Autio said. "He probably heard a few people say behind his back that he'd never succeed. That's not the stuff you say to Finlanders because they will devote the rest of their lives to proving you wrong."

Jarvi vowed to vindicate himself by entering one last duck stamp contest. He spent hundreds of hours making sure he got every single detail right for a painting of lesser scaups. What he needed most was photographic reference material, but he didn't want to use someone else's pictures. He had to witness a bird with his own eyes, capture it on film, and then use it as a tool to help him paint as close to the truth of reality as possible.

Missing, however, was a reliable venue for studying his subjects. He found one in the outskirts of Floodwood in the form of a cattail-encircled holding pond where flows were channeled from the local water-treatment plant. The water, slightly warmer, became an attractive pit-stop area for ducks and geese flying north in the spring prior to when ice-out reaches the lakes. Hundreds, sometimes thousands, of ducks and geese congregated there. The water was bone chilling and the air even more frigid.

Jarvi knew the ducks wouldn't let him get close, so he devised a special blind, modeled ingeniously after the look of a muskrat house. He pushed it through the muck toward the feeding areas. It was so convincing that while positioned inside his blind he was confronted by real muskrats looking to adopt it as their abode. He had to shoo them away.

Encased in multiple layers of long underwear and rubber chest waders, he set out in the darkness before dawn morning after morning and stood alone, camera in hand. On many days, he turned so hypothermic he could barely move, having to stand in a hot shower to help bring the feeling back to his limbs.

Autio remembered Jarvi's determination. "He has always been a perfectionist," he said. "We were on the high school basketball team together and after practice was over he would insist that we stay behind in the gym and keep running plays until we got them right. When we learned that Brian

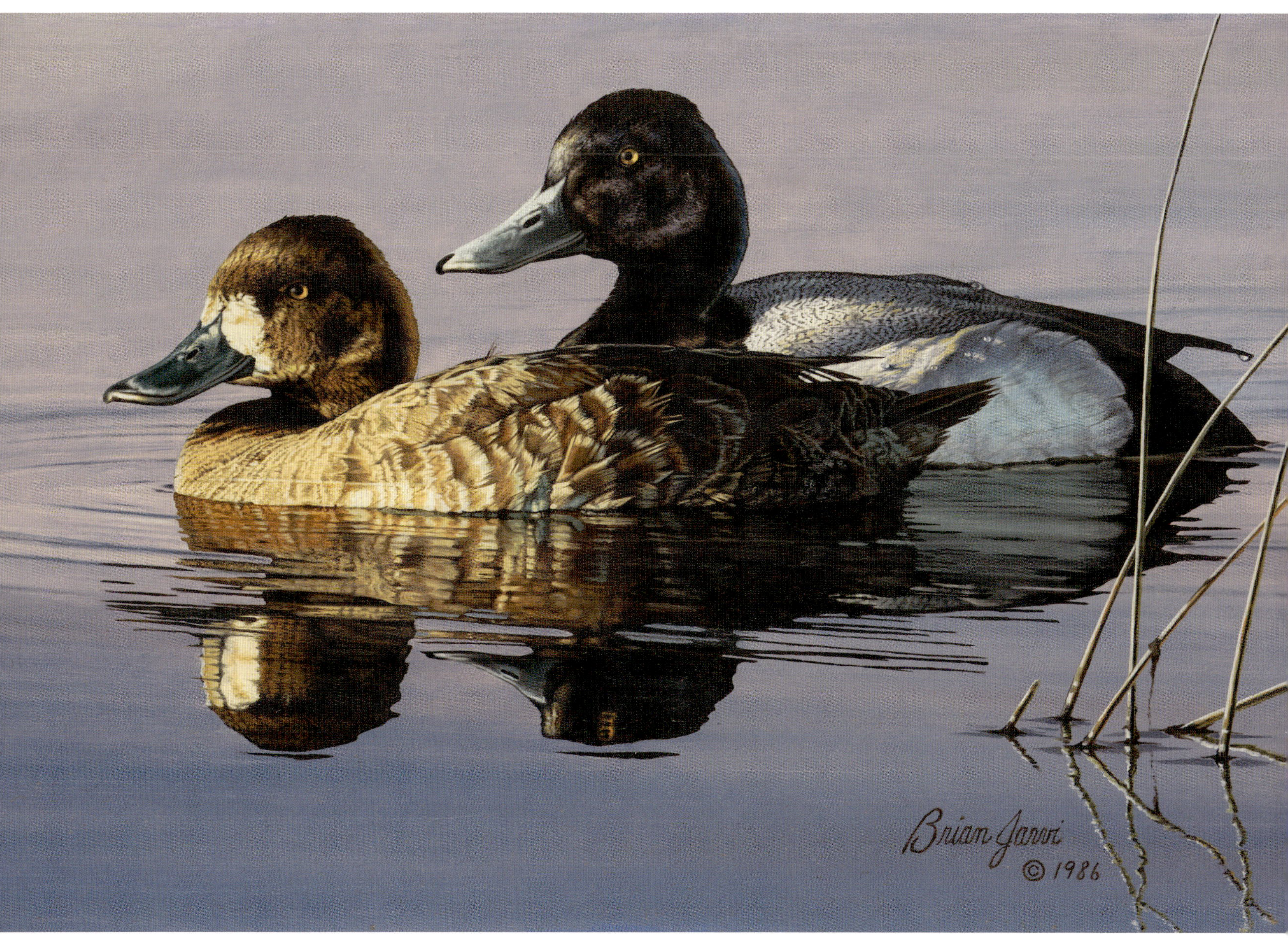

was spending his mornings in the sewage lagoon watching ducks, we knew he had a good reason for being out there, as unpleasant as the thought was."

In their Scandihoovian noir film *Fargo*, Minnesota-born directors Joel and Ethan Coen delivered a lampooning nod to wildlife art in their home state. The movie's heroine, Marge Gunderson, is married to an aspiring duck stamp artist battling against the vaunted Hautman Brothers. The brothers exist in real life, and there has never been a band of siblings that has so dominated the Federal Duck Stamp Contest. They triumphed more than a dozen times. But Jim, Bob, and Joe Hautman, who are friends of Jarvi, weren't alone.

1986 Minnesota Waterfowl Habitat Stamp, 1985
Lesser Scaup
Oil on Panel
6½ x 9"
Collection of Brian and Raelene Jarvi

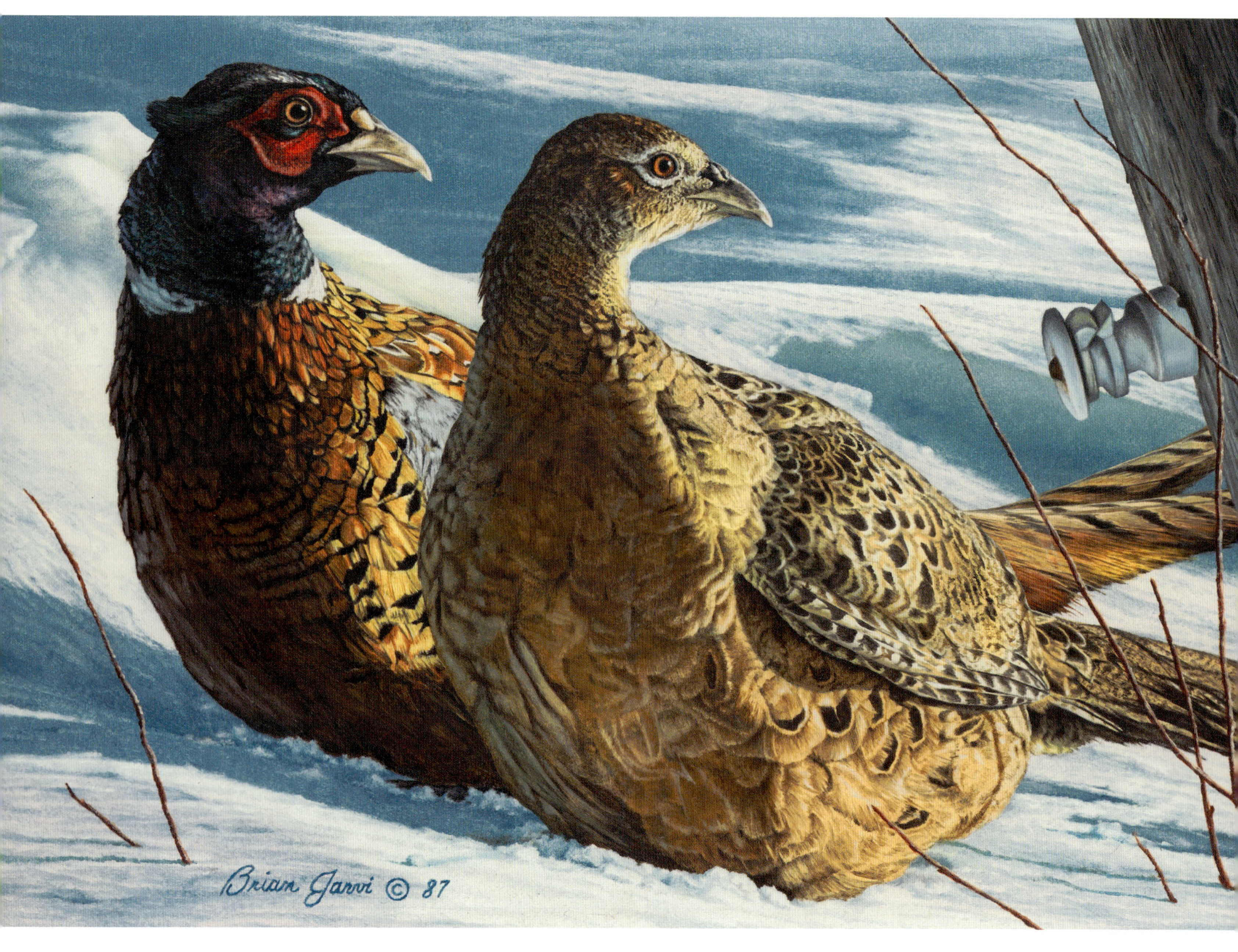

1987 Minnesota Pheasant Habitat Stamp, 1986
Pheasant
Oil on Panel
6½ x 9"
Collection of Brian and Raelene Jarvi

Minnesota is known for having produced a high concentration of duck stamp champions and easel painters distinguished for their (sometimes kitschy) pictures of white-tailed deer, moose, wolves, and fish. A giant among the lot, Francis Lee Jaques grew up in the bogs around Aitkin, where he hunted, fished, and earned spending money by selling the furs of muskrats, mink, and foxes. Then fate intervened. He got drafted in 1918, was sent to France, and, while there, bided his time drawing and making watercolors. After coming home, he met Impressionist Clarence Rosenkranz in Duluth. Soon after, he gained confidence to give painting a try. He met a woman, a passionate conservationist writer, and married her. She encouraged him to submit his work to the vaunted American Museum of Natural History in New York City. The museum hired him to paint some momentous diorama backgrounds that today are considered American classics. That 25-year stint led to Jaques being invited to paint some acclaimed dioramas at the James Ford Bell Museum of Natural History in St. Paul. Together, Jaques and his wife worked alongside the great conservationist Sigurd Olson in getting protection for the Boundary Waters Canoe Area Wilderness—think of it as Minnesota's roadless equivalent of Yellowstone National Park.

Of course, Jarvi was familiar with Jaques, who represented an untouchable enigma. There must be something in the water of the northern lakes region that fuels a fearless stubbornness, a belief that anything is possible for those who are willing to try to defy others who sell them short.

Wracked with self-doubt following his previous attempt, Jarvi agonized that his portrayal of a drake and hen lesser scaup for the state duck stamp contest wouldn't measure up to the discriminating whims of the judges. He toiled on it nightly for weeks on end. Throughout history, such a search for perfection has been a crazy maker for many artists. When the day of open judging arrived in the Twin Cities, Jarvi knew he had given it his best shot. His sister Kim, who was also there in the room, described the scene: "They started with half a dozen finalists and worked their way down from fifth place to fourth place until the top winner was announced. Finally, it came down to two artists. When Brian's name wasn't called as a runner-up, he lowered his head, realizing what had happened." Jarvi isn't the kind of person to spike the ball in the end zone or slide across the ice playing air guitar after scoring the winning goal. After he won, he almost wanted to disappear as a group of well-wishers swarmed him. He would rather stand quietly at the side of a room and observe people looking at his art.

To an outsider, winning a state duck stamp contest might seem insignificant, even fleeting, but it represented a breakthrough for Jarvi's self-confidence. In fact, his devotion to exacting detail gave rise to a term that some duck stamp artists, including the Hautman Brothers, use when they want to punch up a painting area to command more attention. It's called "Jarvi-cizing," and it's a compliment. The win also cemented the foundation for him to complete a project a quarter-century later that many believed was too impossibly complex.

The next year, 1986, thousands of Minnesota waterfowl hunters carried a reproduction of Jarvi's portrayal of lesser scaup on their license. Jarvi didn't become rich on the sale of associated collectible art prints, but it did enable him to quit the railroad job and try to make a go of it.

Jean Melloch, the oldest of the Jarvi siblings, remembered how eyebrows were raised when Jarvi announced he was going to support his family by painting. "It was a huge leap of faith, gutsy, when he gave up the railroad job," she said. "Our dad was still alive when Brian started to achieve recognition with the stamps, so he got to see the beginning of Brian's climb and he was proud." (Ralph Jarvi died in 1999 and Virginia Jarvi in 2012.)

Painter Bruce Miller, who commanded esteem for winning the Federal Duck Stamp Contest in 1993, has known Jarvi since the early 1980s. He remembered the months after Jarvi placed fourth in the 1982 state contest and then had his entry published in the Minnesota Department of Natural Resources magazine, the *Minnesota Conservation Volunteer*. "I took a look at Brian's portrayal of a northern shoveler and thought it was a photograph. Then I looked closer and realized it was one of the most phenomenal paintings I had seen in my life. Little nuanced details like the bill of the drake being slightly open but hyperrealistic—it was unbelievable," Miller said.

When Jarvi triumphed a few years later, Miller said it wasn't a surprise. In fact, he was one of the judges. "I got to know him in short order after that," he said. "He is one of the funniest guys. He and his cronies from Floodwood are real characters from the Northwoods. There have been times when Brian has had to interact with some of the most refined, effete people in the world, yet he's never forgotten where he came from. That speaks to the fact that he's never lost himself in his fame."

As the range of Jarvi's portfolio grew, so did his reputation. He submitted the winning entry for the 1987 Minnesota state pheasant stamp and is the only artist ever to win the state waterfowl and pheasant stamp contests back to back. In the two decades that followed, he racked up annual accolades as artist of the year for conservation organizations that cumulatively had millions of members. At art shows, his individual works often earned merit awards from judges and fellow painters.

Jarvi was on a roll. He added large mammals to his body of work—white-tailed deer, timber wolves, bears, and game species of the West such as elk. During his inaugural showing at Safari Club International, the response was instantaneous. But success in a single year does not guarantee longevity.

Bruce Ogle met Jarvi in 1987 at a banquet for the Ruffed Grouse Society in Grand Rapids, where the artist had donated a painting. "I bid on it but didn't get the piece because there were others with a lot more money who wanted it more," Ogle explained. "I was told not to worry because Brian donated a painting to the local banquet every year." It took five years before Ogle was able to land a piece, a Jarvi portrayal of a drumming grouse. "That was my first Jarvi and still is my favorite," he said. In 1994, he had an epiphany when he realized how rapidly Jarvi's stature had grown. Ogle attended the Safari Club International convention in Reno, Nevada, and saw how Jarvi's work compared to other living painters. "Wow, that was the eye-opener for me," he said. "His works were so much more realistic and showed more detail than others'. Plus, there were people at his booth at all times seeking him out. By the end of the show, he had sold all his paintings."

FIRST TRIP TO AFRICA

In 1989, Jarvi made his first trip to Africa. He was invited by conservationist, actor, and entrepreneur Christopher Law and his friend (and artist) Dennis Curry. Along with Jarvi and Curry, the group comprised Daniel Smith—whose portrayal of a lesser snow goose in 1988 won the Federal Duck Stamp Contest—Gary Moss, Al Agnew, and Rick Kelley. The idea was that the sixsome would visit several national parks in Kenya and Tanzania and then, upon their return, produce high-quality limited-edition prints to raise money for conservation efforts. "I know this sounds like a cliché, but Brian drank it all in," said Law. "He got hooked on Africa in a way the others did not. We could easily have had six artists painting six elephants, but Brian went his own way."

Jarvi told his guide he wanted to encounter Africa's rawest edge; he was hell-bent on not doing something predictable. At Ngorongoro Crater, they found a waterhole encircled by numerous tracks of predators and prey that converged in the mud. "I had a feeling something might happen," Jarvi remembered. As they waited, the bush turned still, and then out came a crashing Cape buffalo followed

The Lioness, 2001
Lion
Oil on Belgian Linen
16 x 12"
Private Collection

OVERLEAF
African Requiem, 1990
Lion and Cape Buffalo
Original Lithograph
17 x 24"
Private Collection

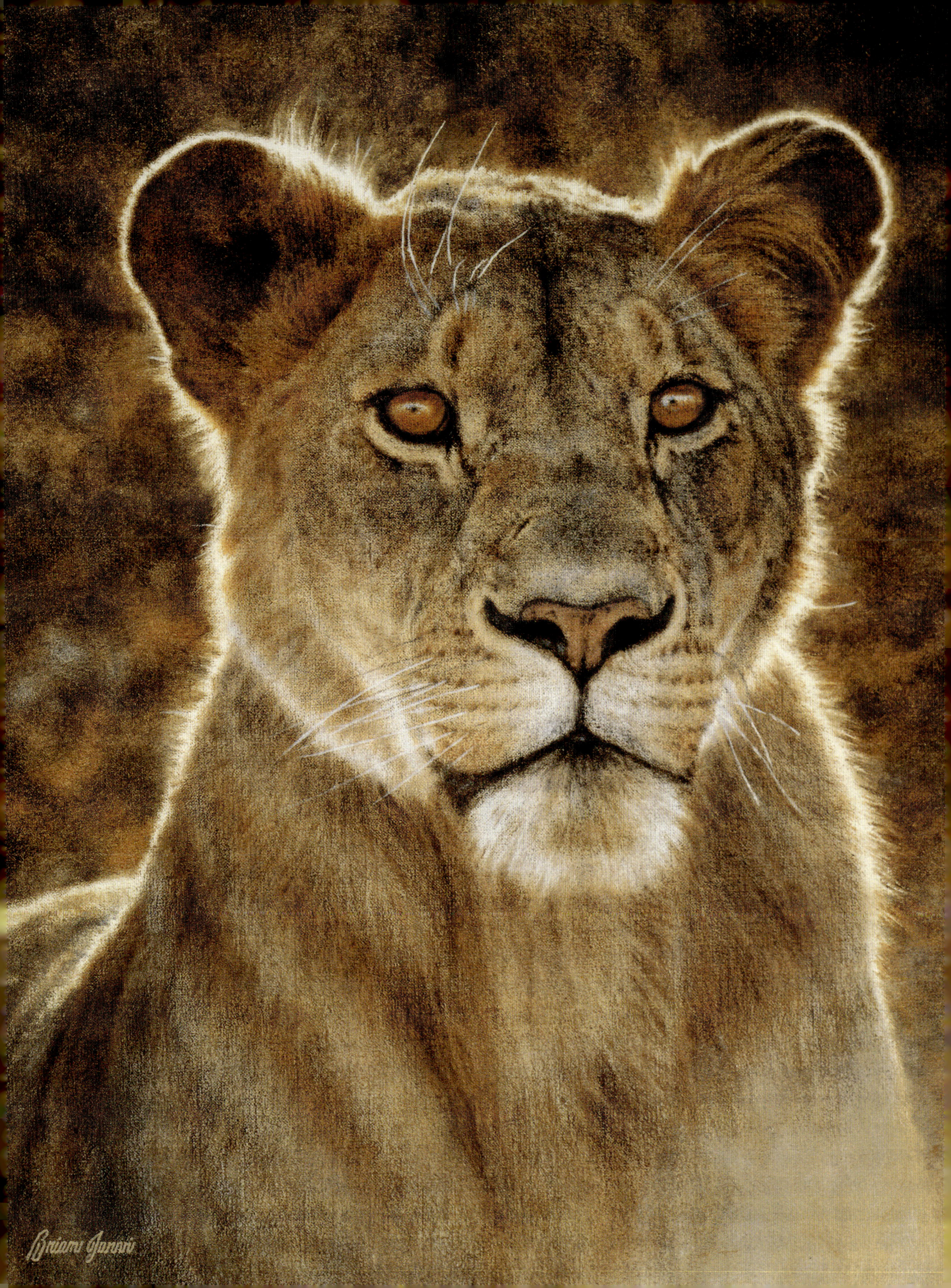
Brian Jarvi

by two lions. The body-to-body encounter, involving grunts and roars, curved horns to claws and canid fangs, resulted in the big cats taking the buffalo down. "It was the most exciting wild experience I had ever had. This wasn't some nature documentary on TV; it was real," said Jarvi.

An original lithograph emerged, *African Requiem*, which was not only brilliantly executed and marked an auspicious debut of Jarvi the painter of African wildlife, but also spoke to the artist's work ethic. "Even after he saw the kill, he didn't quit. To make sure he understood the habitat, he went back to Ngorongoro for a week," Law said.

"I know how difficult it can be in general to make a living as an artist, but when Brian got back from Africa he knew what he was going to do," his friend Jay Autio said. "I would ask him, 'Who in the hell is going to buy African art in North America?' And his reply was, 'It isn't about the subject; it's about how well you paint it. If the art is good enough, people will be drawn to it and put it on their wall.' He was right. His art makes a strong impression."

After the Africa trip, Jarvi worked with Curry, a master printmaker, at his studio in Cambria, California, to design the limited-edition print. "Brian rendered a male and female lion after they had taken down a Cape buffalo," Curry said. "His painting style was very much like drawing, the way he worked with values. He is a master of predator and prey scenes. Looking at what Brian did with that experience made the whole project seem worthwhile." As a result of the artist's involvement, more than $150,000 was donated to grassroots conservation efforts being carried out by friends of Dr. Richard Leakey and Jorie Butler Kent of the luxury safari company Abercrombie and Kent.

Within a few years, Jarvi had gallery representation in London. Saudi princes were buying his works for six figures. "I think the way he emerged, a lot of people were asking, 'Who is this Jarvi guy?'" Law said.

Jarvi has a unique painting style that is immediately recognizable. "Brian is a master of illusion," said Daniel Smith, one of North America's premier wildlife artists and a good friend of Jarvi. "He knows how much information to include and what to edit out. It's a delicate dance that allows the viewer's eye to fill in the blanks. His work is mesmerizing and full of spontaneous life. I'm not aware of any other wildlife art that has the focus and dedication to African subjects that Brian has."

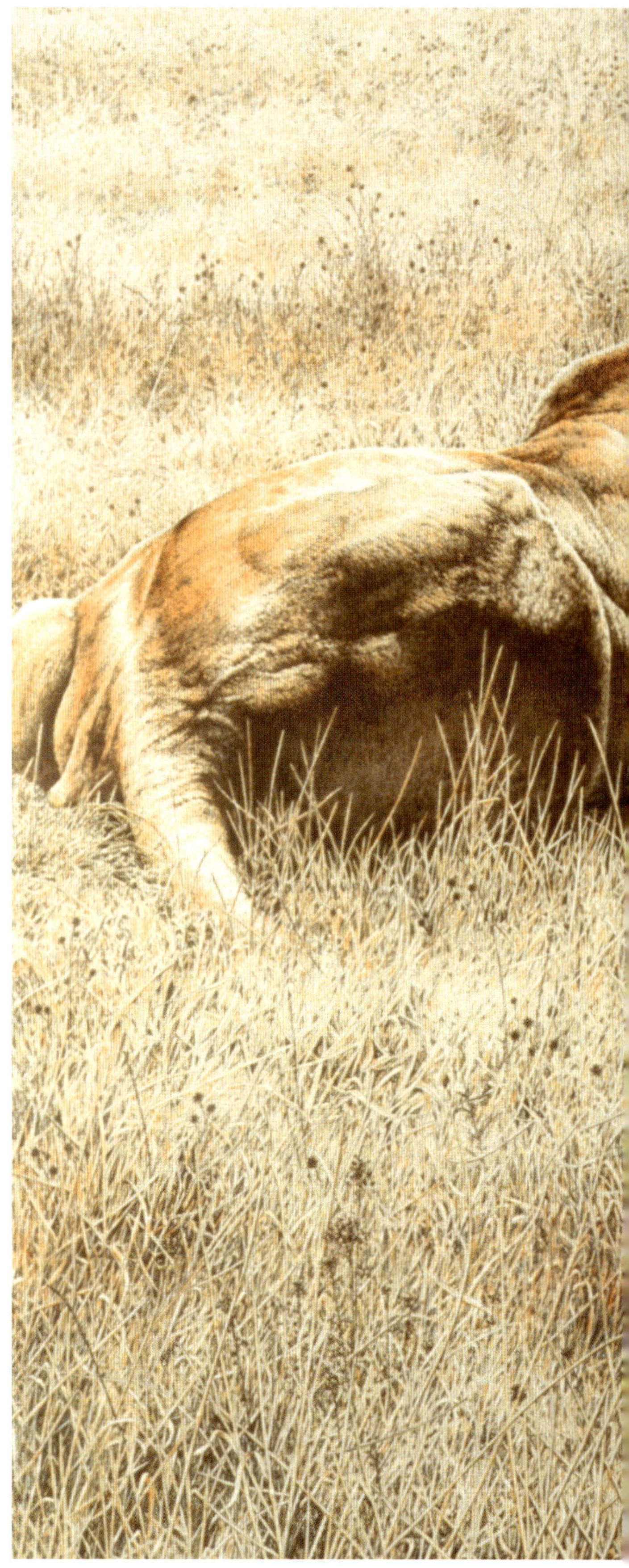

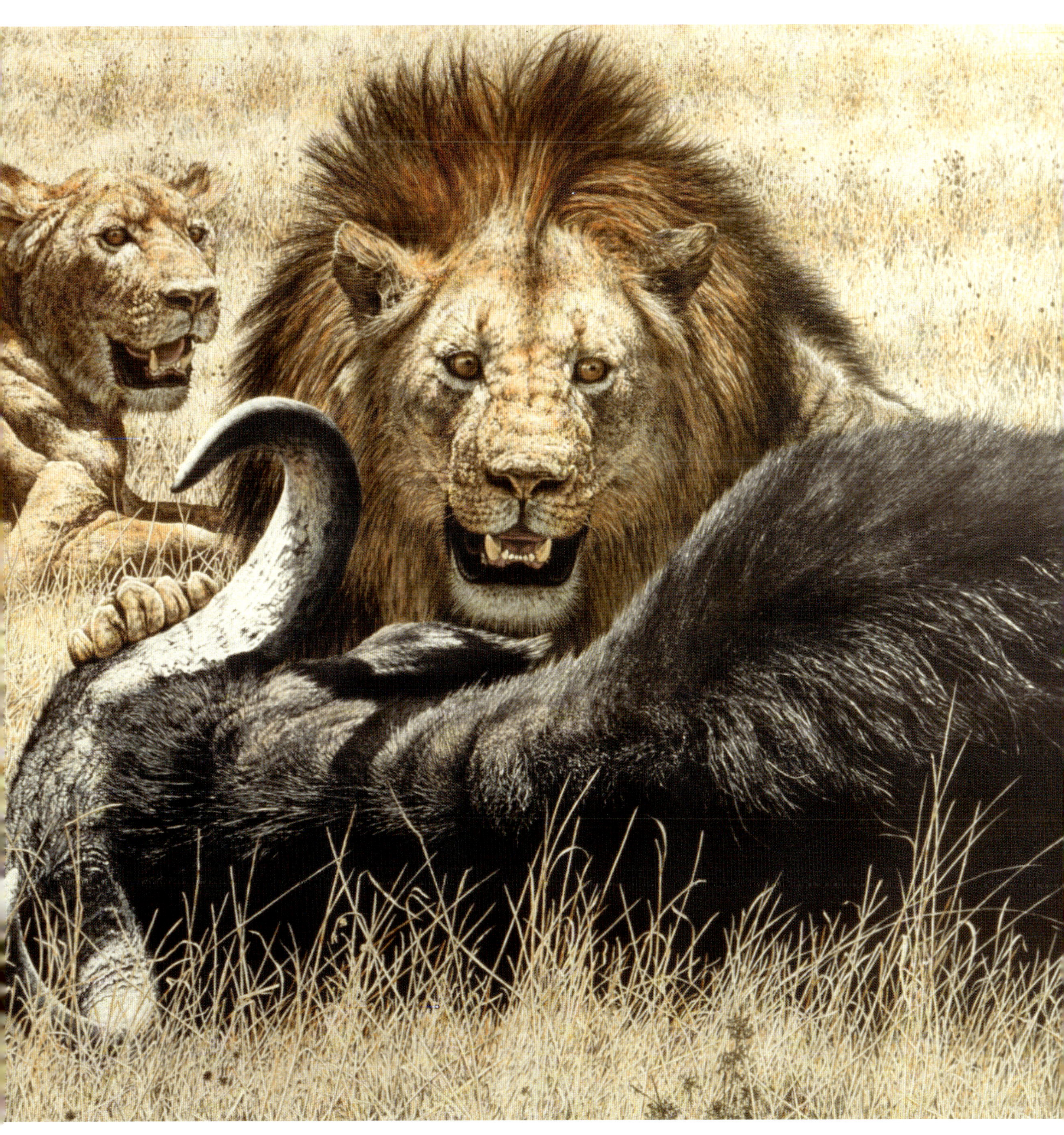

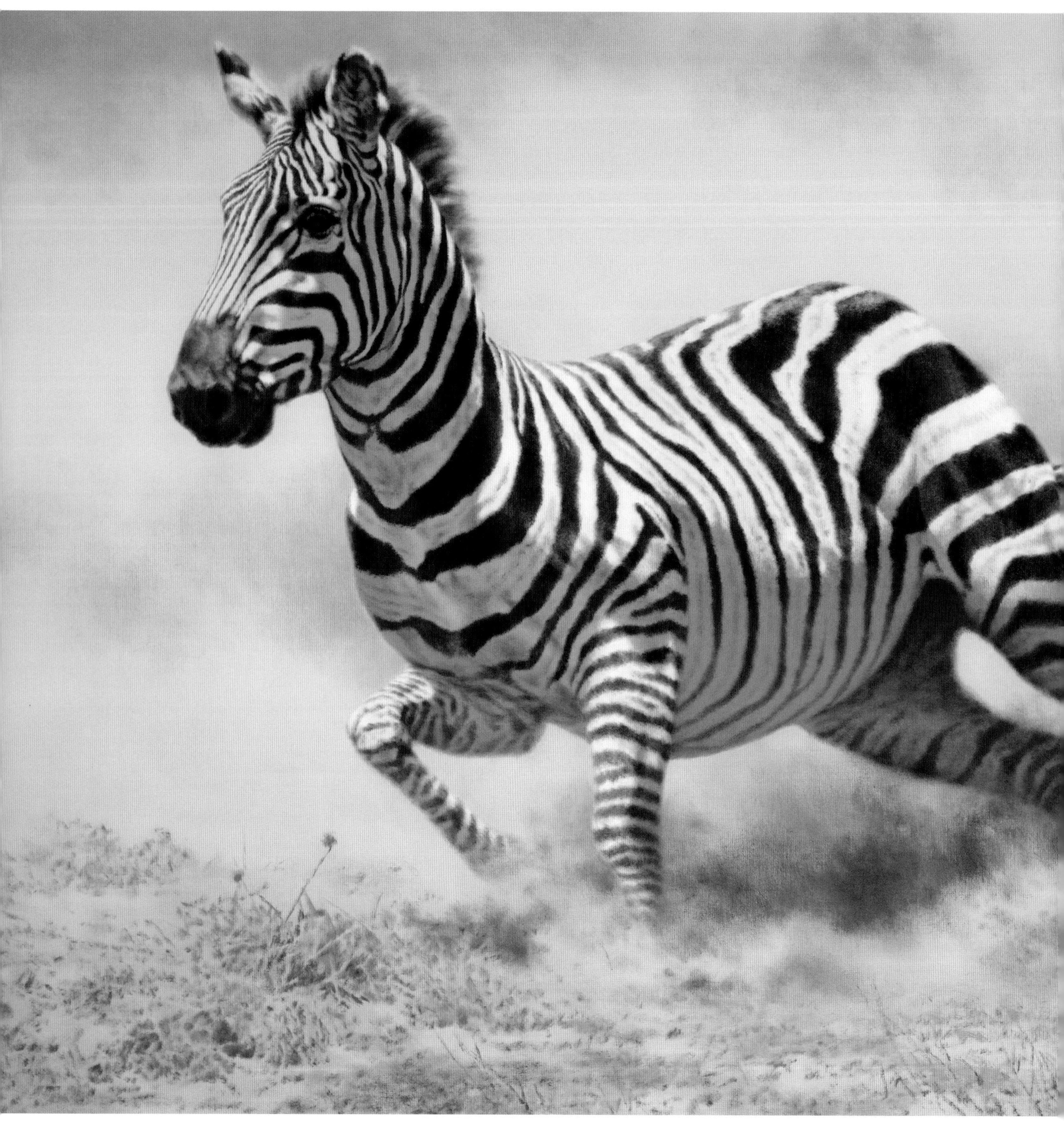

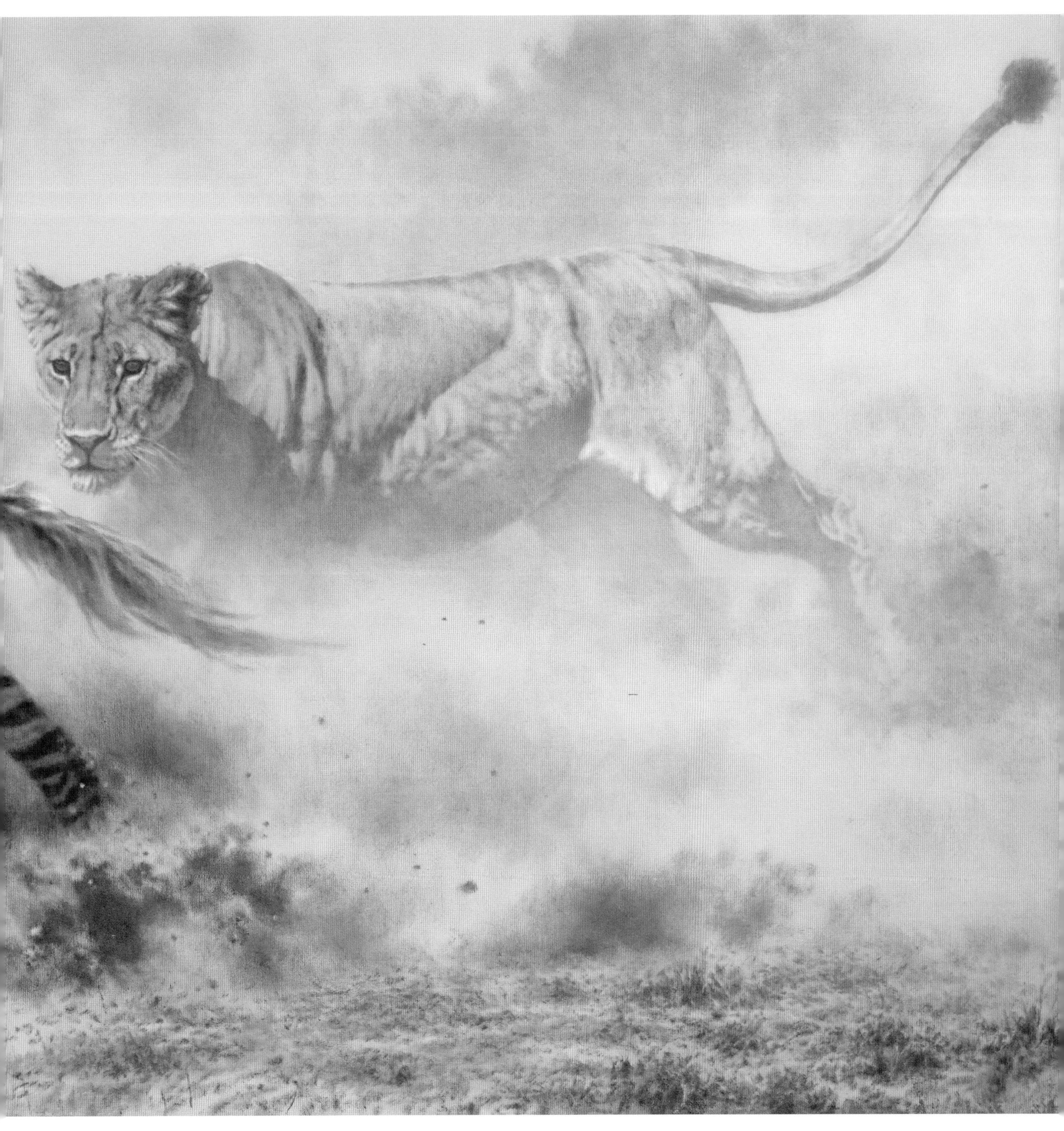

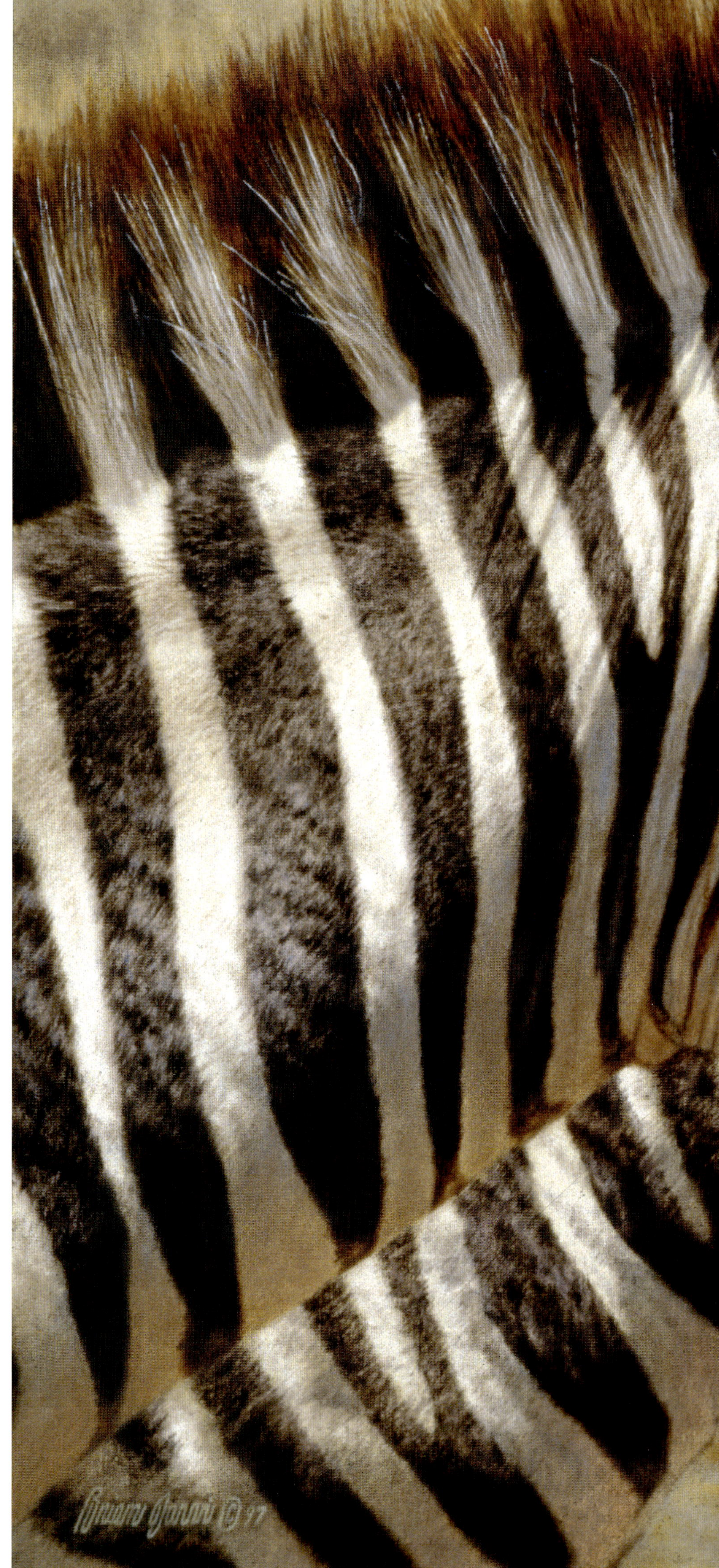

PREVIOUS SPREAD

Razor's Edge, 2006
Lion and Zebra
Oil on Belgian Linen
48 x 96"
Private Collection

Under an African Sun, 1997
Zebra
Oil on Belgian Linen
15 x 18"
Private Collection

OVERLEAF

The Savage Land, 2010
Lion and Cape Buffalo
Oil on Belgian Linen
36 x 72"
Private Collection

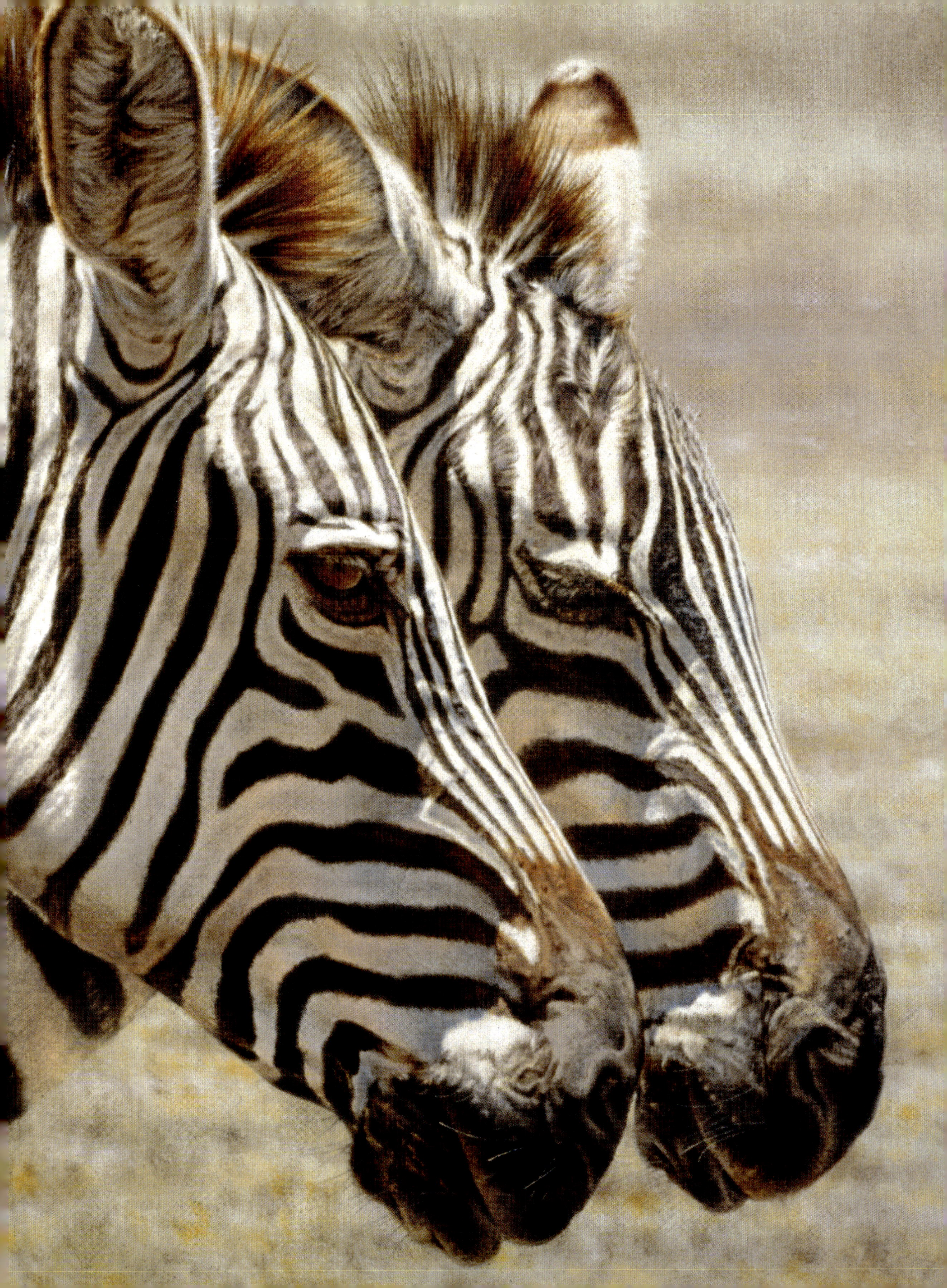

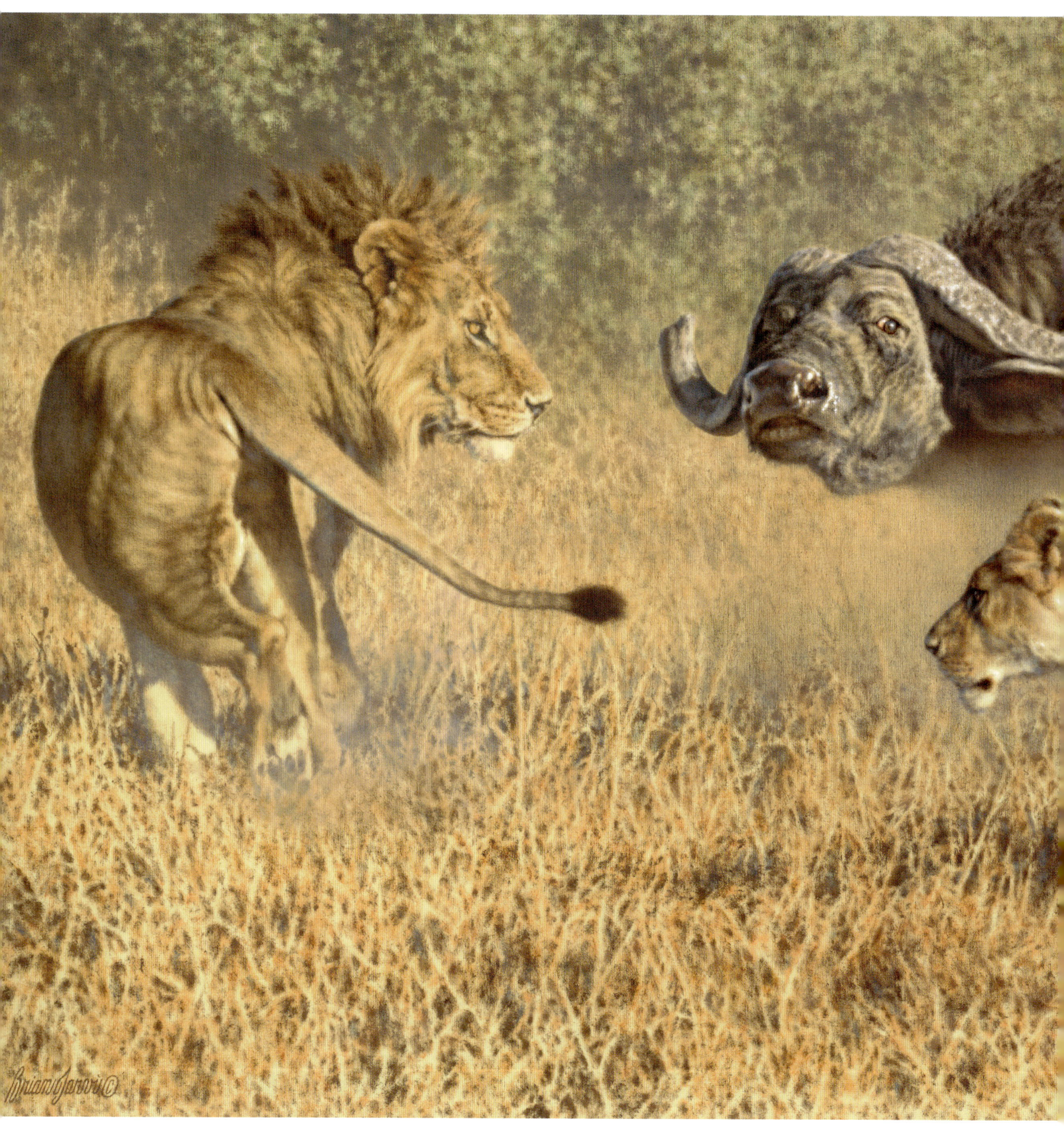
Brian Jarvis©

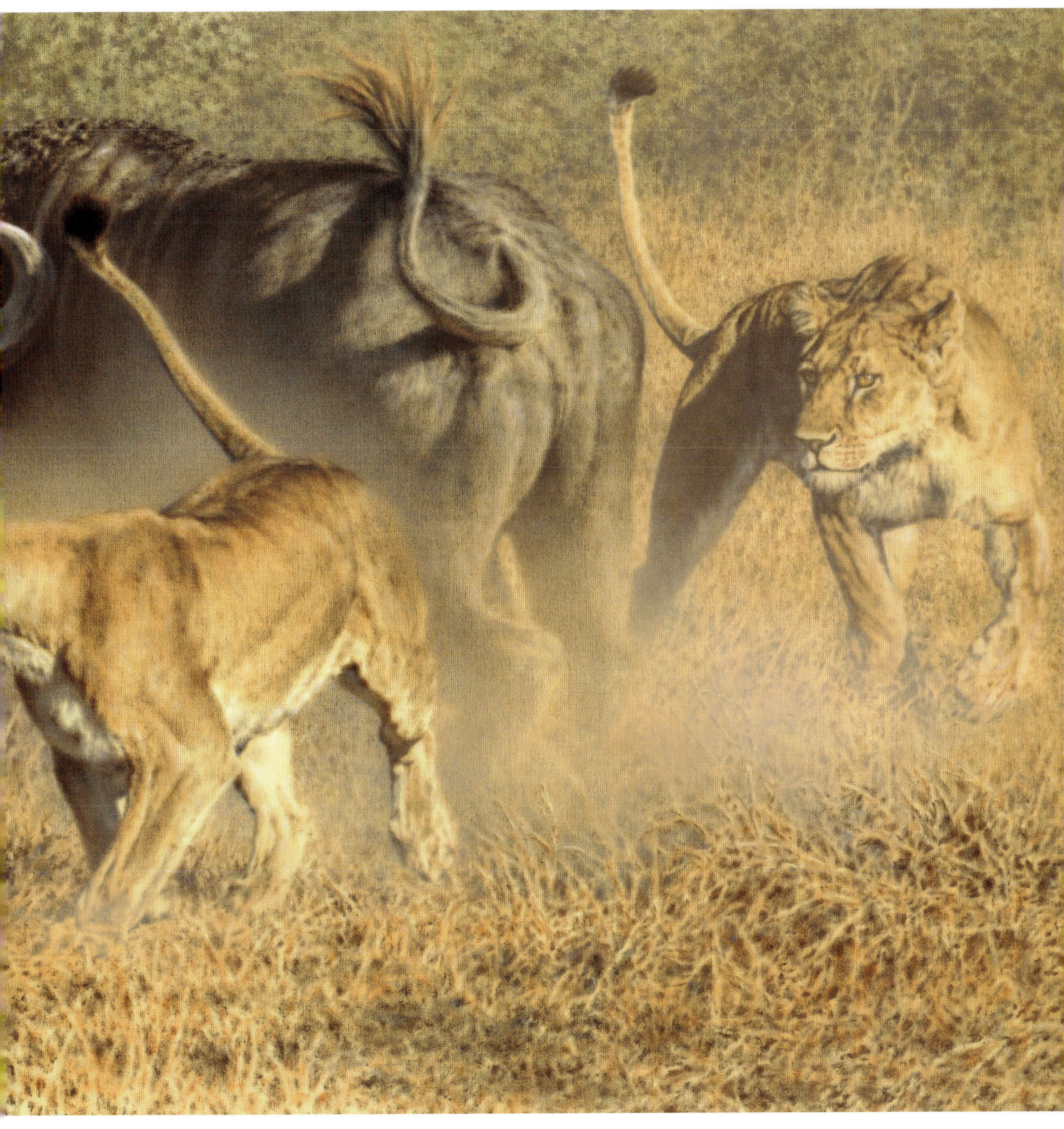

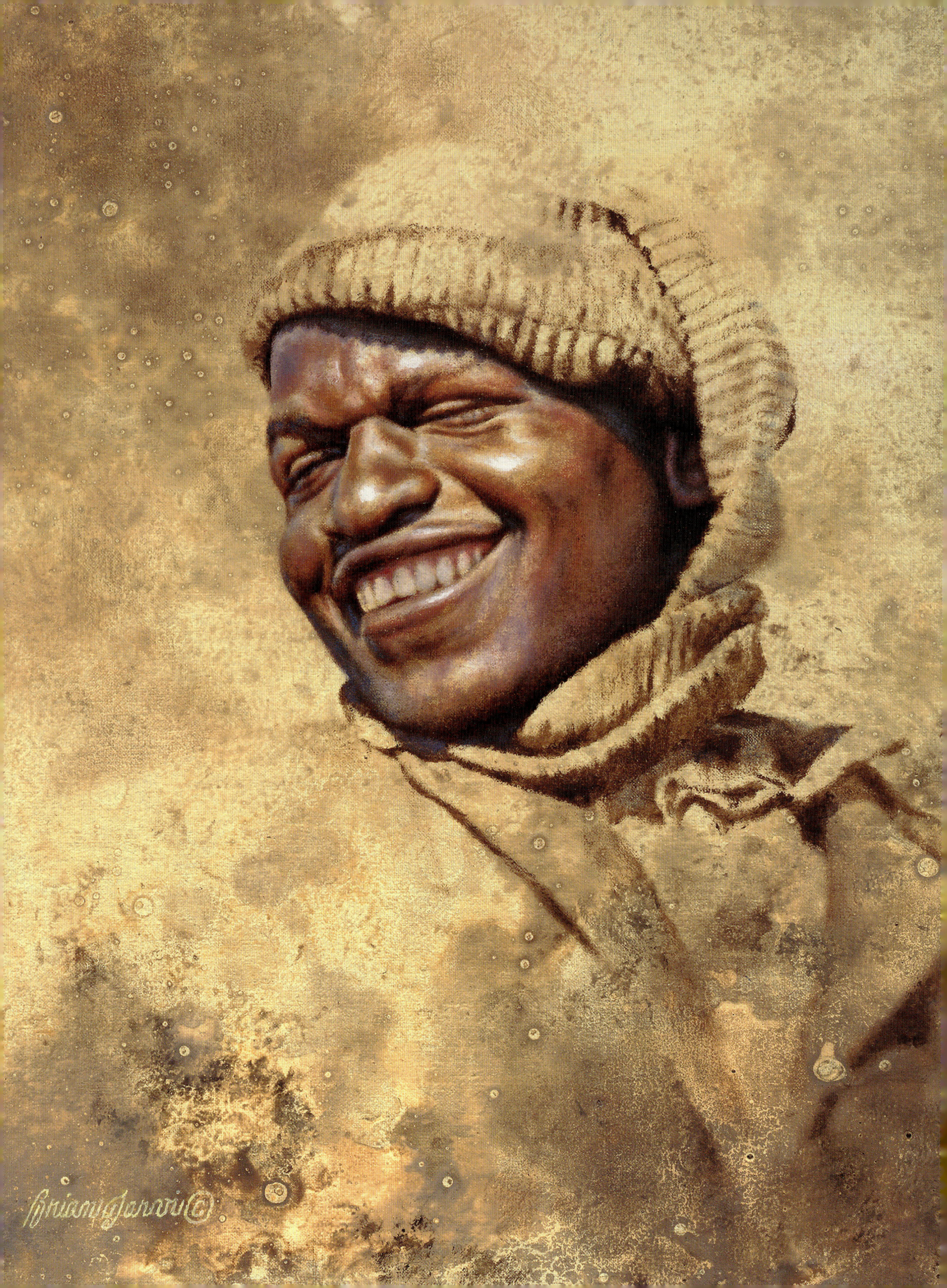

During the decades of gestation as the concept of *African Menagerie* solidified, Jarvi amassed an impressive corpus of individual paintings, works that now hang in private collections around the world. He became known for his up-close-and-personal action scenes. "The excitement lies in transferring situations I have witnessed on safari onto canvas," Jarvi once told an interviewer. "When choosing the themes and subjects of my paintings, I like to draw on personal experience—for example, a bull elephant making a charge or lions attacking a Cape buffalo. The tremendous diversity, the sheer numbers, and the often-sobering confrontations of wildlife have spawned more ideas than I can ever hope to translate."

What put him on the map were pieces like *The Savage Land*, a lion and Cape buffalo scene; *Razor's Edge*, a heart-palpitating predator and prey scene of a lion and zebra; *Rogue* and *Rush*, engaging portrayals of bull elephants; *Reticulated*, a portrayal of giraffes; *Return of the Wildebeest*, a depiction of a lion pride waiting for the great Serengeti migration; and a series of human figurative pieces celebrating the Masai tribe.

The piece *Yellow-Billed Storks*, which feels like a 16th-century Flemish painting, was featured in the 2005 Leigh Yawkey Woodson Art Museum's *Birds in Art* exhibition. The work hinted at the classical painterliness that would emerge fully fledged in *African Menagerie* and the attendant studies. All the while, Jarvi gave away hundreds of prints made from his works to help raise hundreds of thousands of dollars for conservation causes. He even collaborated with actor Ted Danson on a print of orca whales titled *Orca-stration* as a fund raiser for ocean species conservation.

M. Stephen Doherty pointed to Jarvi's painting *The Last Gladiators*, a spectacular portrayal of two bull elephants jousting for supremacy in the bushveld. Each painting was the product of studies and, while appearing spontaneous and effortless, only emerged from toil. If Jarvi didn't think a piece was up to snuff, he tossed it in the trash bin. But the studies he completed for *The Last Gladiators* were gems in their own right. "Those studies were wiped out, repainted, wiped out again, and adjusted until Jarvi knew he had a basic image that had pictorial impact and was accurate in terms of the elephants' anatomy, perspective, and attitude," Doherty explained. "Only then did he go through the long process of bringing the image to life with layers of oil colors."

The Tracker, 2007
Homo Sapiens
Oil on Belgian Linen
16 x 12"
Private Collection

OVERLEAF, LEFT
Samburu Warrior, 2006
Homo Sapiens
Oil on Belgian Linen
28 x 16"
Private Collection

OVERLEAF, RIGHT
Masai Mother and Child, 2005
Homo Sapiens
Oil on Belgian Linen
28 x 16"
Private Collection

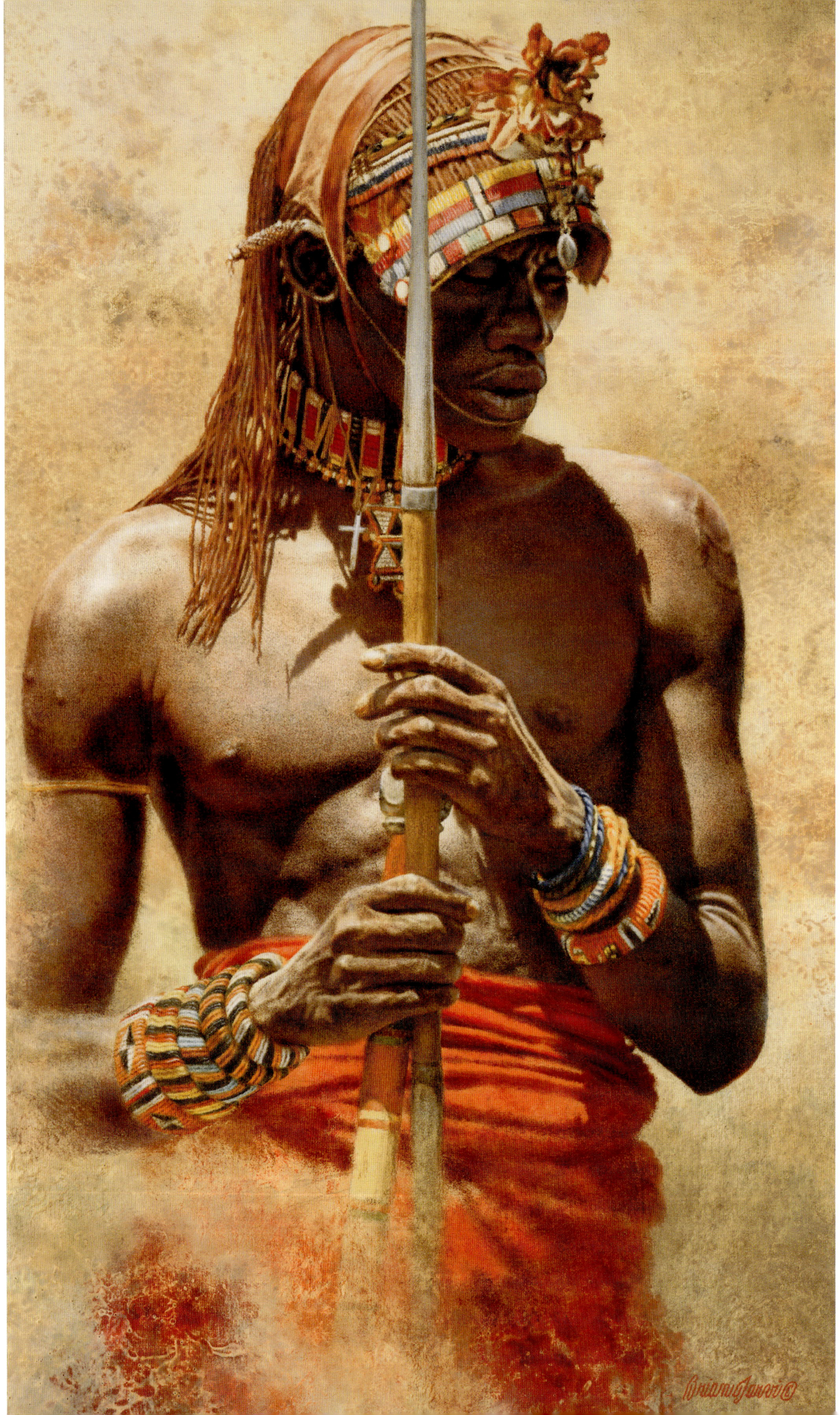

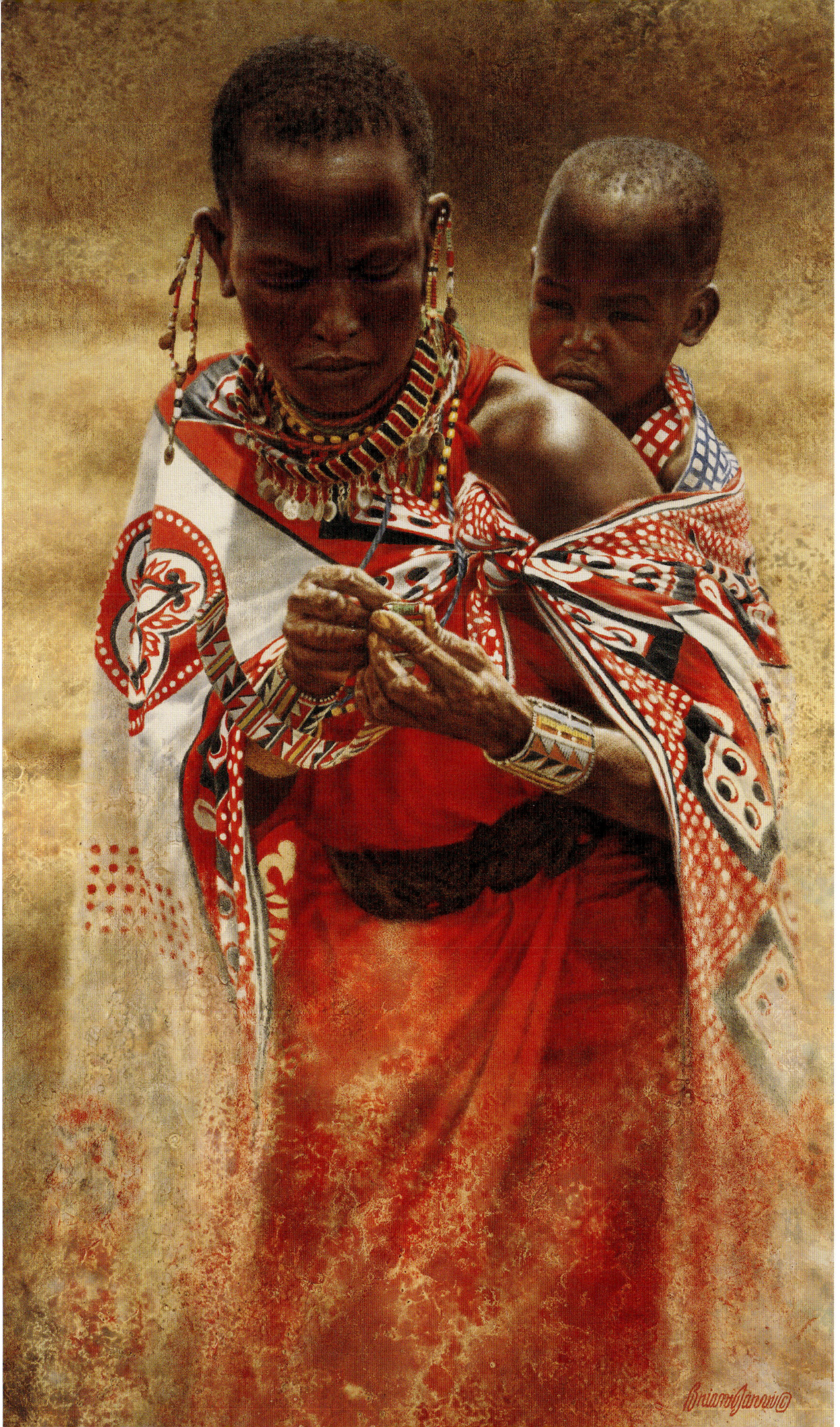

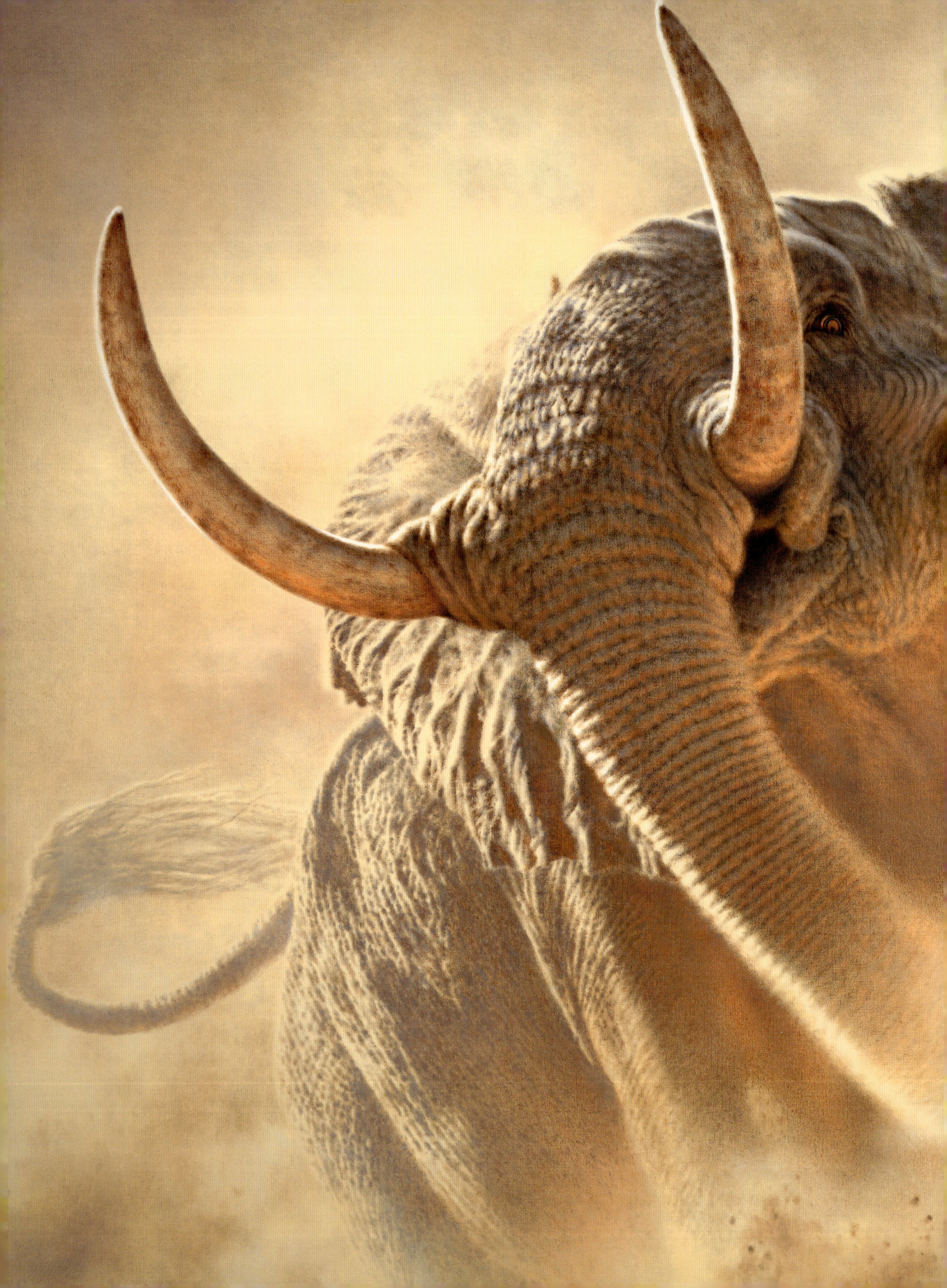

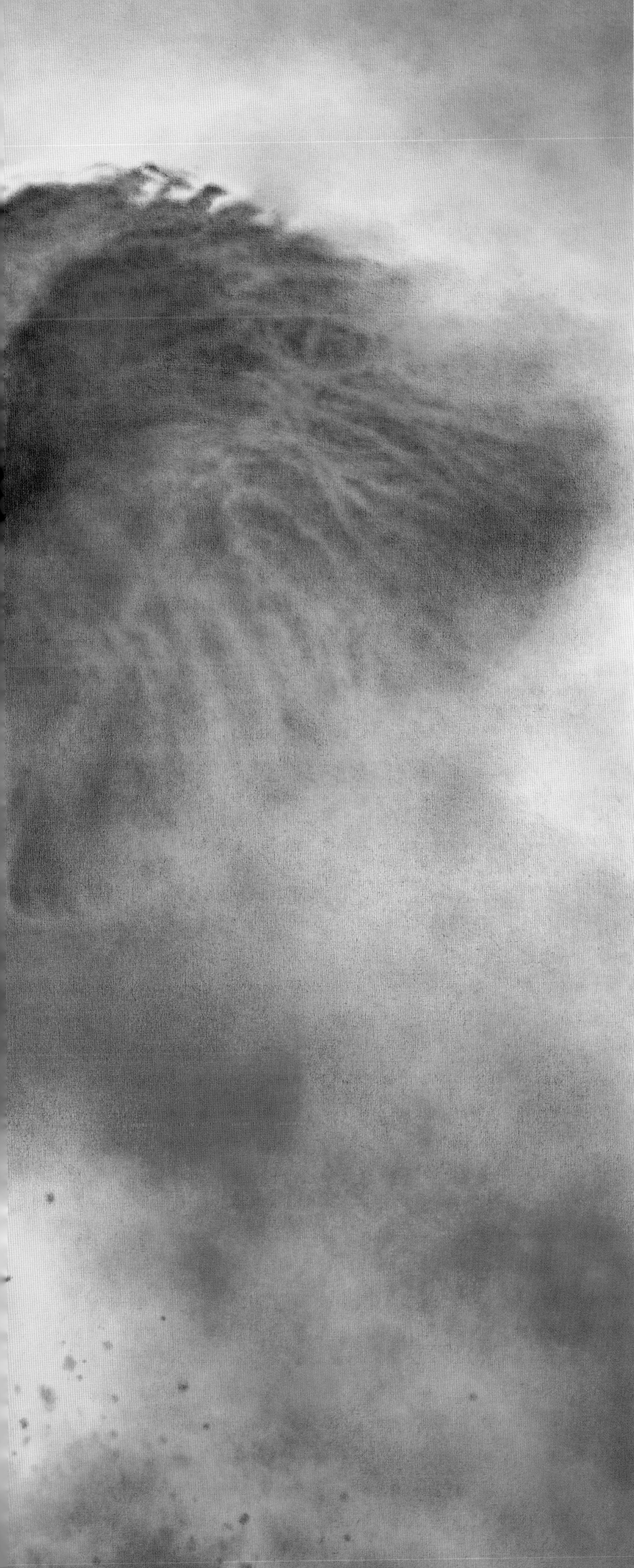

Rogue, 2008
African Elephant
Oil on Belgian Linen
72 x 96"
Private Collection

Prior to completing *African Menagerie*, *Rogue* was the largest painting Jarvi had ever done.

OVERLEAF, LEFT
Reticulated, 2006
Giraffe
Oil on Belgian Linen
30 x 15"
Private Collection

In 2006, this piece was juried into the Society of Animal Artists' *Art and the Animal* exhibition.

OVERLEAF, RIGHT
Rising Son, 2000
White Rhinoceros
Oil on Belgian Linen
36 x 24"
Private Collection

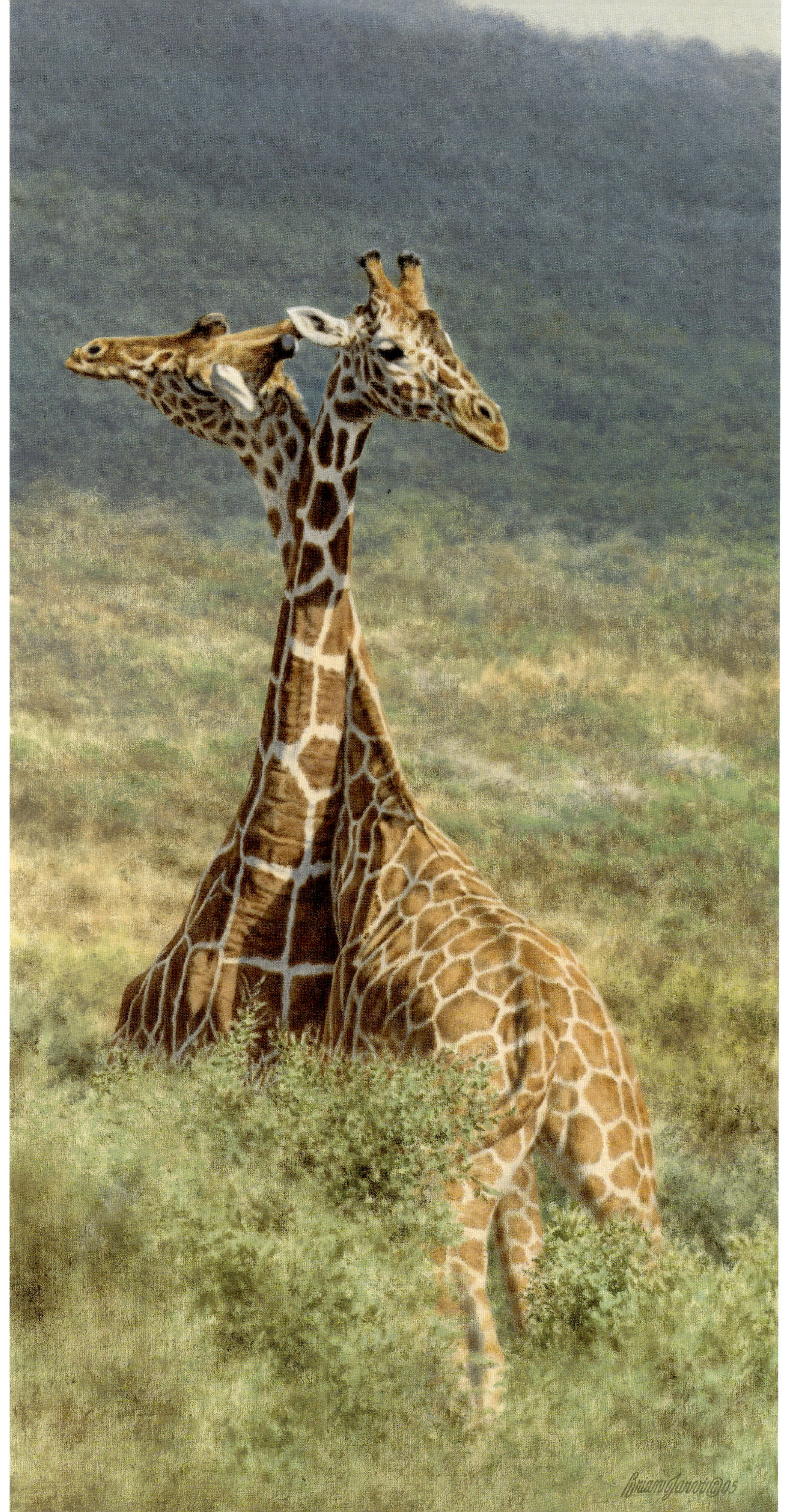

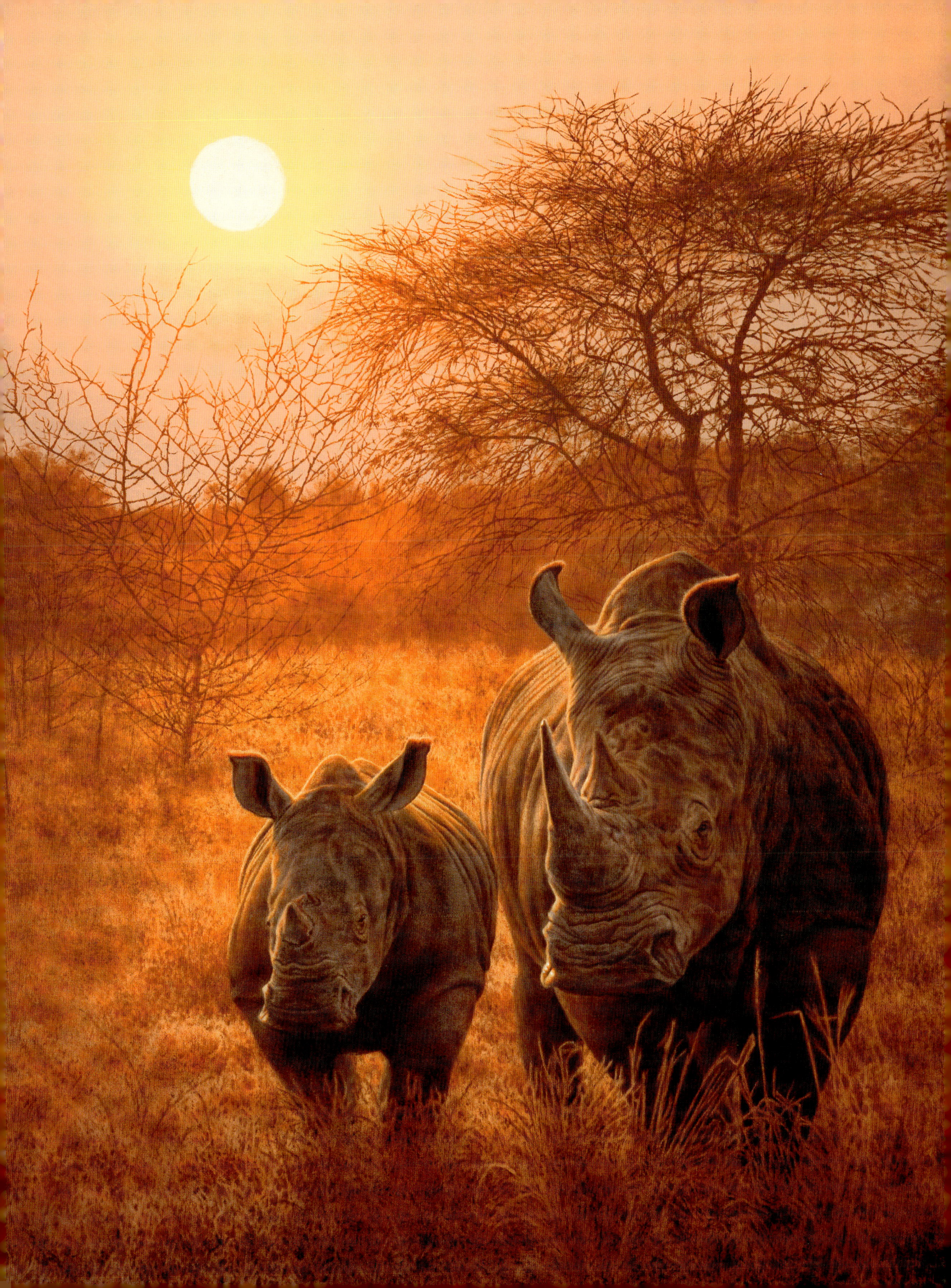

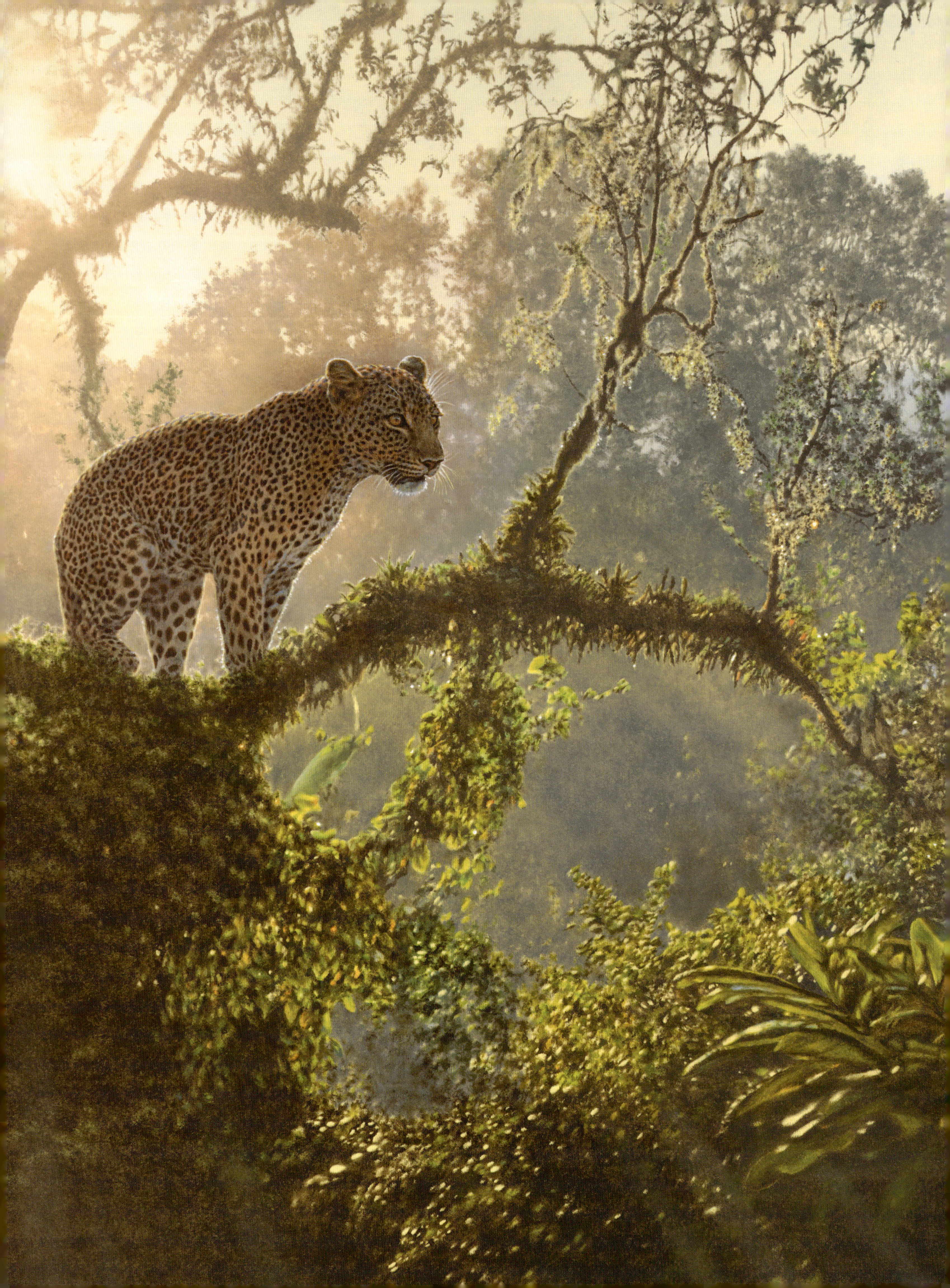

Doherty also mentioned Jarvi's leopard painting *The Overlord*. "About 25 years ago, I saw a leopard perched in a tree with sunlight filtering through the lush vegetation," said Jarvi. "It was such a captivating scene that it stuck with me all those years even though it only lasted a few moments. I started making sketches of how I might re-create those moments that said so much to me about the African environment and the life of the animal. I dug back to find some of the slides I had taken at the time, played with some oil sketches, and gradually developed an image that captured the impact and meaning of those distant memories."

A MONUMENTAL NEW PROJECT

Jarvi has been to Africa 13 times. In 1998, he went on a safari with Daniel Smith and Bruce Miller. Each member of the threesome was in the midst of a personal transformation, actually trying to shed the label of being "duck stamp artists." In some ways, they almost felt as if they were typecast, like an actor who plays one role and is only associated with that part. "The limited-edition wildlife print market phenomenon had run its course," Miller said. "It used to be a compliment whenever someone would say, 'Oh, your painting looks just like a photograph,' but after a while it ended up becoming demeaning."

According to Miller, after painting tightly and aspiring to render a photographic-like painting, it's difficult to change into a looser style that is competent and unmistakably your own. Jarvi continued to evolve, forcing fine art critics to sit up and take notice. Even today Miller is somewhat perplexed that Jarvi isn't better known as an American painter—not as an American wildlife artist, but as a fine artist who just happens to paint animals. "If he's not the best wildlife artist in the world, then he ranks within the top three and I don't know who the other two are," he said. "The caliber of animal painter he is shows how damned fine a painter he is in general."

For Jarvi's studies (which are interrelated to the giant *African Menagerie* narrative), the painter chose titles that drew on a diverse range of touchstones, ranging from biblical scripture and mythology to science fiction, classical literature, and scientific reference points, including Darwin's theory of evolution and Swedish zoologist Carl Linnaeus's binomial nomenclature, the system he devised for identifying and naming species. (Linnaeus also coined the term *Homo sapiens* in 1758.)

The Overlord, 2013
Leopard
Oil on Belgian Linen
36 x 24"
Private Collection

OVERLEAF
Orca-stration, 1991
Orca
Oil on Panel
20 x 30"
Private Collection

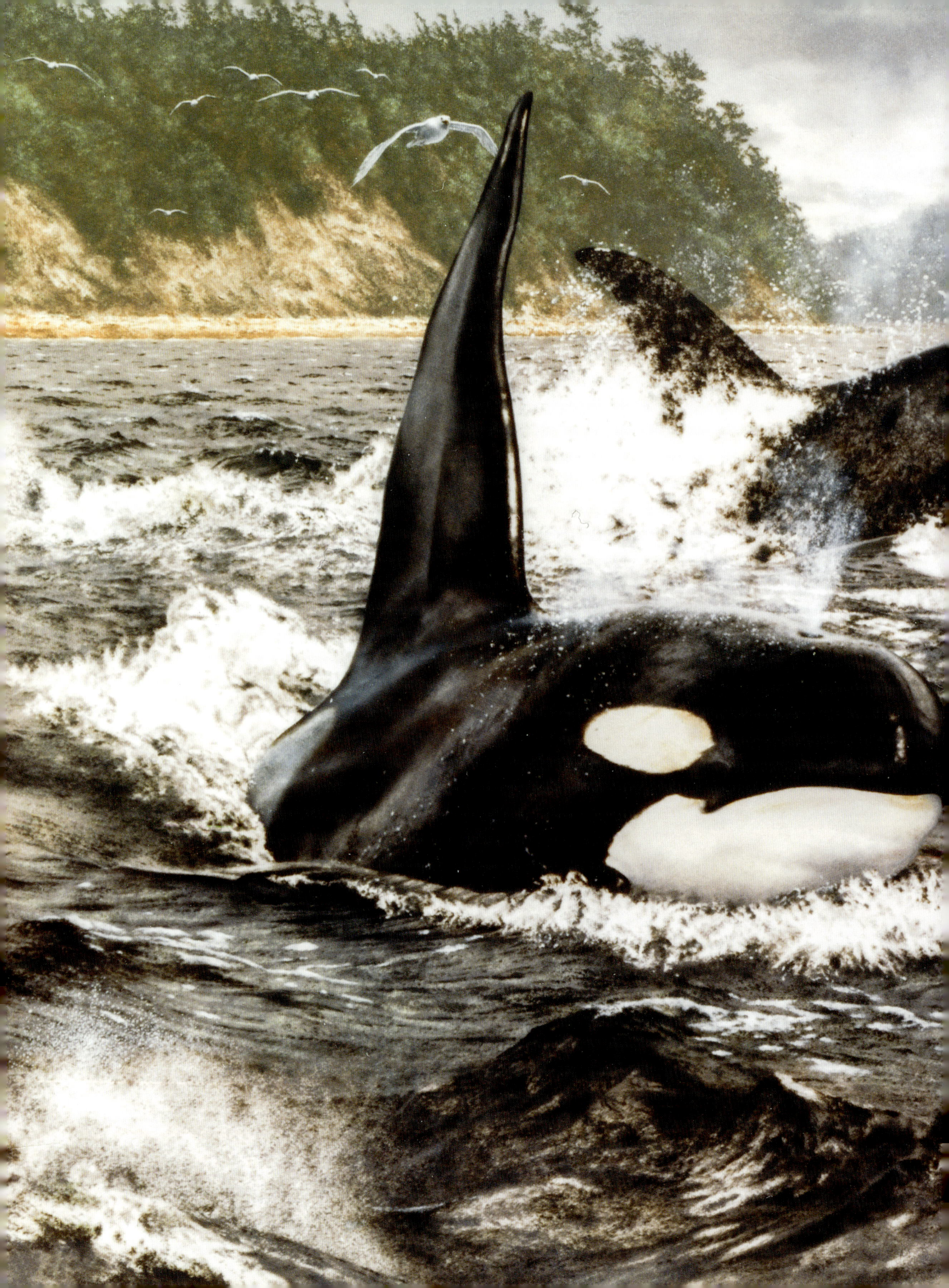

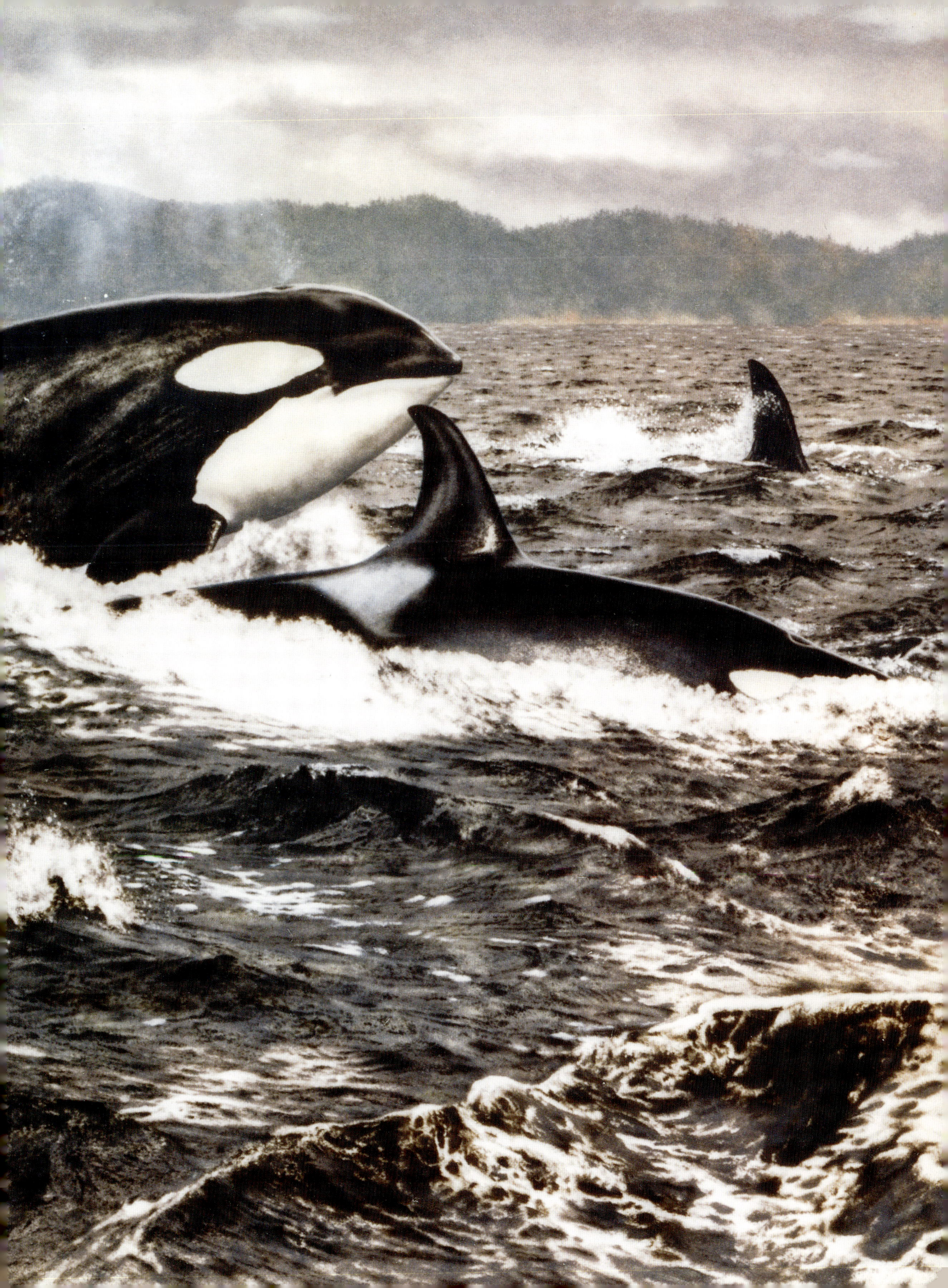

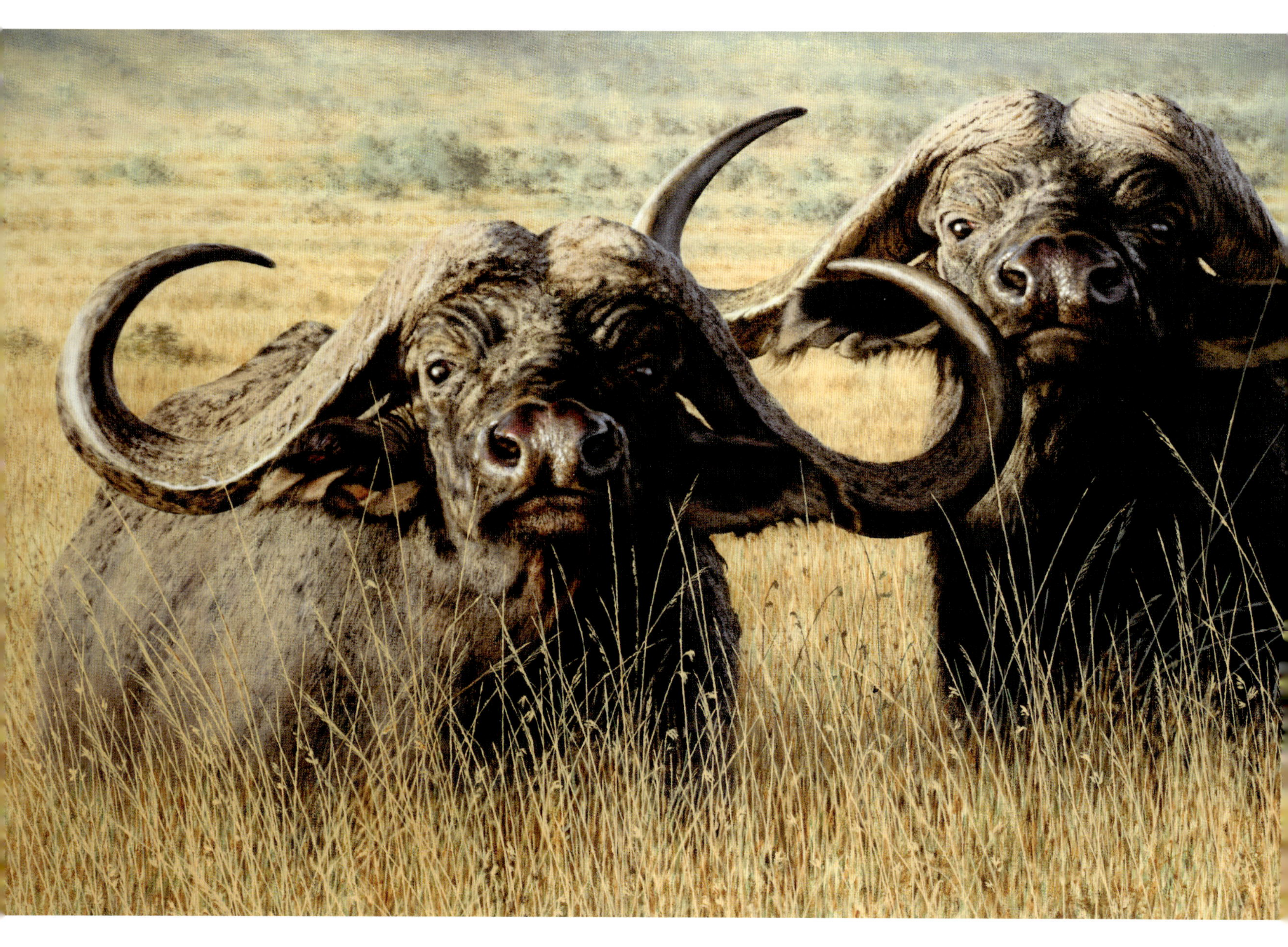

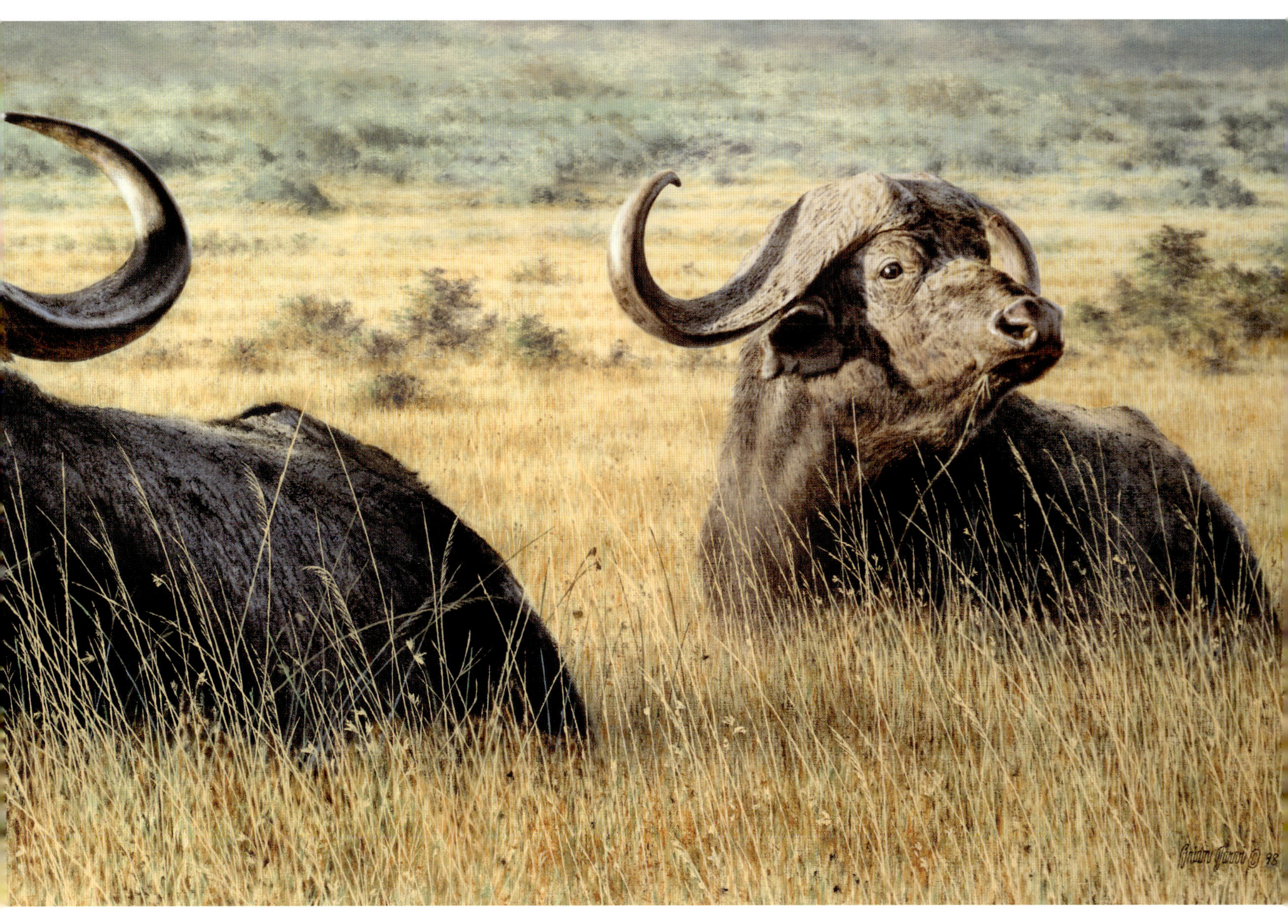

Lion Scent, Cape Buffalo, 1998
Cape Buffalo
Oil on Belgian Linen
20 x 60"
Private Collection

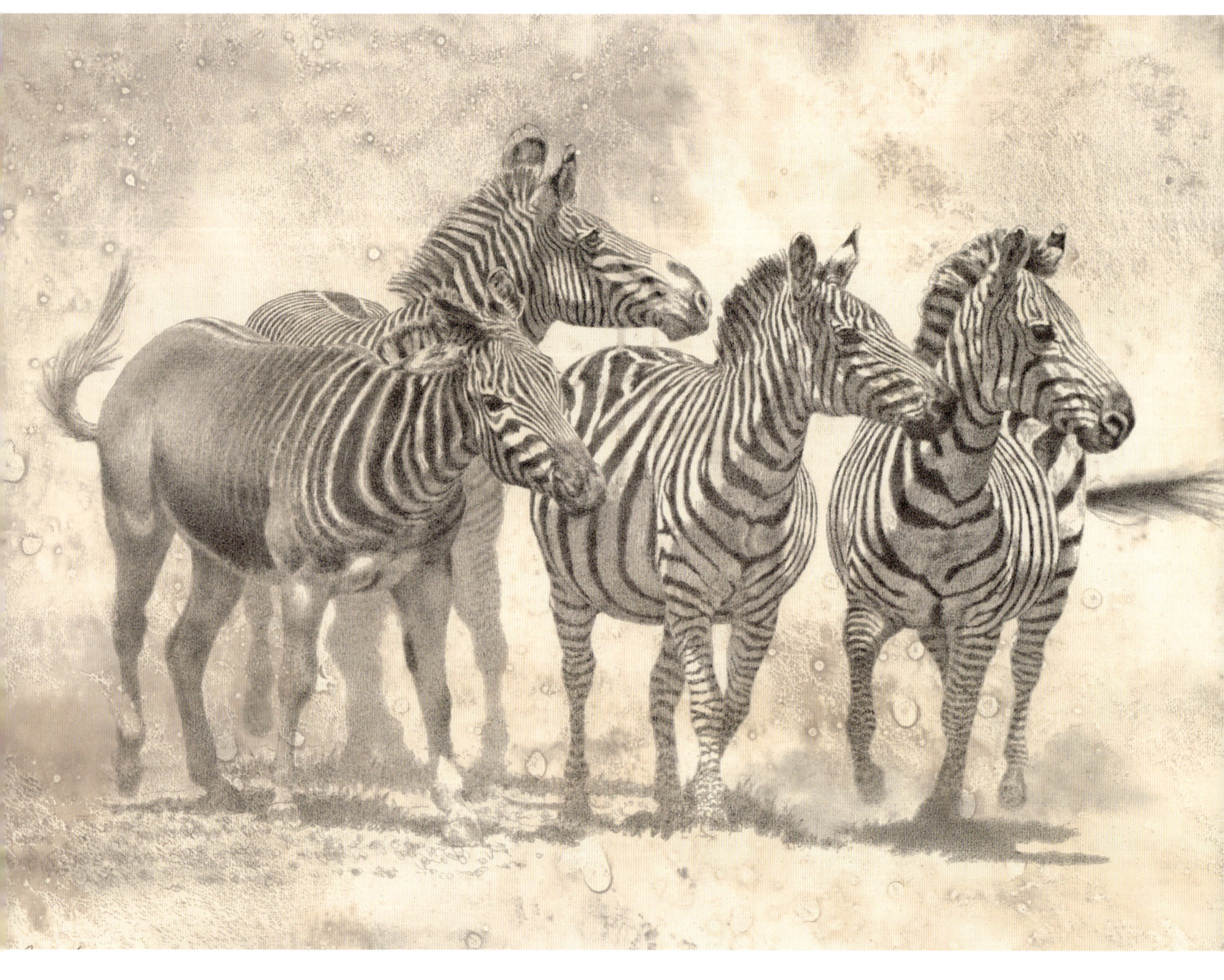

The Four Horses, Study I, 2015
Zebra Species
Pencil
18 x 24"
Collection of David and Kathleen Chesness

In *Predatoria*, for example—a paean to those at the top of their respective food chains—Jarvi featured a 450-pound African lion and a banded mongoose, a midsize carnivore, along with a leopard, spotted hyena, honey badger, cheetah, and, of course, a human. In *African Menagerie*, they are all grouped together.

With *The Last Quagga*, a detailed full-color study, he invited viewers to reflect on what was lost with the extinction of the quagga zebra, a cousin of several zebra subspecies lost to meat and sport hunters. In *The Four Horses*—a nod to the Four Horses of the Apocalypse—Jarvi featured a mountain zebra, Grevy's zebra, Burchell's zebra, and quagga, issued as a warning that we must not allow the three persisting zebra subspecies to follow the same fate as the quagga.

In *Twelve Monkeys*, he clustered a convergence of primates, many of which share a close genetic lineage with *Homo sapiens*—the western lowland gorilla, yellow baboon, Wolf's mona monkey, mandrill, colobus monkey, bonobo, and chimpanzee—that in many places are critically imperiled.

In a sprawling study that would make John James Audubon proud, Jarvi created *Silent Song, The Birds*—four by three feet—that focused on 22 of Africa's most iconic avians, including the ostrich, shoebill, crowned crane, lilac-breasted roller, bataleur eagle, and lesser flamingo.

So stunning were these inaugural studies that the first six were juried into major exhibitions, including those held by the Society of Animal Artists, Artists for Conservation, and the Leigh Yawkey Woodson Art Museum's prestigious *Birds in Art* show. Kathy Foley, the museum's director, said Jarvi has an uncommon touch with birds and his art is reminiscent of the spare atmospherics in the great Flemish paintings between the Renaissance and Baroque eras.

Through the studies, Jarvi gamely presented an illusion of outlines that are embedded in our primitive psyches, dating to when our ancestors began walking upright and slowly making the trek toward achieving a landscape of emotion and the ability to think, reason, have empathy, and even put altruism ahead of self-interest. And yet here was an omega man who was being humbled, parlaying with other entities written off as lesser life-forms. And what he discovered and brought to our attention in *African Menagerie* is that in spite of all the accrued human brainpower—the ability to adapt, innovate, and command superiority—we do not possess the means to manufacture a species, let alone a larger community of species, once they are gone.

One might also think of *African Menagerie* as a mosaic that prevents us from contemplating the plight of individual species in isolation from each other, a way of rationalizing the fraying of a few threads slowly over time until we no longer notice the tapestry is only a pale gossamer of its former grandeur. Jarvi will have none of that; now that he has wooed us in with his brushstrokes, where from a few inches away we can appreciate the nearly microscopic artistic decisions he has made with paint, so too are we encouraged to behold the outdoor world once we leave the gallery. In this way, *African Menagerie* functions like an entire art collection presented before us in one fell swoop.

Jarvi subscribes to the ideals of two-time Pulitzer Prize–winning scientist E. O. Wilson, "the father of modern thinking about biodiversity," who agrees with another writer, Richard Louv, that society suffers from "nature-deficit disorder." Kids aren't getting out in the woods as generations before them did; it isn't their fault but that of their elders. According to Wilson, human attitudes and outlooks on life can be positively influenced if they are infected with a heavy dose of biophilia or love of nature. "Fortunately, that is a condition for which there is no cure," he said. "Once you bring nature into your soul, you are forever changed and you are changed in a way that does not allow you to never care about all other life-forms."

In this age of virtual reality, in which iPhones can deliver us digitally to a far-flung corner of the world, gadgetry is anesthetizing us. Jarvi believes art should be an experience that fires all of our senses, and he hopes it ignites a spiritual response as well. His only regret about *African Menagerie* is that the experience doesn't come replete with a scratch-and-sniff kit that would allow viewers to take in the smells of the wild bush, to viscerally feel the back of their neck hair stand on end at the approach of a predator.

ADMIRERS AND COLLECTORS

The Jarvis' neighbors at Sugar Lake, Jan Gunderson and David Cress, have insight into the painter's work habits that many don't. They spent many nights sitting around a campfire or on the dock talking about Jarvi's dream. "And what a project it became," Gunderson said. "It grew in scope and complexity as the project took form. Neither Dave nor I had ever known a painter before—and both of us found his work amazing. We often got to watch his paintings take shape. His left-handed, upside-down, mirror-reflection painting technique wowed us."

To explain their own appreciation for Jarvi's rarefied ability to translate Africa, they held up a safari in Kenya for which the painter had served as a guide and chaperone. "We had an incredible adventure traveling to three Kenya park reserves with Brian. Up before dawn every day for two weeks, we left camp in a Range Rover with a Masai guide and returned in the dark every evening. We saw incredible sights—and we lifted our cameras whenever Brian raised his," Gunderson said. "The commission painting Brian did for us of Mount Kilimanjaro and the safari animals we saw while in Amboseli National Park is really dear to us. Our safari experiences are memorialized in paint."

Todd Nelson, president of Kalahari Resorts, features Jarvi's work at his vacation destinations. "I was immediately hooked when I saw his lion piece. It is incredibly difficult to capture cats in drawings, and Brian has created the most authentic renderings of lions that I have ever seen," he said. Of the pieces in his personal collection, Nelson said it is hard to identify a favorite. "But if I had to single out one, it would be the bongo painting that was part of the series of studies he did for *African Menagerie*. That striking gaze captured me."

Will Beecherl, a fine art collector from Texas, bought his first Jarvi piece at a Safari Club International auction. "As I've collected art, I've become ever more impressed with the caliber of Brian Jarvi's work," he said. "What he does—his ability to paint animals as true and realistic—is exceptional." Jarvi also painted a Rocky Mountain elk for Beecherl and his wife, Kay, on commission, and a few years ago, Beecherl invited Jarvi to go elk hunting with him in Colorado. Jarvi left his gun at home and brought along a camera and notepad. "In his pieces you can sense what the animal is thinking and gain appreciation for its instinct," said Beecherl.

Walter Broich met Jarvi through Safari Club International. He and his wife, Darci, are also proud collectors. Among the three original Jarvis they own is one that immediately brought tears to their eyes upon delivery—a commission portrait Jarvi painted of Jose, their English springer spaniel/black Labrador mix, who they found as a pup in Spain. "It is fabulous," Broich said. "In some paintings of animals, the subjects look stiff and dead. Jarvi transports their energy into your room."

Few of Jarvi's collectors know him as well and have supported his career more than businessman and nationally known sportsman David Sandstrom. He met Jarvi in the 1980s when Sandstrom was president of the Ruffed Grouse Society and also in the process of helping launch another conservation organization called Geese Unlimited. Jarvi was later twice named artist of the year by Geese Unlimited for his fund-raising efforts.

Sandstrom commissioned Jarvi to paint *The Big Five—Lion, Leopard, Rhinoceros, Elephant, and Cape Buffalo*. "The first

Bongo Study, 2016
Bongo
Oil on Belgian Linen
36 x 24"
Collection of Todd and Shari Nelson

In 2017, this piece was juried into the Society of Animal Artists' *Art and the Animal* exhibition and won its prestigious Award of Excellence.

OVERLEAF
Return of the Wildebeest, 2001
Lion
Oil on Belgian Linen
40 x 72"
Private Collection

The Halcyon Gallery in London, a former hub for Jarvi's international collectors, reported that after installing *Return of the Wildebeest* in its street-side display window the painting caused several fender benders.

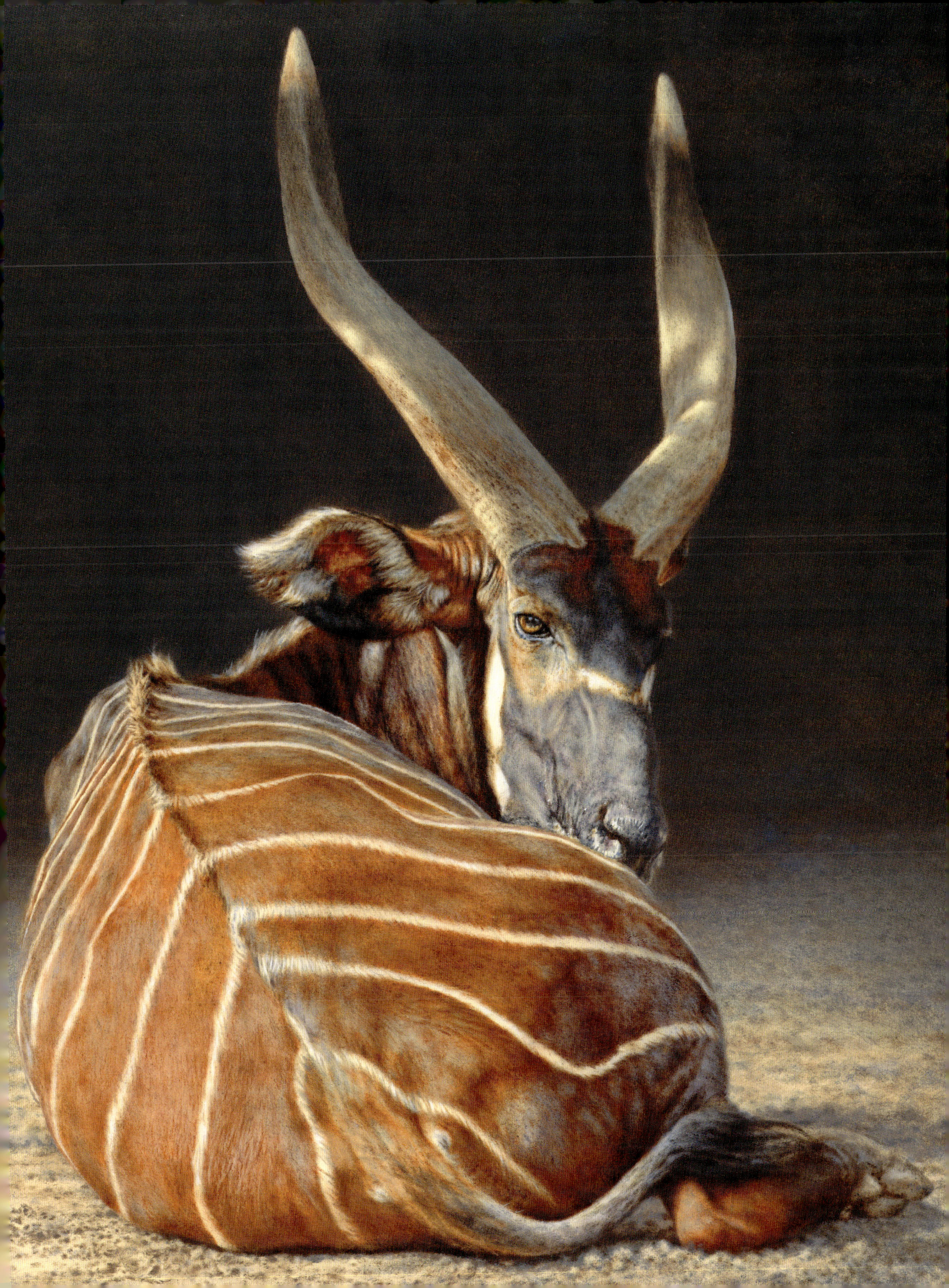

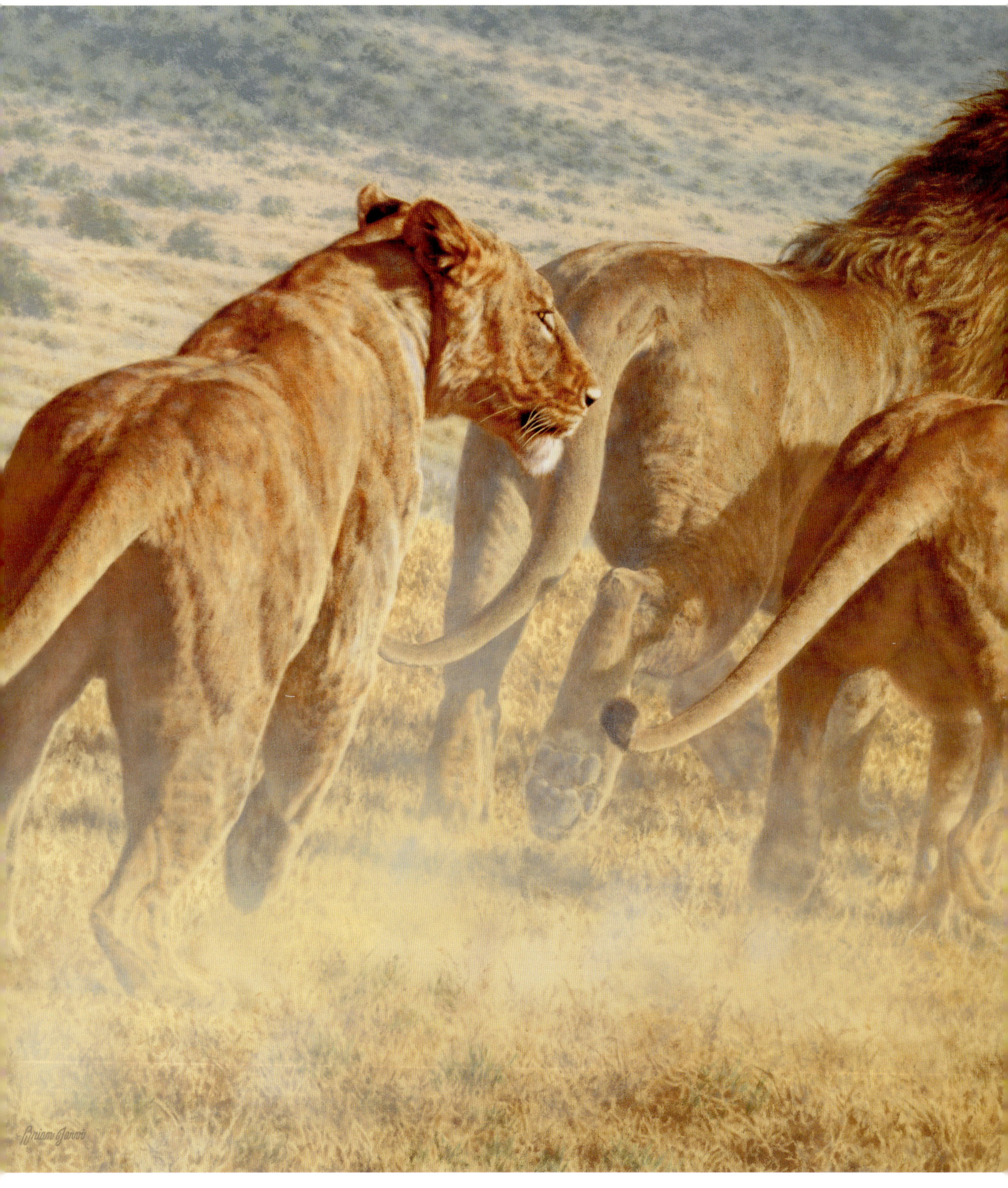
Brian Jarvi

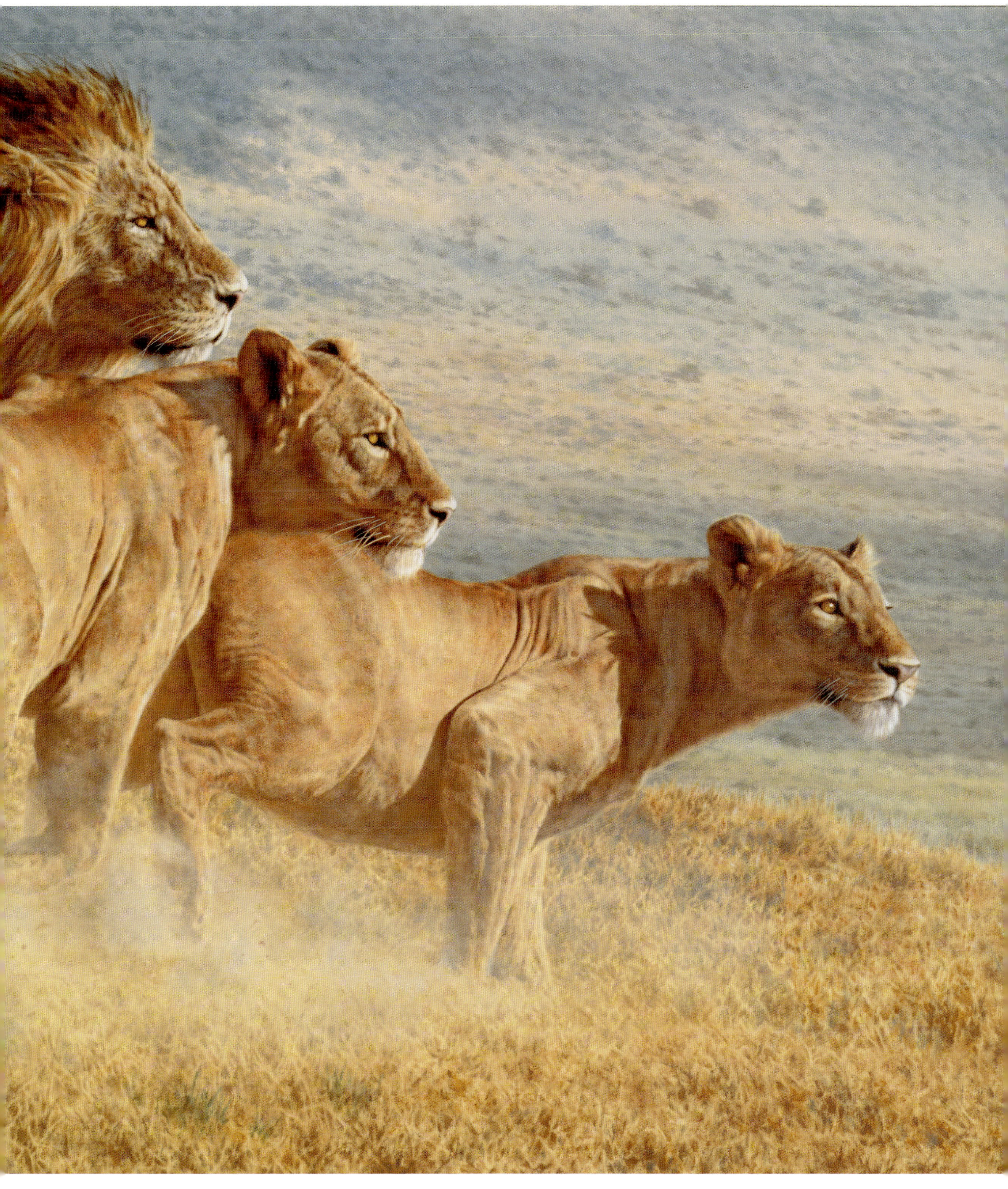

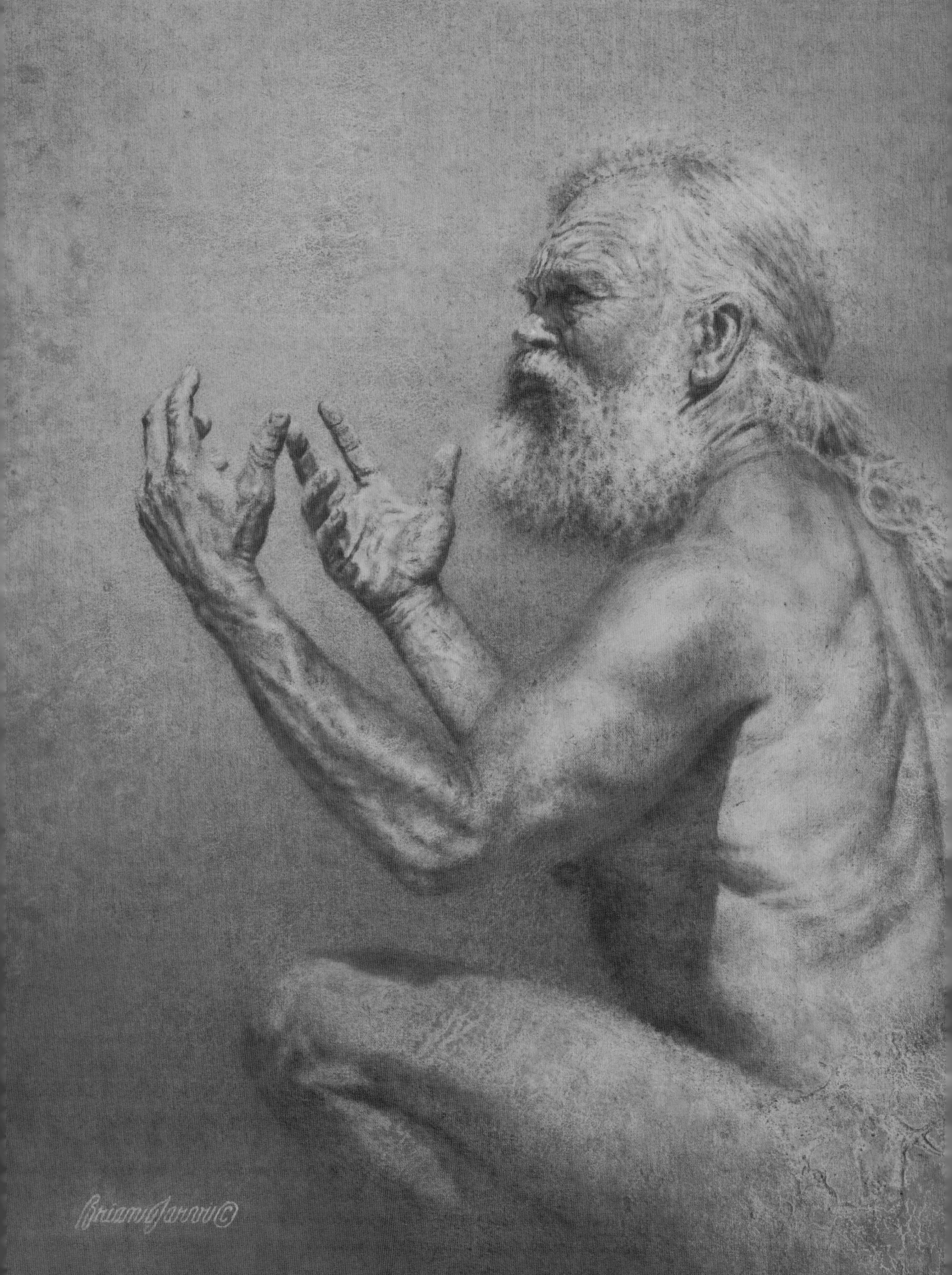
Brian Jarvi ©

two pieces in the series to come off of Brian's easel were the elephant and lion paintings," he said. "Both knocked me out. His portrayals of the animals were exactly as I had seen them."

Over the years, Jarvi has been enlisted to paint several of Sandstrom's hunting dogs and has been a regular visitor to his lodge in the Chippewa National Forest. With more than a dozen major works on his wall, as well as studies created for *African Menagerie*—including a pre-work of *Homo sapiens* titled *The Oracle*—Sandstrom is one of the artist's most prolific collectors. "Our house could be its own Brian Jarvi museum," he said.

Steve Wilcox, Sandstrom's best friend, is also a collector. "I met Brian and Raelene in an interesting way. As a little community banker, I sold them the lake house they had when they first moved to Grand Rapids," he said. The Jarvis and Wilcox became fast friends and they've gone on safari together. Wilcox, who is highly discriminating in judging nature and sporting art, said, "What I appreciate about Brian is that he doesn't paint soft scenes that could wind up on a Hallmark greeting card. He conveys the intense life force, the power and glory and spirit of the animal. David and I have been fortunate enough to be there [the wild African bush] and see up close what Brian is communicating in his art. The animals are so alive. When you see the paintings on the wall, you think they are going to move."

PEER REVIEWS

Within the world of contemporary wildlife painters, few have as much name recognition as John Banovich. Renowned for his portrayals of African species, he commands rock-star appeal among sportsmen and women, especially those art collectors who regularly embark on hunting and photography safaris. Between them, Jarvi and Banovich have won several wildlife artist of the year honors from prominent organizations, and their pieces have commanded six figures at auction. They are admirers of one another, not rivals.

"I'm a big fan of Brian's work and always have been," Banovich, a native of Butte, Montana, said. "We emerged in our careers around the same time. Brian captures the soul and character of Africa as good or better than anyone. The thing I admire most about him is that here he is, in the last third of his career, as I am, and he's taken this unbelievably bold move to create something very few artists in living history have attempted to do on such a scale. I'm really blown away."

The Oracle, 2016
Homo Sapiens
Oil on Belgian Linen
20 x 18"
Collection of David Sandstrom

OVERLEAF
The Last Gladiators, 2009
African Elephant
Oil on Belgian Linen
28 x 42"
Private Collection

For North Americans who take on African subject matter, there's no guarantee—regardless of the ability to paint big-game animals on this continent—that the skill will translate to sub-Saharan subjects in a way that exudes authenticity. "The lines of animal forms and gestures are just so strong in Brian's work; it's as if he has lived in Africa his whole life," Banovich explained. "Normally, an artist is very good at depicting herbivores or carnivores, say elephants or lions. But Brian is capturing multiple species and yet each one is exquisitely composed."

Banovich said he feels a kinship with Jarvi in heeding the call to try and make a difference by using art to stir people's emotions. And they both enthusiastically endorse the North American Wildlife Conservation Model that has revived the prospect of species on this continent and has been applied to Africa. According to Jarvi and Banovich, if one is really dedicated to saving African species, then one cannot be anti-hunting because in many countries it is vital to local economics and shows that animal populations are worth more alive than dead. That's crucial in the fight against black-market forces killing rhinos for their horns and elephants for their ivory tusks.

"I relate to Brian. I think that we artists are compelled by our fascination with nature. We're not trying to tell the world how to think, but when you spend hours and weeks and months collectively in special places as we have, it is your hope that your art might stir the viewer," Banovich said. "Africa is changing so rapidly. The fall of some of these ecosystems is happening on a scale I never could have imagined when I first set foot in them three decades ago. We are at the stage where it's all hands on deck. The most important thing is that we stand up and take notice. This is where art comes in."

Ross Parker was raised on a farm in rural Zimbabwe. During his childhood in the wild bush, he hunted, fished, tracked big game, collected bird eggs, attended the prestigious Plumtree School, and dabbled in art. Those experiences informed his thinking today as a conservationist and owner of two fine art galleries in Florida—Call of Africa's Native Visions Galleries—which specialize in African wildlife and landscape art.

Zimbabwe gave rise to one of the most talented groups of modern nature painters in history. People who were raised there understand the nuances of color, the way the morning air is clear, and how afternoon storms bring a

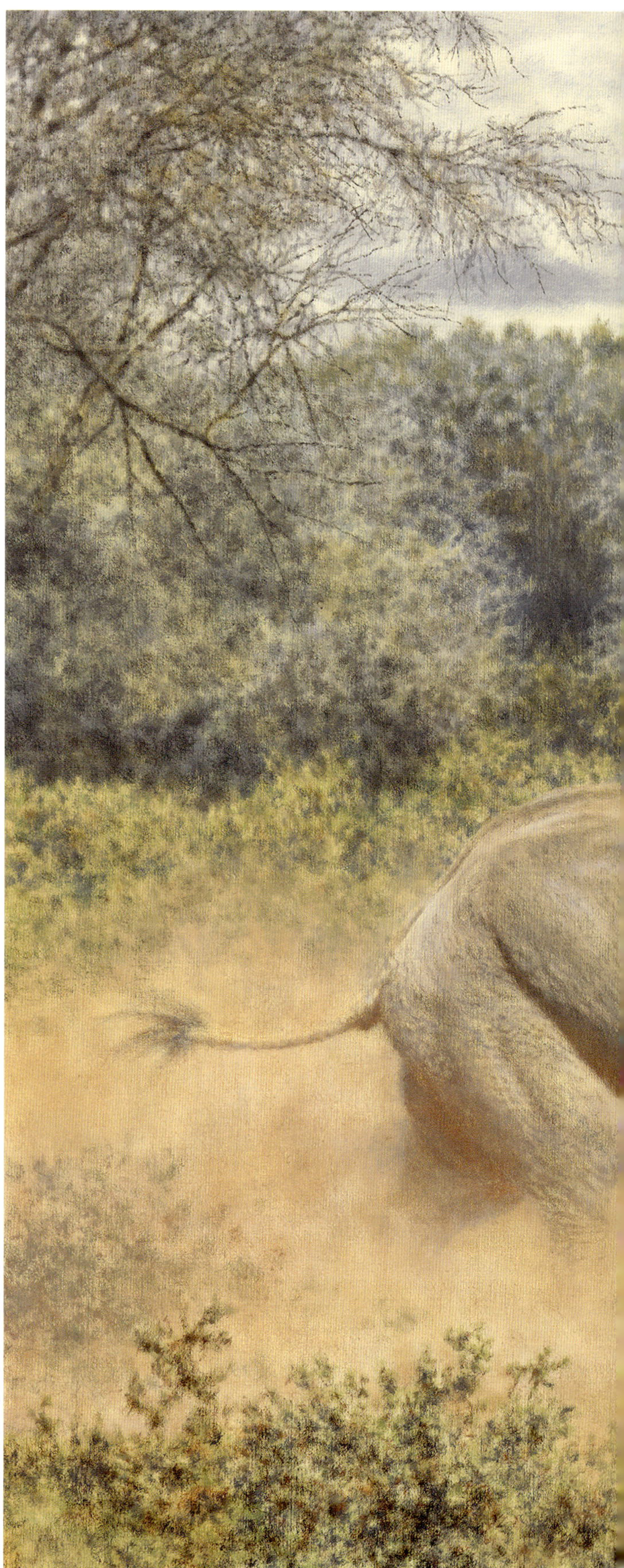

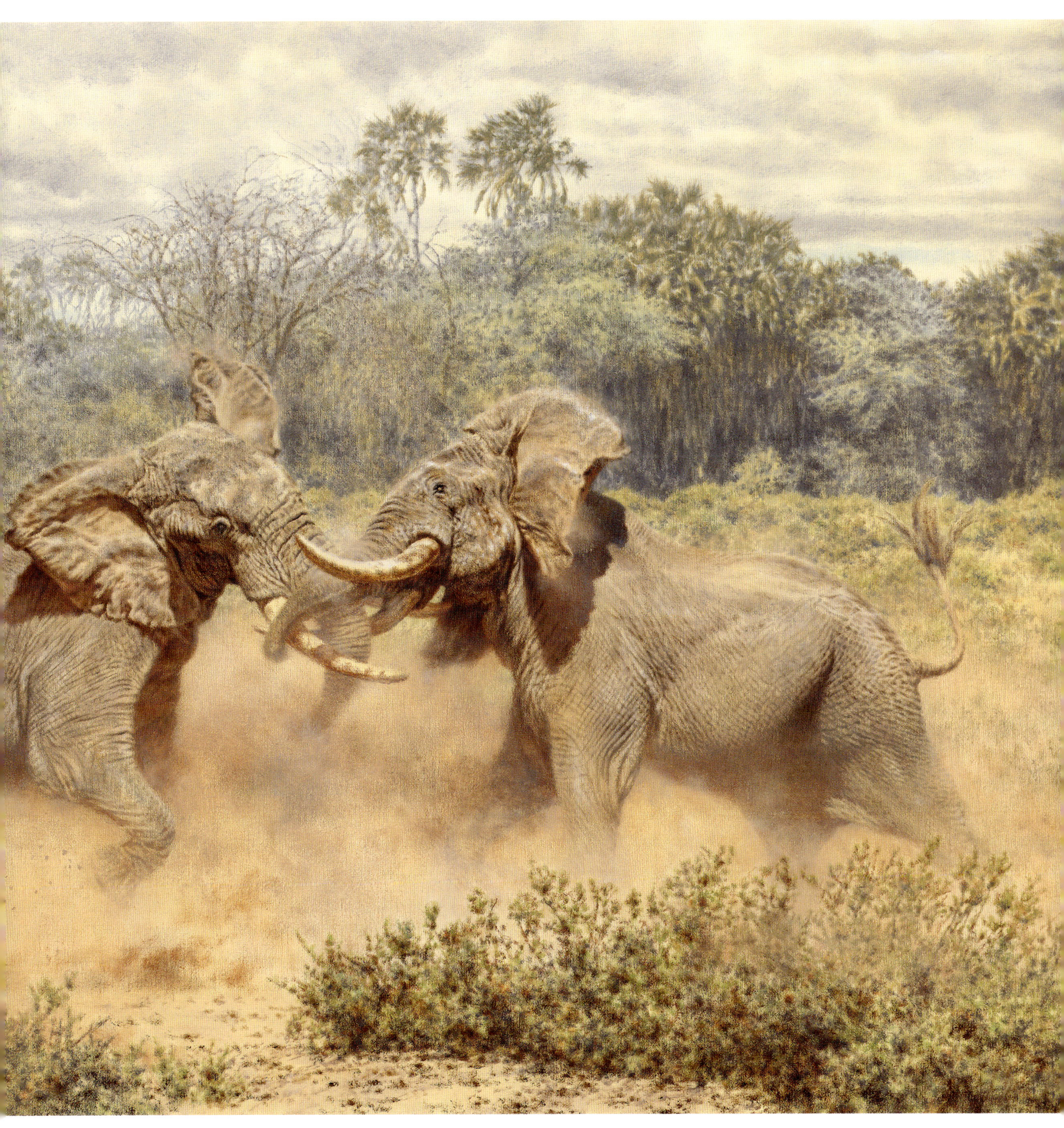

mixture of diffuse atmosphere effects. Reddish and orangey hues float on the horizon as the last rays of sunlight push through thin layers of dust. A lot of artists from outside Africa aren't able to bring that essence to life. "You can't imagine how rare it is for an American to make paintings that leave you feeling as if you're coming home," Parker said. "Jarvi has absorbed Africa via osmosis, soaked in it at the soul level, and is able to articulate it, as if it were always innate to his being."

"The thing I like most about Brian is that he's an artist who has evolved over time and his work continually gets better," said Parker. "The evolution happened with his palette. His early work was kind of monochromatic, but he's become a colorist. His design and composition is really great and the anatomical accuracy is excellent. Much as I tout my African-born artists, Brian holds his own with the best of them."

Shirley Greene is one of those heralded Zimbabweans Parker mentioned. Her impressionistic portrayals have won numerous accolades. According to Greene, many North American wildlife painters seem to work in a rather cool palette, which is not normally representative of the very warm, bright African light, usually made even warmer by being filtered through atmospheric conditions such as dust, smoke, or haze. Other problems arise simply from a lack of working visual knowledge of the animals or birds portrayed.

Jarvi, she said, has an uncanny feel for Africa. "He sensitively captures African light even in very nuanced ways; for example, suggesting its strength not only by warmth and brilliance, but also as it reflects off the ground surface onto the animal's underside, which we would usually think of as being in dark shadow. His palettes are never garish or oversaturated, but unified and sophisticated, which I find very evocative and pleasing to the eye."

Greene also admires Jarvi because it's obvious he worked hard at understanding animal movement. It's said that it takes ten thousand hours of practice before one moves to a superior level of competency, but Jarvi and Greene paid their dues times 10. "With anatomical accuracy, Brian just doesn't miss a beat. Try picking apart the bongo antelope in the foreground of *African Menagerie*. You can't do it. The skin, musculature, facial detail, the all-important eye—he gets it all, and it's a delight."

Of course, the coup de grace of *African Menagerie*, as Greene and others have pointed out, is that Jarvi brings wry

Mara Ballet, 2001
Ostrich
Oil on Belgian Linen
16 x 12"
Private Collection

OVERLEAF
Giraffes and Lala Palms, 2004
Giraffe
Oil on Belgian Linen
12 x 16"
Private Collection

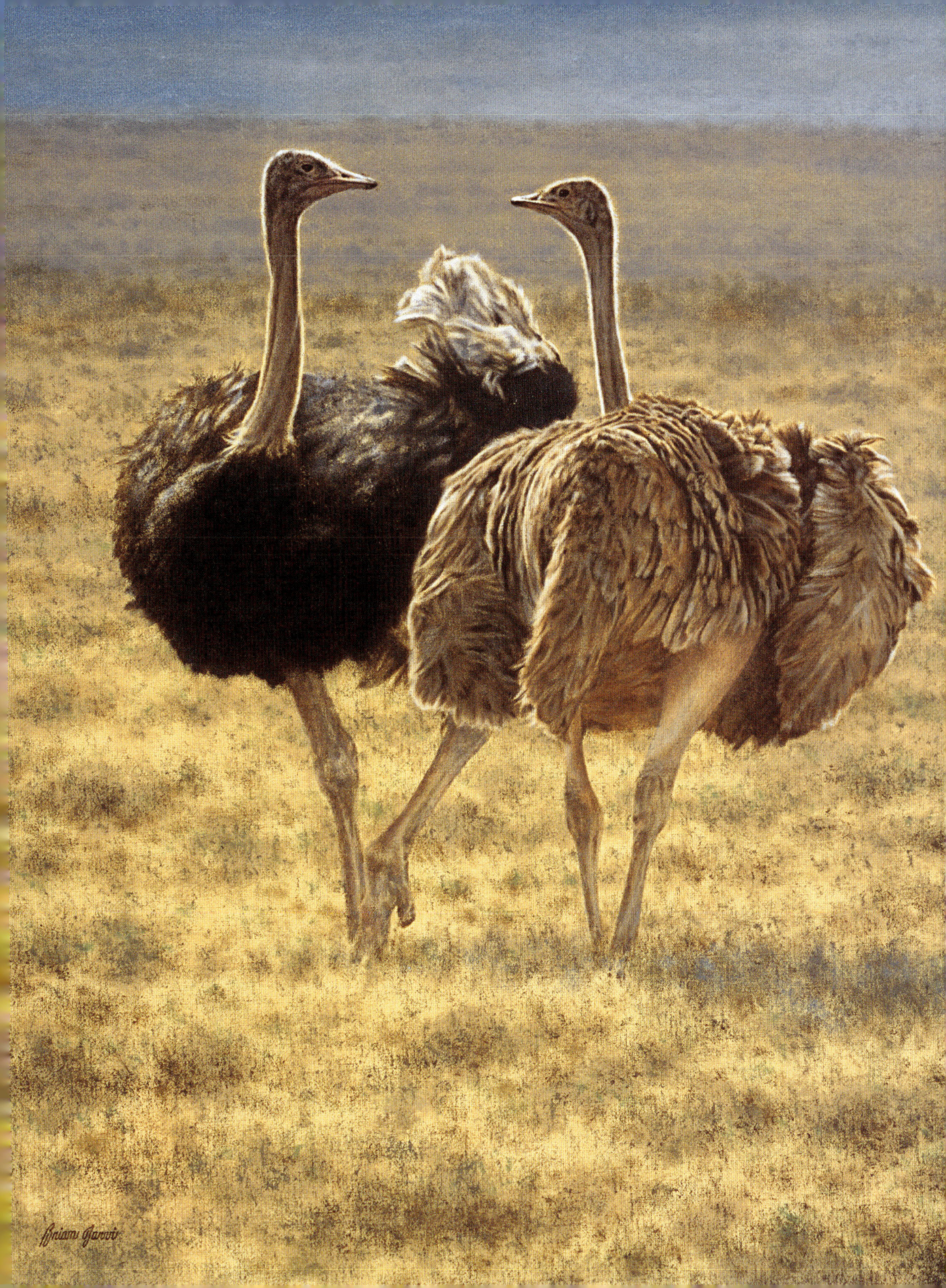

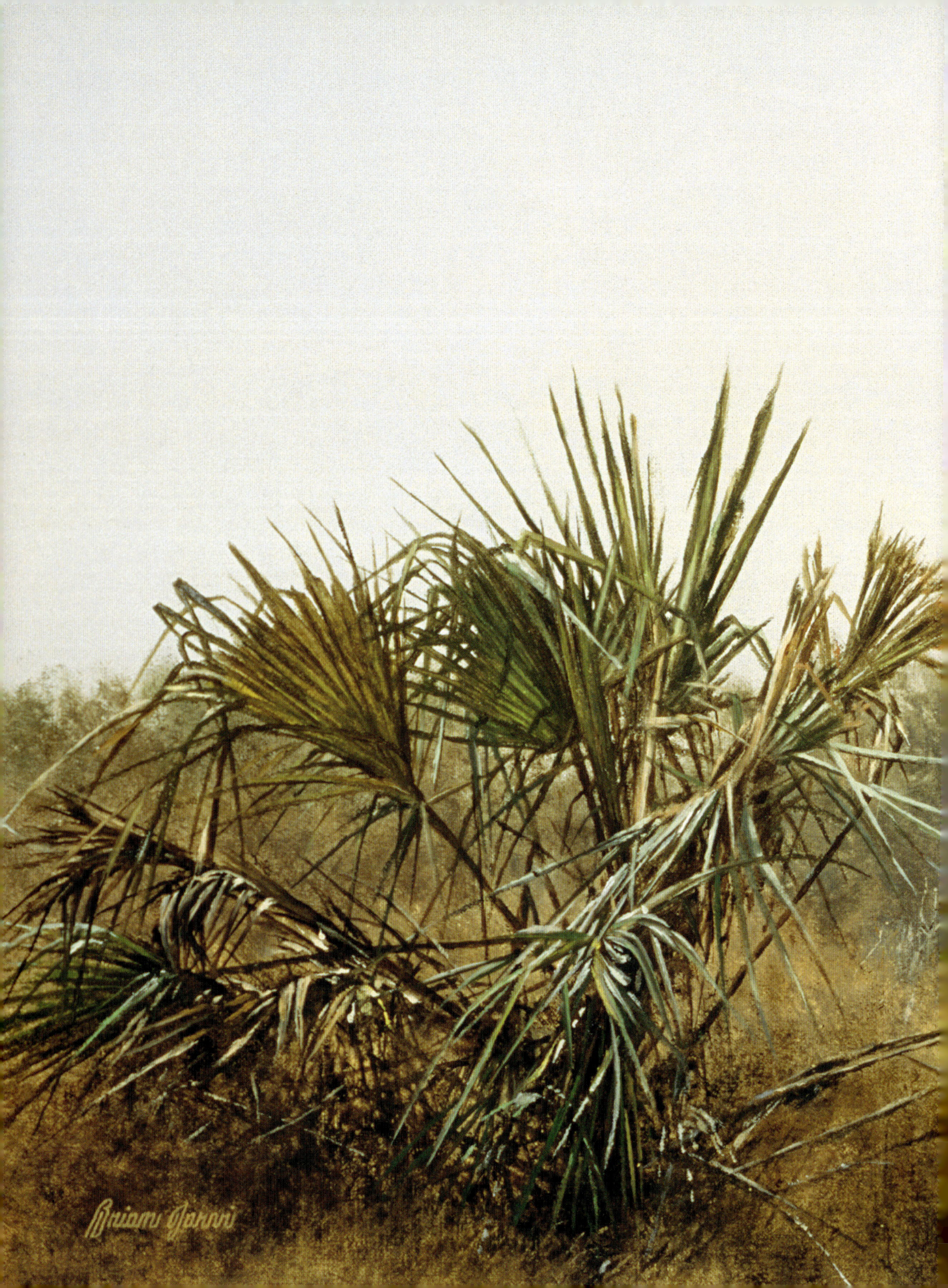

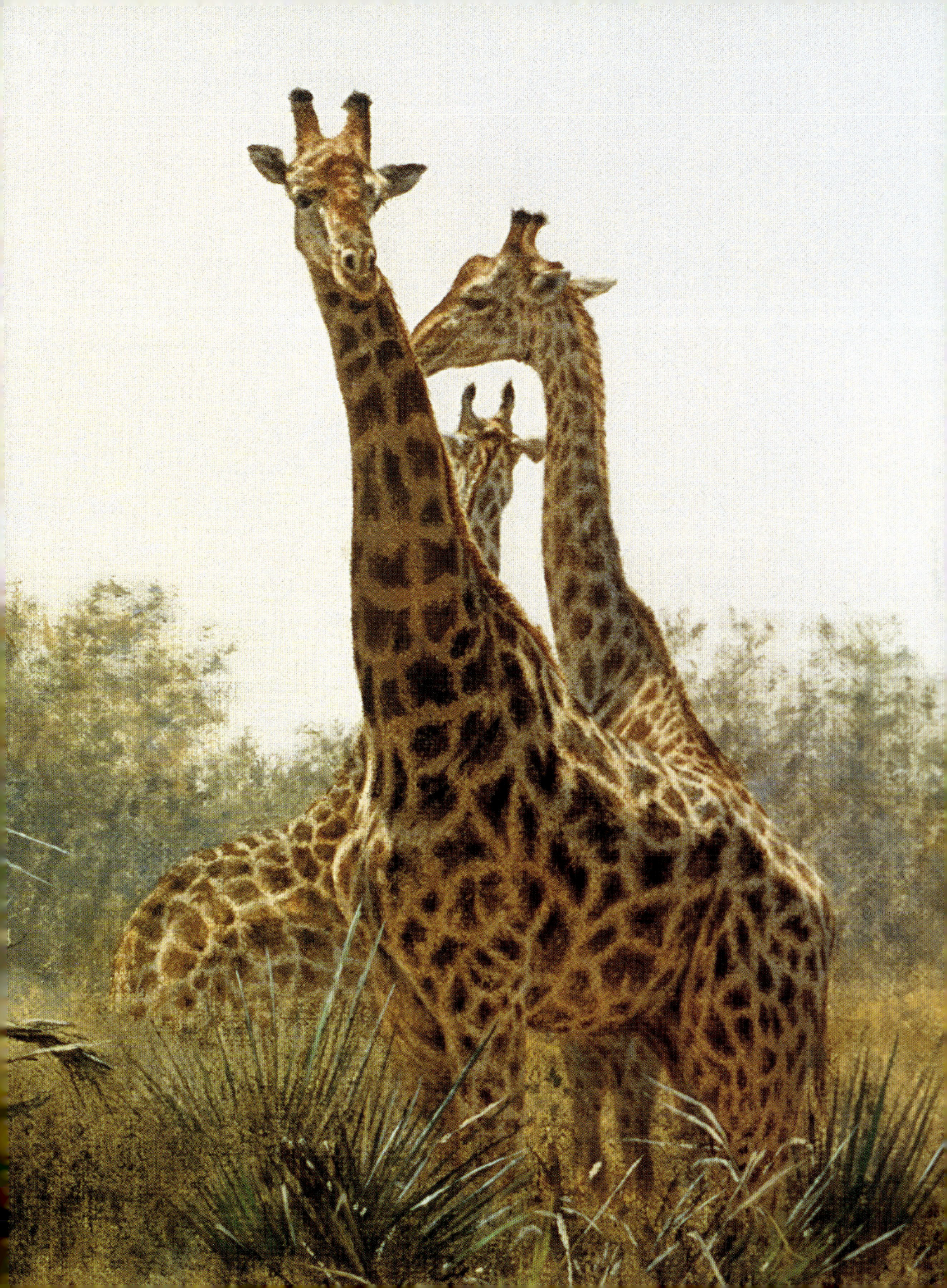

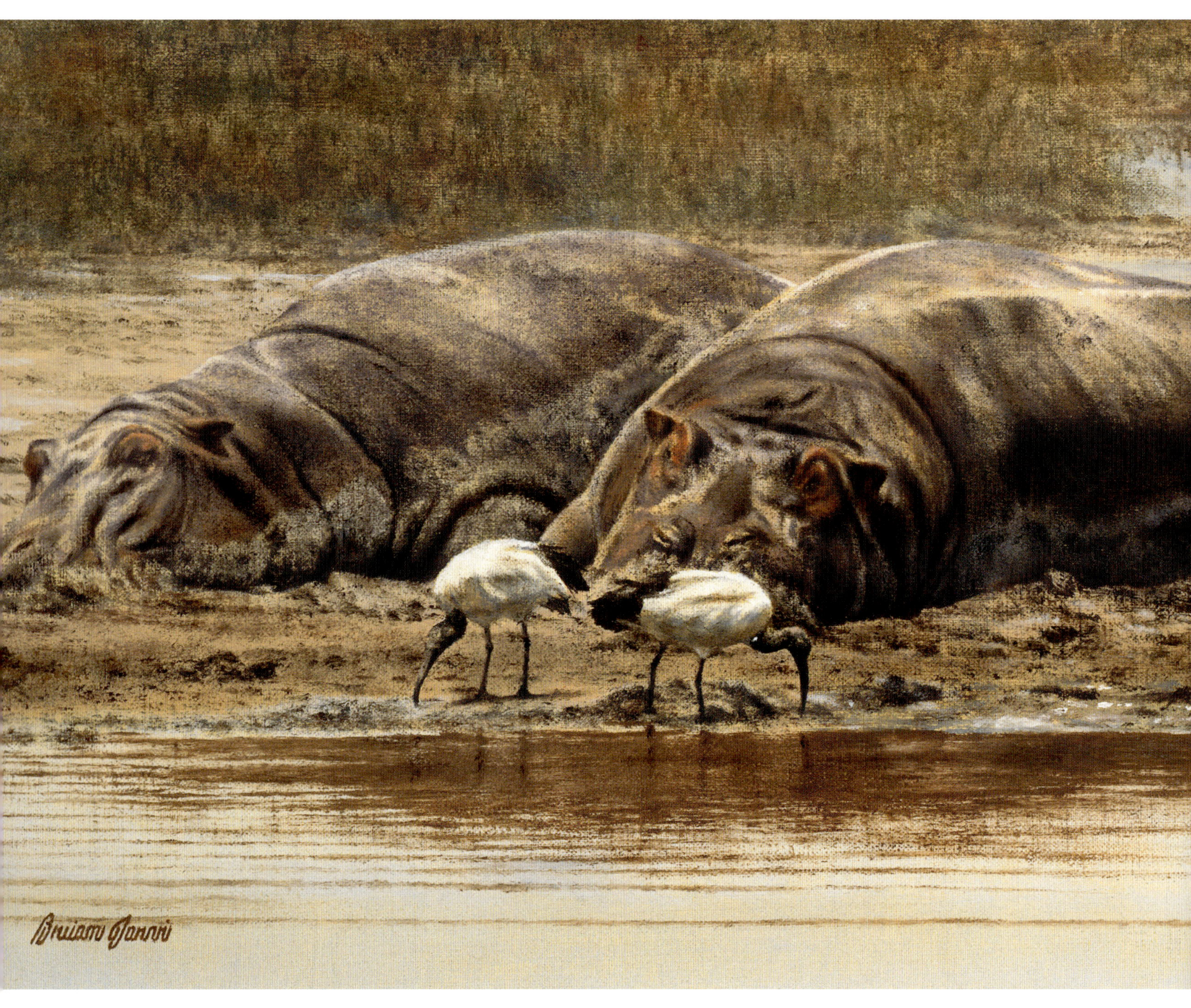

On the Flats, 2003
Hippopotamus
Oil on Belgian Linen
8 x 20"
Private Collection

humor to test whether we are paying attention. "He does this with lovely touches," said Greene. "The scene suggests that this group of beasts and birds have gotten together spontaneously for a meeting in a curiously believable fashion, a remarkable feat on its own, for it doesn't feel contrived. The underlying poignancy to the scene is of course that many of the species depicted are endangered or nearing extinction through habitat loss and poaching."

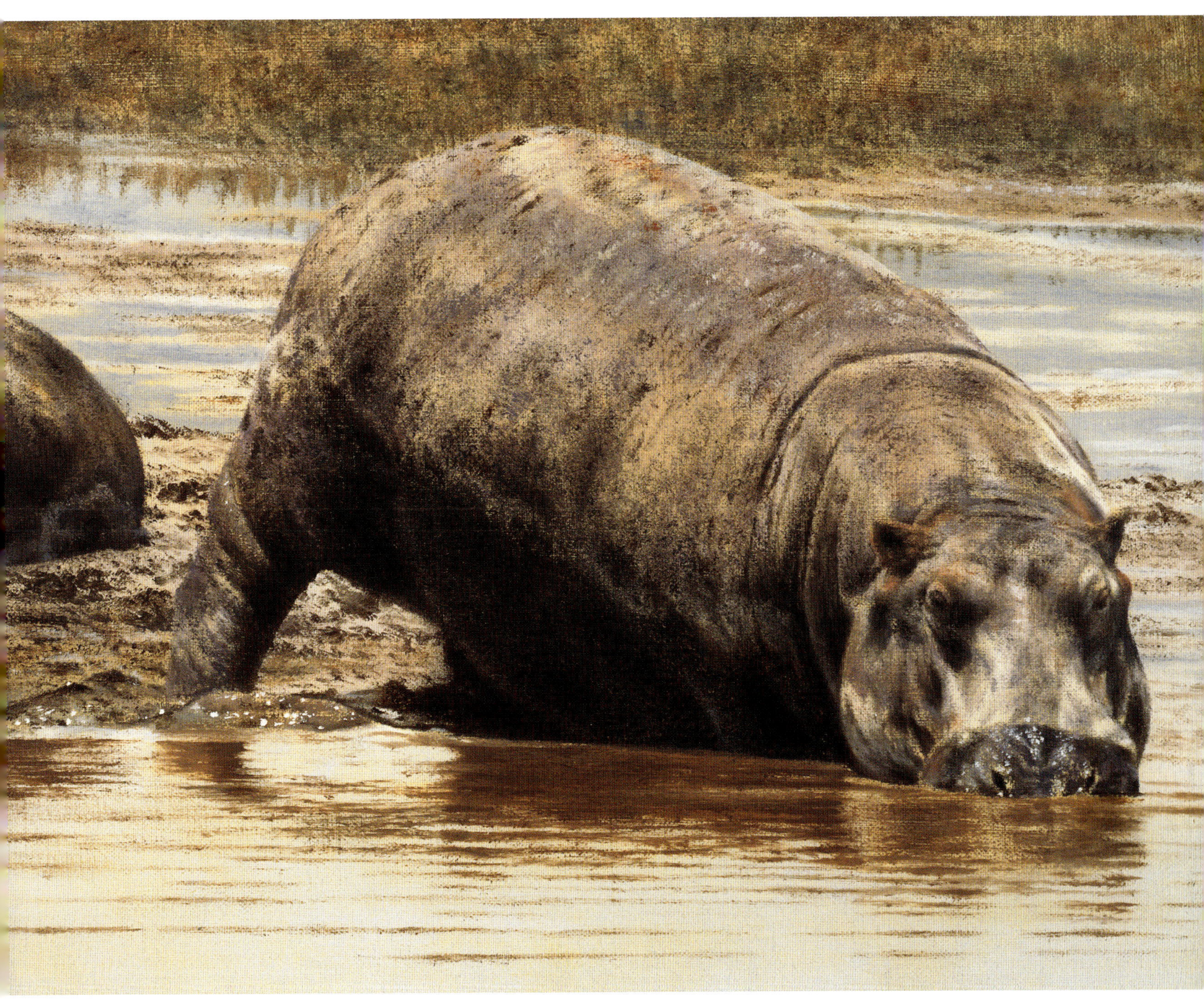

The painter Joseph Sulkowski, counted among the greatest living sporting artists in America, said Jarvi has created something historic. "In the 21st century, the roots of the classic traditions remain alive in Brian's work," he said. "In *African Menagerie*, he has brought all the elements of the painter's art along with a deep knowledge of his subject together in a brilliant and uniquely masterful style that transcends the constraints of time and space."

THE ARTIST'S INTENT

While we all have busy lives, tending to the needs of our own families, making ends meet, often not pushing ourselves out of our comfort zones, perhaps resigned to believing there is nothing we can do to make a difference, Jarvi's hope for *African Menagerie* is that it represents a reason to assemble before beauty.

"I didn't create *African Menagerie* to be a piece that could be interpreted only as decoration," Jarvi said. "I want to engage the viewer. I want you to be inspired. I want you to feel wonder. I want you to lose your breath and squirm when the reality hits home that many of these species are imminently threatened. These animals don't belong to any one people or country. They are part of our common heritage. I want viewers to realize that if we want the richness of biodiversity to continue for future generations to enjoy we need to become involved. I hope this painting helps spark a vow from viewers who will say they refuse to let extinction happen on their watch."

How the fate of Africa's treasure trove of creatures is decided has implications for wildlife everywhere else. In one sense, *African Menagerie* is an homage to wildlife, an investigation of the web of life to which we are all connected and intertwined, and a rallying cry. Art historians also say that *African Menagerie* is a historic yardstick laid down at our feet that functions not only as a wake-up call for contemporary viewers, but also as a timeless lodestar for humans to come. As its subtitle—*An Inquisition*—suggests, it is also an opportunity to hold a mirror up to ourselves.

"I want to inspire thought, not shove information down the throat of the viewer," he said. "It would be easy to overwhelm viewers with details alone and lose sight of the larger theme, which is the convergence of a community of species interwoven together, each representing vital strands in the web of life."

"I've known Brian since he was a duck stamp artist and I understand how Africa infected him because it did the same to me," said Jarvi's friend, the American wildlife artist Jan Martin McGuire. "It's scary because once you begin studying your subjects and realize the prospects of their survival is tenuous, you can't go back to where you were and pretend you live in ignorance. You feel compelled to do something about it. The problem is too many humans using up too many resources and pushing animals into smaller and smaller

Bonobo Study I, 2011
Bonobo
Oil on Belgian Linen
14 x 11"
Collection of Annette Bishop

This piece was juried into the 2011 Artists for Conservation international exhibition.

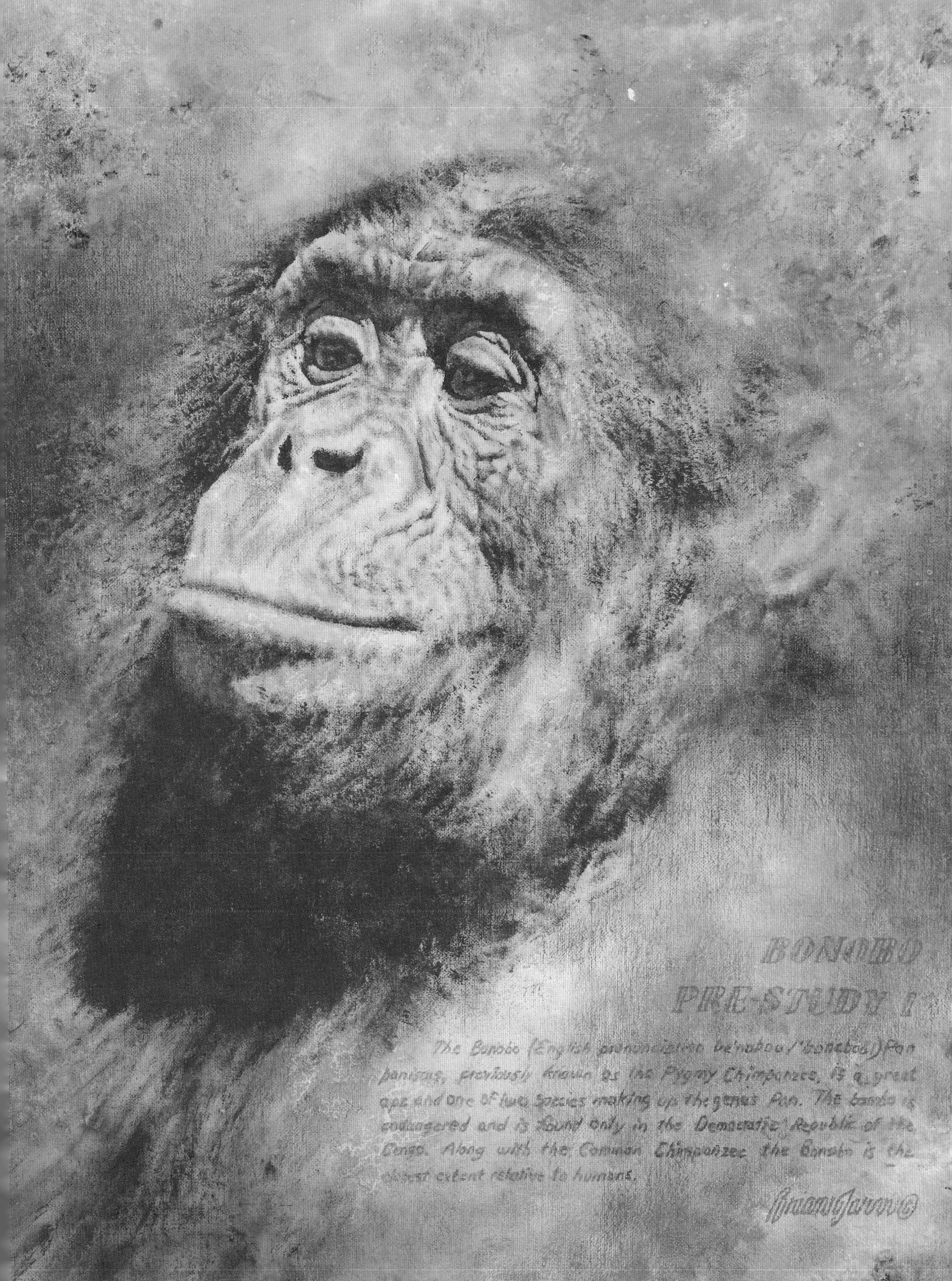
BONOBO
PRE-STUDY I
paniscus, previously known as the Pygmy Chimpanzee, is a great
ape and one of two species making up the genus Pan. The bonobo is
endangered and is found only in the Democratic Republic of the
Congo. Along with the Common Chimpanzee the Bonobo is the
closest extant relative to humans.

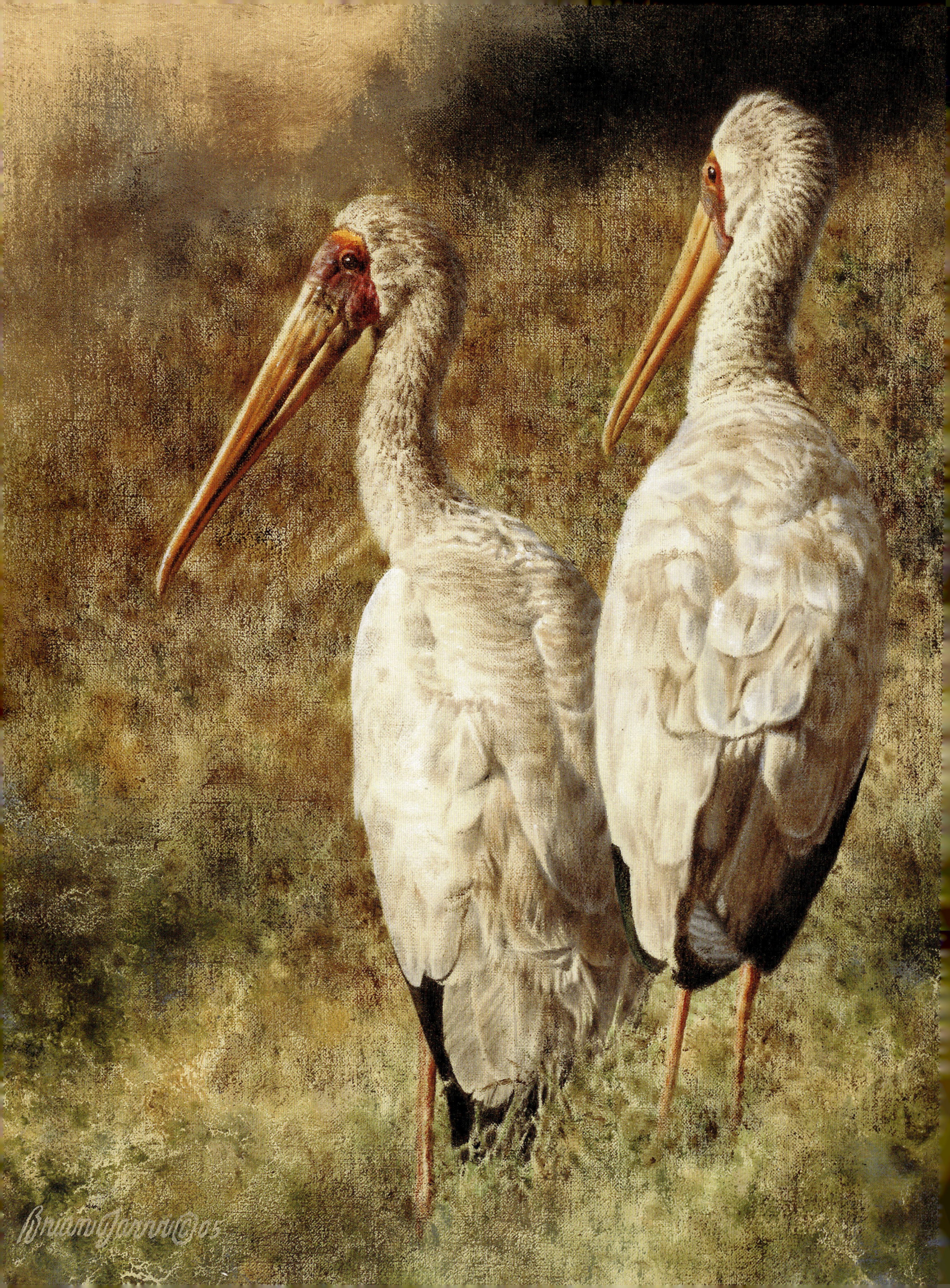
Brian Jarvi ©05

places. What can art do? It can make the case for protecting the habitat nature needs before it's too late."

The iconography of wildlife has never been more topical. As the human population rises and more wild places are being whittled away, wildlife is a reminder of what's at stake. At the same time that *African Menagerie* was traveling the country on its museum tour, Jarvi had a replica made and debuted it at the 2018 Safari Club International convention. It attracted attention from individuals who have some of the most prestigious sporting art collections in the world.

CREATING A MASTERPIECE

African Menagerie is the culmination of a once amorphous dream involving a revelation that Jarvi carried forward for decades. Grand Rapids businessman Steve Gilbertson and his wife, Cindy, were among several citizens who stepped forward at a crucial point and enabled the painter to convert the outlandish idea into a tangible reality. They formed a patron group, affording Jarvi the necessary time he needed for research and compositional execution. "What instantly attracted me was hearing Brian's vision," Gilbertson said. "In actual fact, he had been painting it in his head for years, and boy was he thinking big."

While Jarvi had no doubt he could pull off his opus, it wasn't until *African Menagerie* was unveiled—after countless all-nighters and years of toiling in the studio—that anyone realized its monumental impact. "There's no better place than Grand Rapids to rally behind a cause that might ordinarily be written off as impossible. The community spirit we have here is special," Gilbertson said. "Brian is one of our own and we are proud to be sharing his greatest achievement with the rest of the world."

Jarvi worked on the painting over the course of three years. The first phase involved sketches, the second an array of animal studies serving as preparatory exercises for the master composition, the third roughing in the shapes, and the final the application of paint. Once he deepened the journey into phase three, there was no turning back. The perfectionist Jarvi was constantly wrestling with the Jarvi who was letting go of the insecurity he had and learning to believe in himself.

In his loft studio above his garage, surrounded by his tools and reference points—skulls and hides and tactile trinkets he has collected during his travels—Jarvi spent months

Yellow-Billed Storks, 2005
Yellow-Billed Storks
Oil on Belgian Linen
15 x 12"
Private Collection

In 2005, this piece was juried into the Leigh Yawkey Woodson Art Museum's *Birds in Art* exhibition.

using mathematical measurements to methodically ensure the figures in *African Menagerie* were precisely scaled. He created shading to serve as reference points of value and tonality in his sketches. He continually applied layers of paint to different surfaces to perfect a grand vision of interwoven aesthetics.

As to the genesis of *African Menagerie*, it originally started to gel in Jarvi's mind only as a rough artistic concept. He was, frankly, tired out by the American wildlife art scene, which he said had become formulaic. "I reached a point where I did not want to do what had been done before," he said. Endeavoring to even attempt *African Menagerie* required that he enter a different state of mind, retreating into seclusion, vanishing from the known path that had propelled his career forward.

Something else, however, sent Jarvi into a period of deep introspection in which he actually lost his focus on painting. As his mystique continued to surge, he received devastating news in 2006. His wife, Raelene, was diagnosed with advanced-stage cancer, with treatment options considered longshot hopes at best. With Raelene receiving treatment at the Mayo Clinic and struggling to stay alive, the couple was worried about the impact on their two daughters, Morgan and Haley.

Jeanne Nicklason, a close family friend and mother of a child the same age as Morgan, tried to help buffer the shock. Jarvi had painting commissions to complete, but his thoughts were elsewhere. Raelene kept on him to keep working, knowing it was the only thing that might keep him centered. Nicklason said the whole town of Grand Rapids came together for the family, which is one of the reasons Jarvi decided to hold the opening for *African Menagerie* there.

"We prayed for a miracle and we got one," Jarvi said. "Had I lost Raelene, I don't know if I could ever have picked up a brush again." The painter could barely talk about the possibility of losing his wife without choking up. It was, after all, Raelene who encouraged him to paint with heart and purpose, to not live up to the expectations of others, or, worse, to create art only for money.

"When Raelene was ill, her wonderful girlfriends rallied around her and our family—making meals, housecleaning, bussing our children, doing laundry, driving her down to Mayo, whatever was required," Jarvi said. "As a small measure of our gratitude, we created a special edition of the painting *Yellow-Billed Storks* and presented it to her girlfriends. Raelene often said the image reminded her that she never felt alone; there was always someone at her side."

In many ways, Jarvi considers the completion of *African Menagerie* a gift to his wife and daughters. The night before the unveiling in Grand Rapids, Jarvi, severely sleep deprived, sat in his studio holding a dram of scotch in one hand and a celebratory cigar in the other. And then in drifted Raelene and the couple's now-grown daughters. They gathered around Jarvi in an embrace. He cherished the moment and began to tear up. Once upon a time, Jarvi's siblings had chided him for not finishing his artwork. Before the three people who meant the most to him, he let his emotion pour out and didn't say a word. The artwork spoke for itself, accentuated by the smell of fresh paint still drying.

The next day at the Reif Performing Arts Center, while the seven paintings were being assembled, Haley climbed on stage with her dad, and the two of them shared an emotional moment. "It is such a privilege to call this extraordinary man not only my most valued mentor, but also my father," she said later. "I am incredibly inspired by his talent, drive, and success. There is no one more deserving of dreams coming true. I am thrilled to see his hard work and talent flourish into this monumental masterpiece."

"It's not every day you hear someone say, 'My dad is an artist.' It can be a challenging career to take on, and he had the perseverance to build it on his own. I'm so proud he was able to accomplish *African Menagerie*; it was a lot of work and I was astonished by the outcome," Morgan said. "He's passionate about what he does, not just through his painting, but in how he explains the story behind each piece of artwork."

Some may wonder what it is that gives Jarvi the greatest joy. It's when he watches people spend time in front of a painting—pointing to it, stepping closer in contemplation, walking away, and coming back. He derives satisfaction from witnessing people who are moved by the illusion and wondering how he did it.

"I feel so proud and honored to have witnessed the making of *African Menagerie*," Raelene said. "Brian's ability to make subjects come to life on canvas is truly amazing. It was an immense undertaking and remarkable to follow from inception to fruition."

African Menagerie is a celebration of life on earth. There are, after all, no African elephants or lions on Mars, no rhinos or Cape buffalos on Venus, no pink flamingos, leopards, hippos, or gorillas on the moon. Our planet is a miraculous place. If a microbe were proved to exist on those neighboring

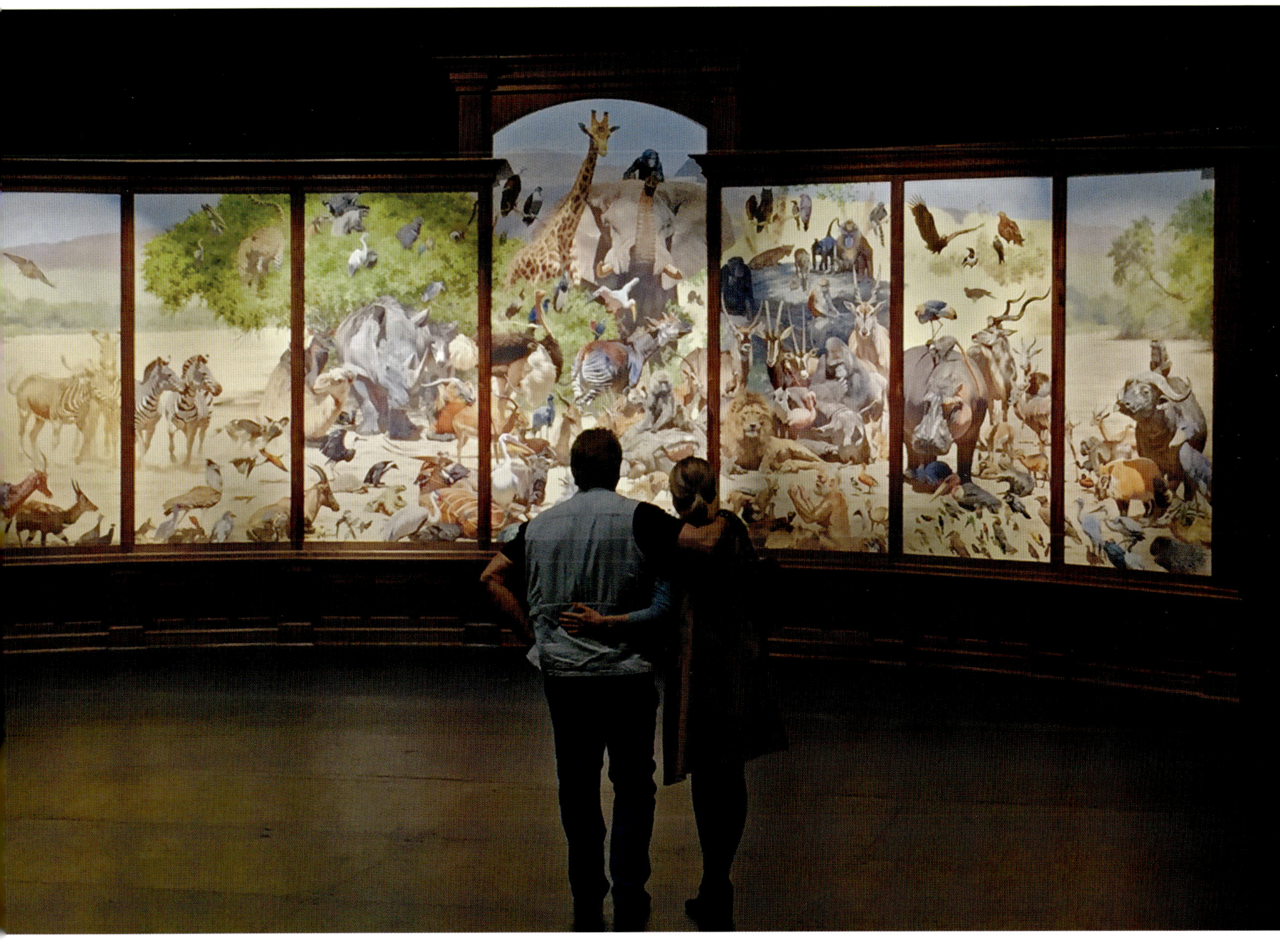

orbs, it would be earth shattering. Instead, higher life-forms are present only here, each one the product of millions of years of temporal engineering. And once a species disappears, like the dodo, it is gone forever.

In these times of social unrest, divisiveness, and uncertainty, it is sad how difficult it is for Americans to rally together and pay attention to the important things that unite us. Jarvi reminds us that it can happen with an art project that seizes our attention with its dazzling beauty, and then inspires us to count what's in front of us and gasp at the thought that much of our living earthly treasure could be lost. *African Menagerie* implores us: let's not let it happen.

Jarvi shares a quiet moment of reflection with his daughter Haley after the final panel is put in place. It is the first time *African Menagerie* has been fully assembled, just hours before the formal unveiling at the Reif Performing Arts Center.

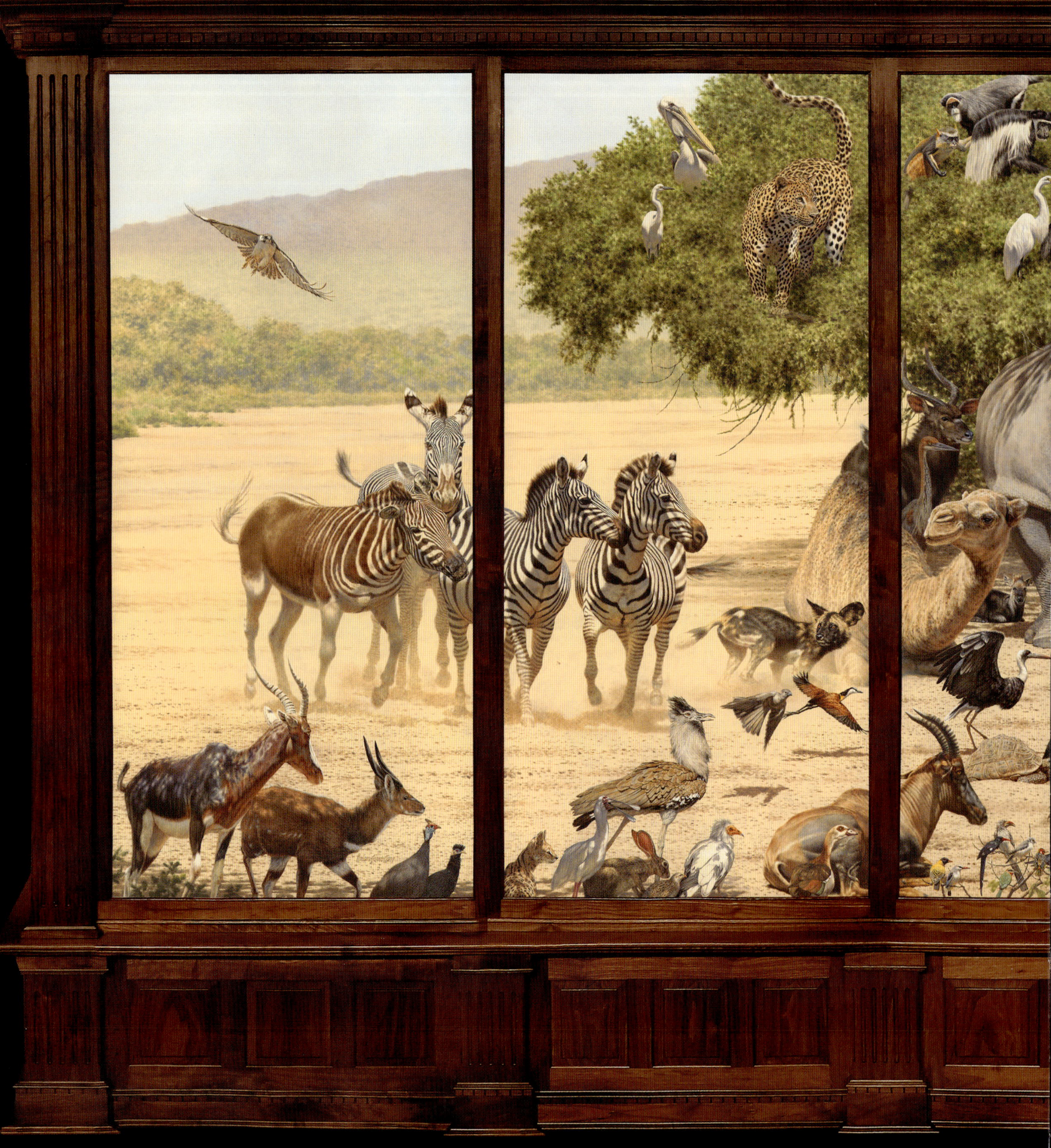

AFRICAN MENAGERIE

A Celebration of Nature

From its conception, *African Menagerie* has experienced a journey of growth, meaning, and evolution. Now it has entered the most exciting phase—completion of the epic vision. Spawned from an early childhood fascination with both living collections of wild animals and paintings from the distant past portraying large groups of exotic creatures, this 28-foot-wide work stretches across seven panels and features more than 200 different African species. • In *African Menagerie*, the gathering of an incredible array of wildlife is set against a grand panoramic backdrop, highlighted by a view of the receding snows of Mount Kilimanjaro. All of Africa's icons—including the elephant, leopard, giraffe, rhino, hippo, and zebra—mingle with such exotic species as the bongo, okapi, and mandrill. More than 100 bird species—ostrich, shoebill, lilac-breasted roller, and sacred ibis among them—are also included. And man, who has come to dominate the earth, has been summoned to this gathering as the natural world seeks answers to the growing issues of survival faced by countless species across the planet. • Allegorical story lines intended to dramatize the urgency of the moment are woven into this idyllic scene. The arrival of the Four Horses on the far left and the Lion and the Lamb, seated directly in front of a Da Vinci-like figure in the foreground, compel the viewer to interact with the scene. • Over the years, the original concept of simply seeking to create art evolved into something far more meaningful: a message to humanity intended to inspire acts of conservation, acts that will save not only the iconic species of Africa but also other wildlife across our fabulously diverse planet.

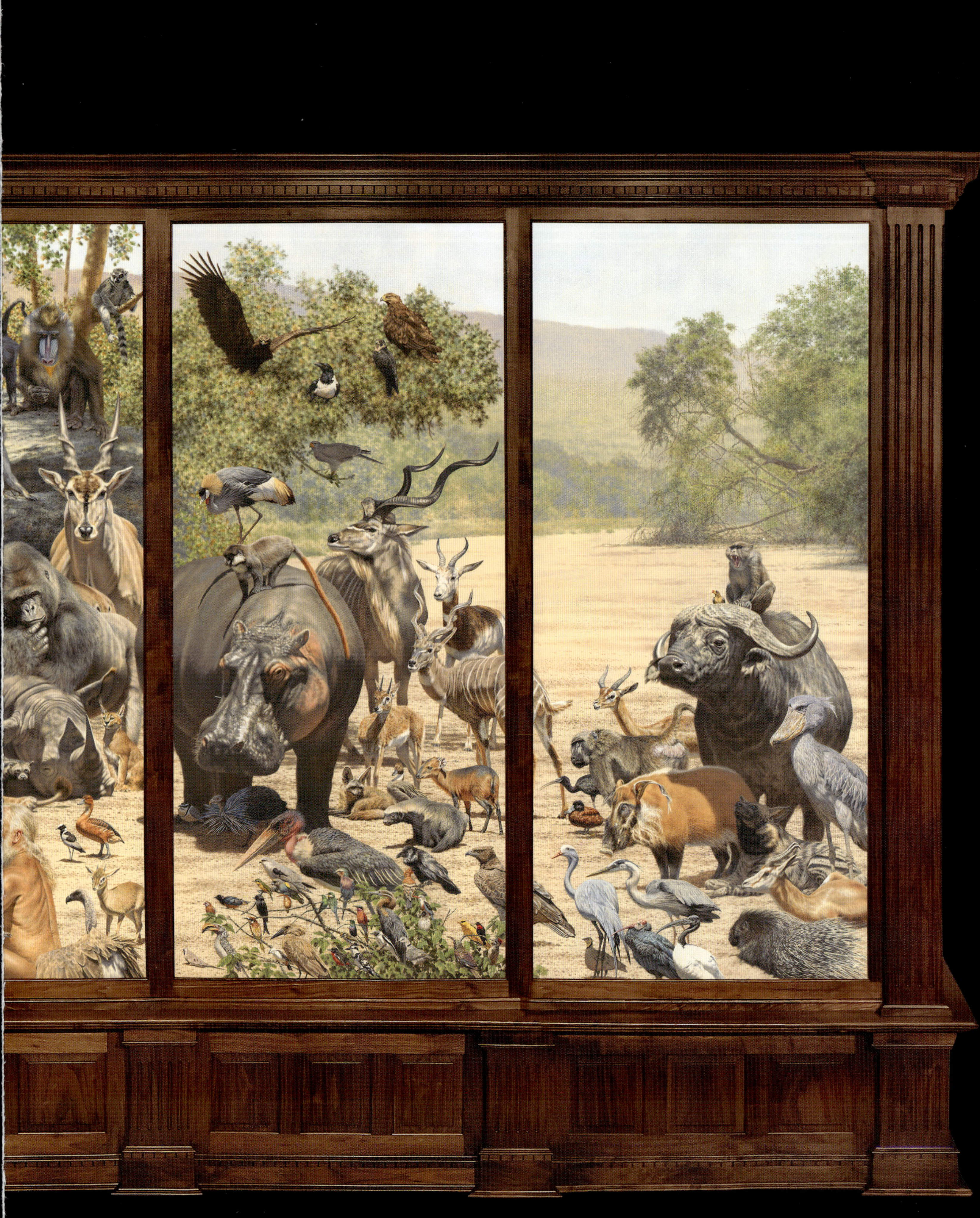

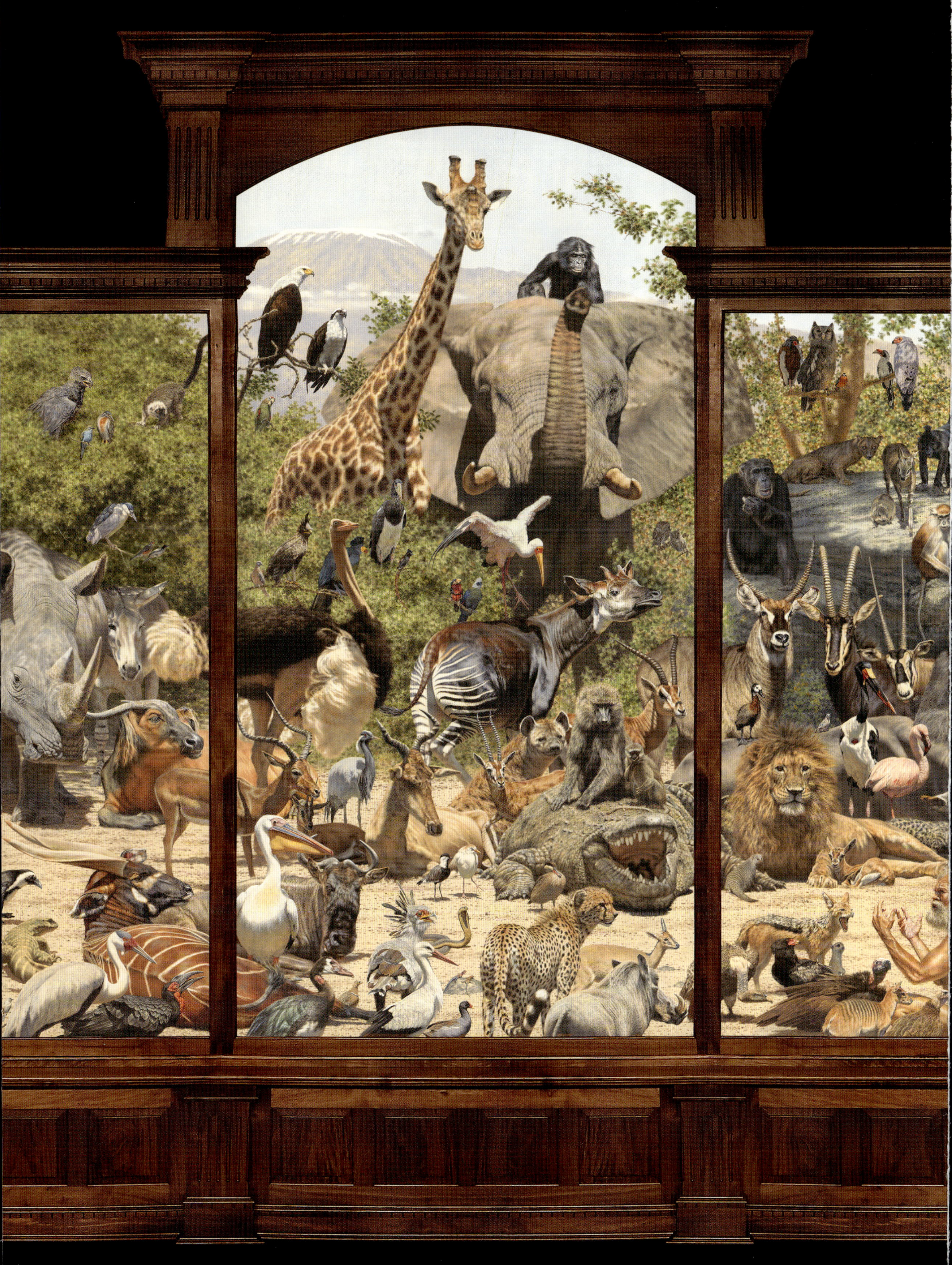

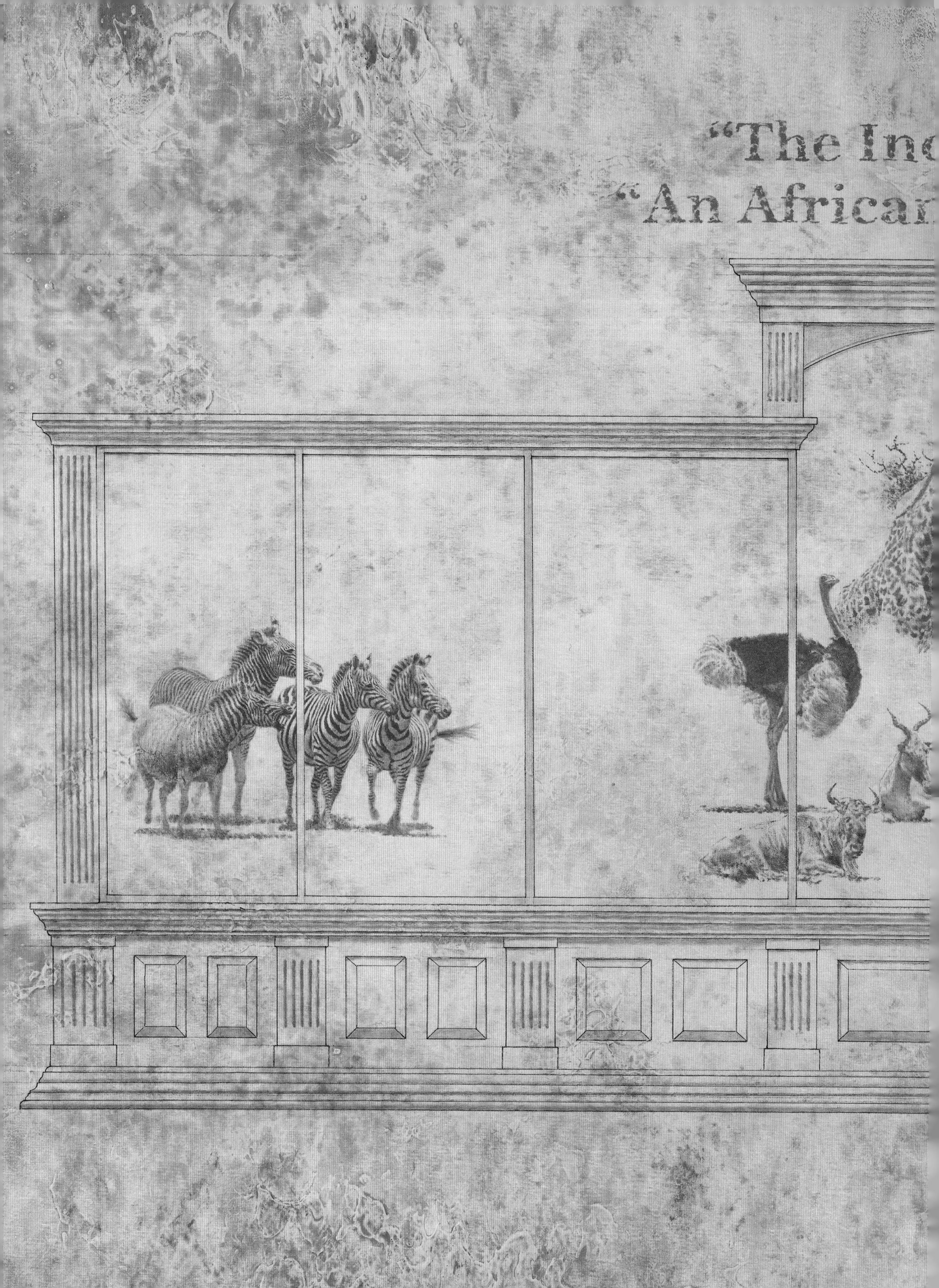

"The In
"An African

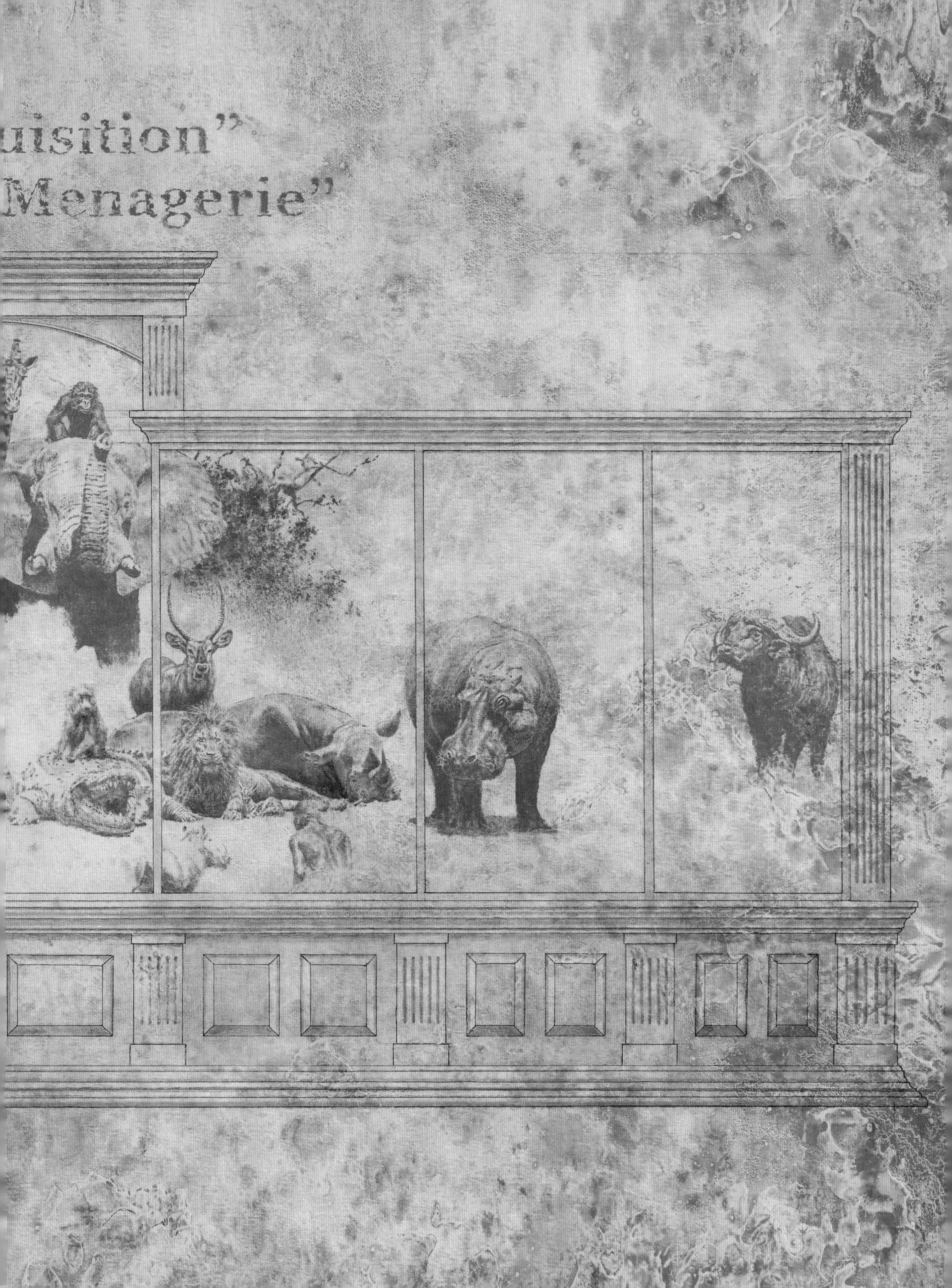
uisition"
Menagerie"

PREVIOUS SPREAD

African Menagerie Prospectus, 2014
African Species
Oil on Belgian Linen
28 x 46"

Jarvi spent years developing the concept for *African Menagerie*—its size, scale, and dimensions. To present vertical animals like a giraffe and elephant at the proper scale, he realized he needed to add an extra foot in height, giving it a reach of nearly one and a half stories tall.

The Last Quagga, 2016
Quagga
Oil on Belgian Linen
36 x 48"
Private Collection

OVERLEAF

African Menagerie, Panels 1 and 2
Detail (Middle)
Grevy's Zebra, Quagga, Burchell's Zebra, Mountain Zebra

One of Jarvi's earliest objectives was ensuring that *African Menagerie* visually read as a classical, timeless painting rather than the replication of a photograph that had come to dominate modern wildlife art. One of the many subthemes is *Four Horses*, represented by four zebras and intended to evoke thoughts of the potential apocalypse of extinction. Jarvi introduced them on the far left, where they enter the scene and set forth a flowing ripple of motion that carries across the painting with other species.

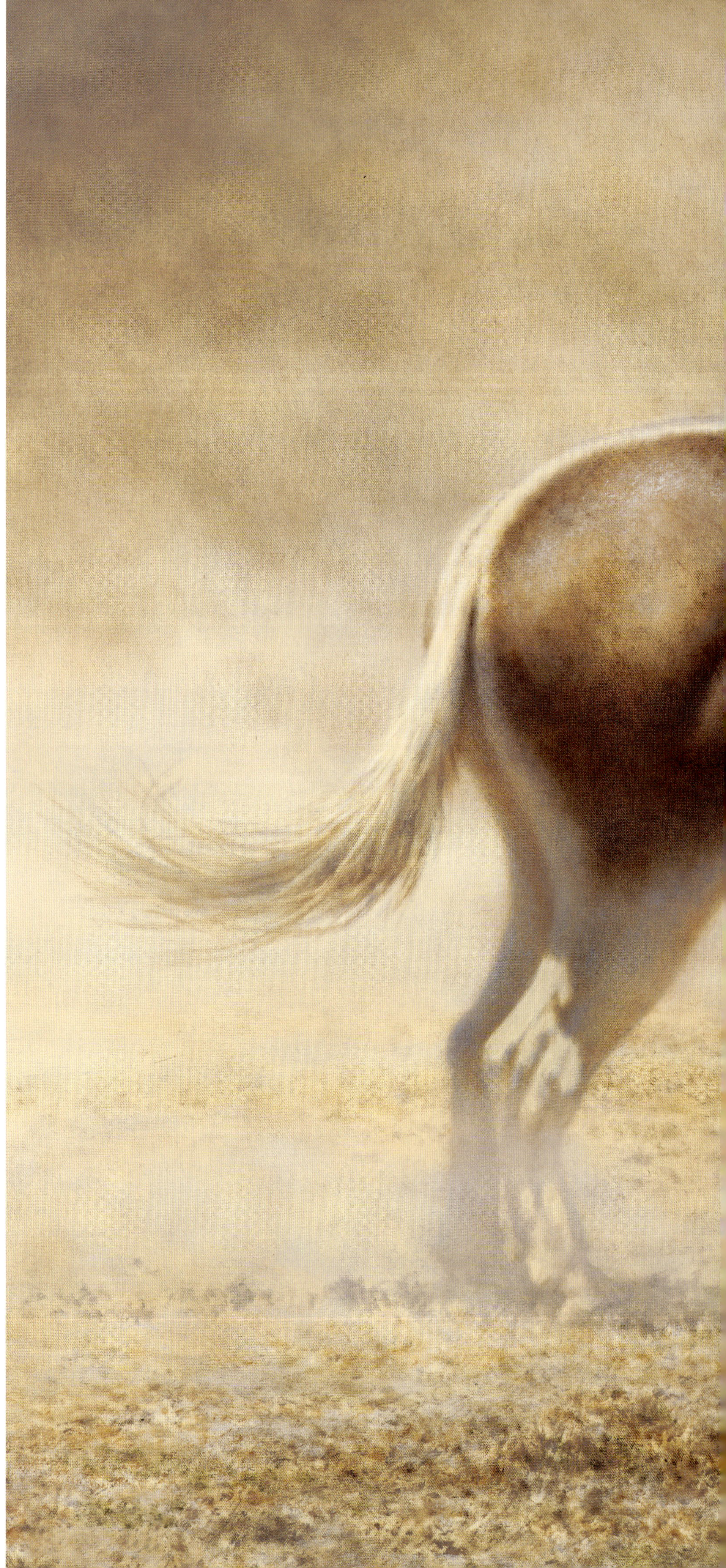

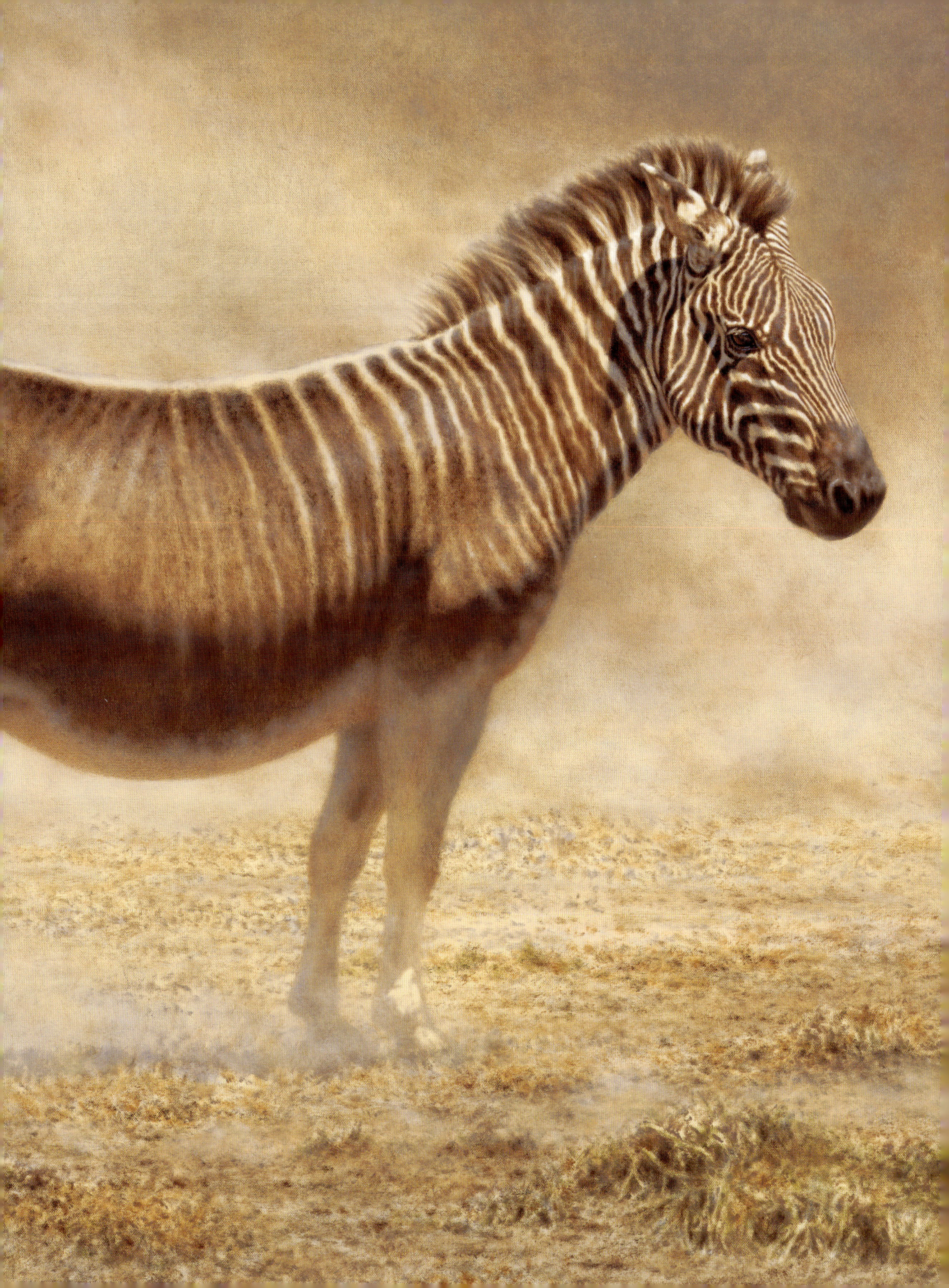

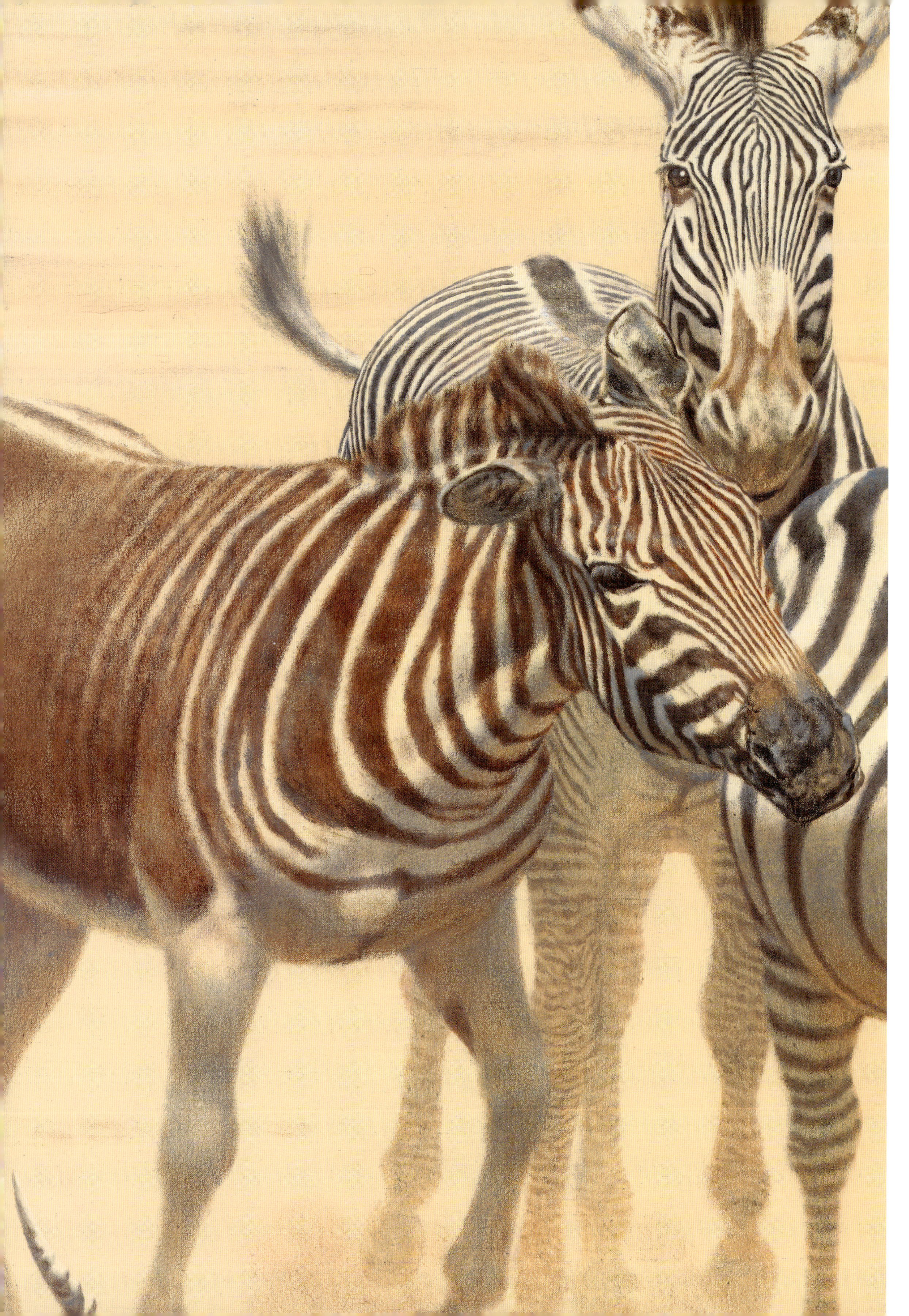

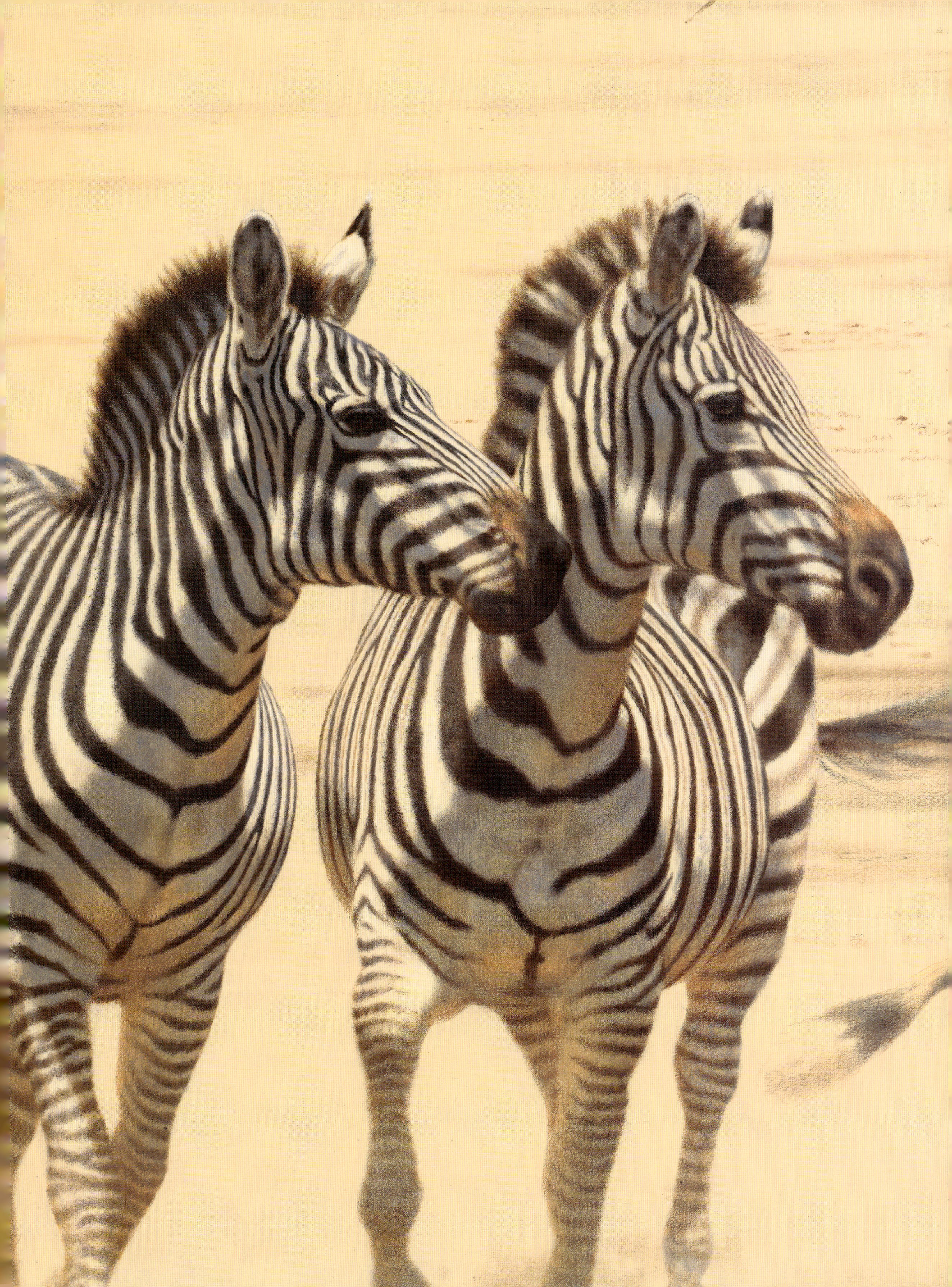

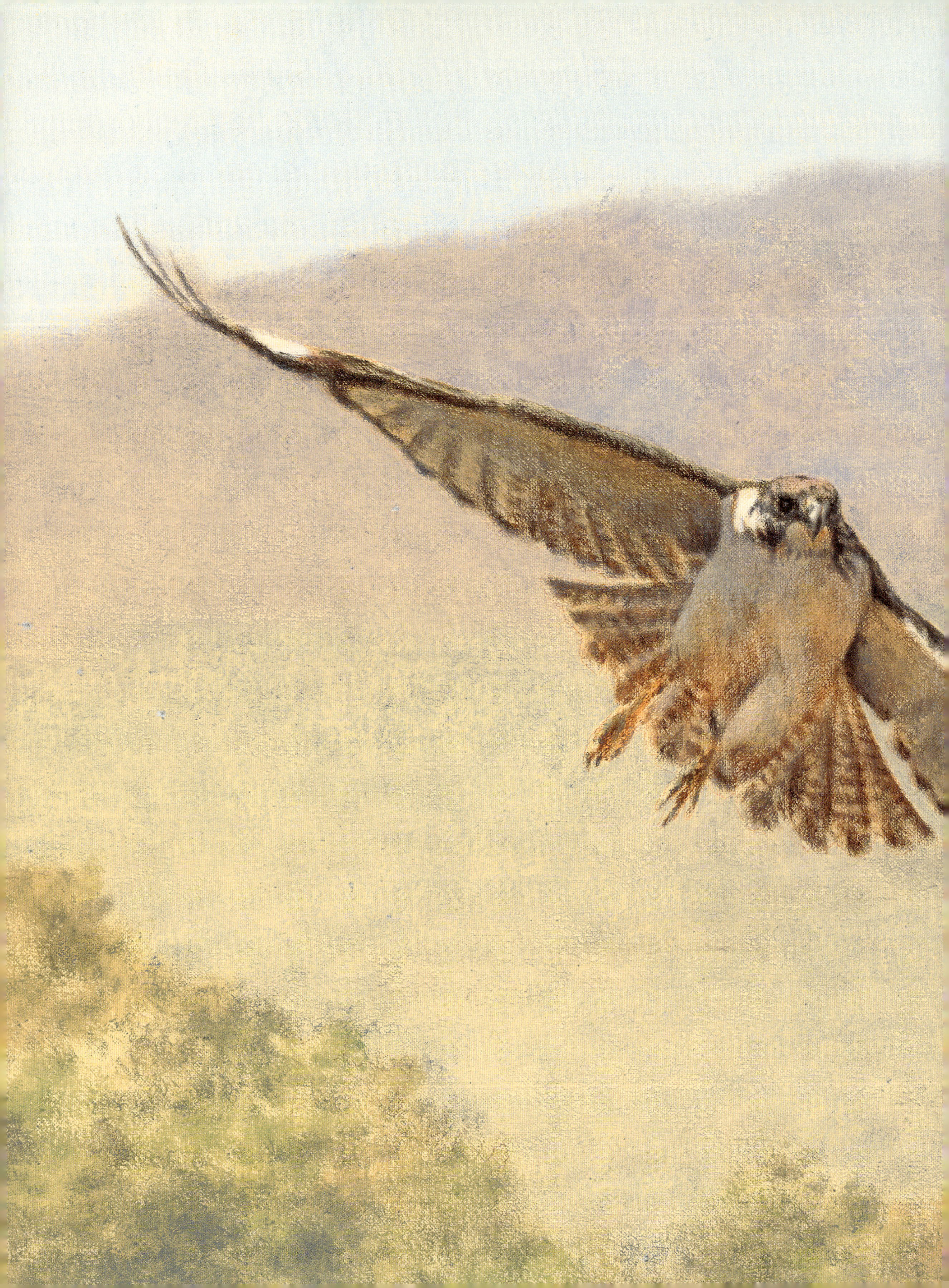

African Menagerie, Panel 1
Detail (Top)
Augur Buzzard

OVERLEAF
African Menagerie, Panel 1
Detail (Bottom)
Bontebok, Bushbuck,
Helmeted Guineafowl,
Crested Guineafowl

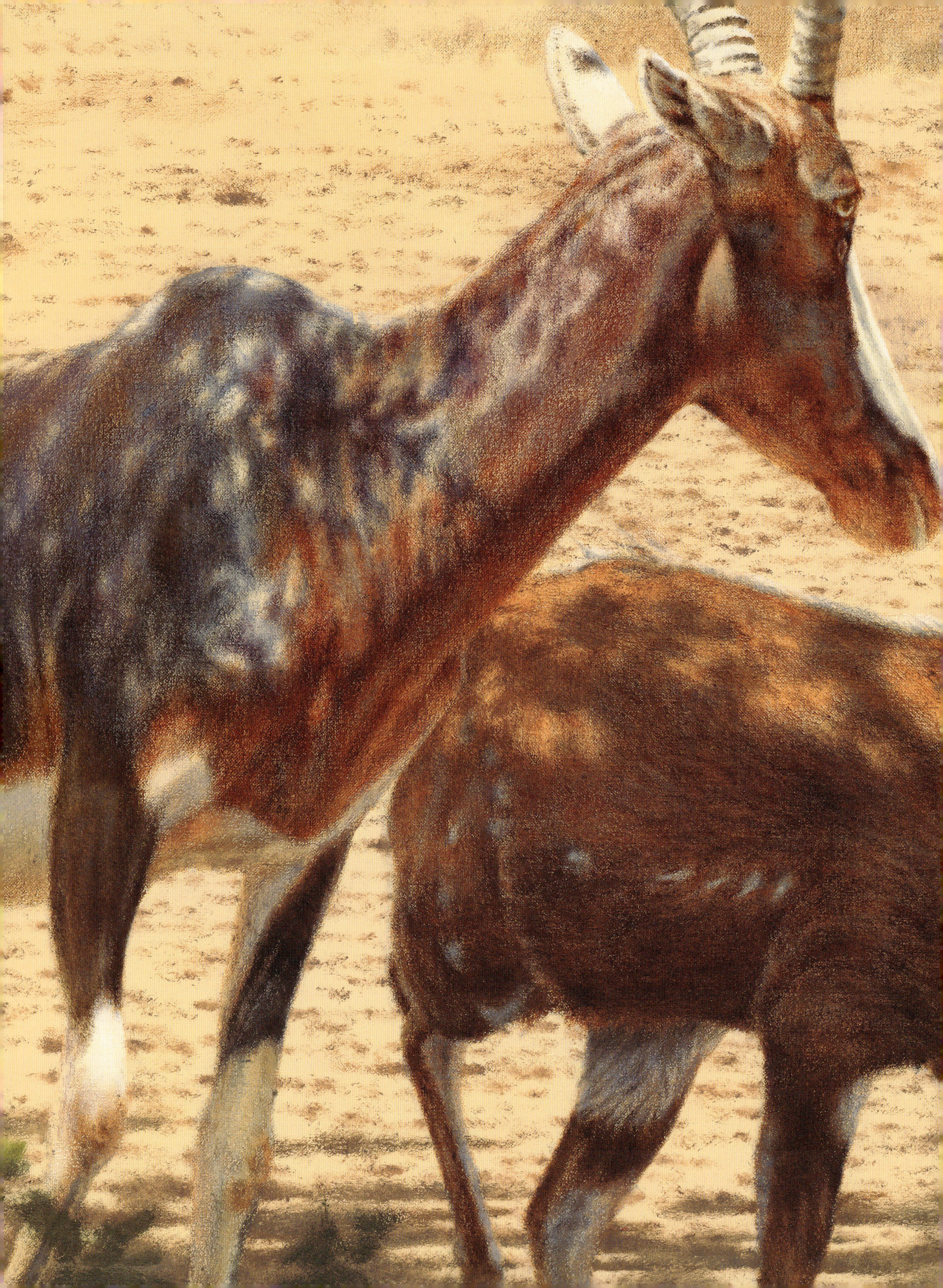

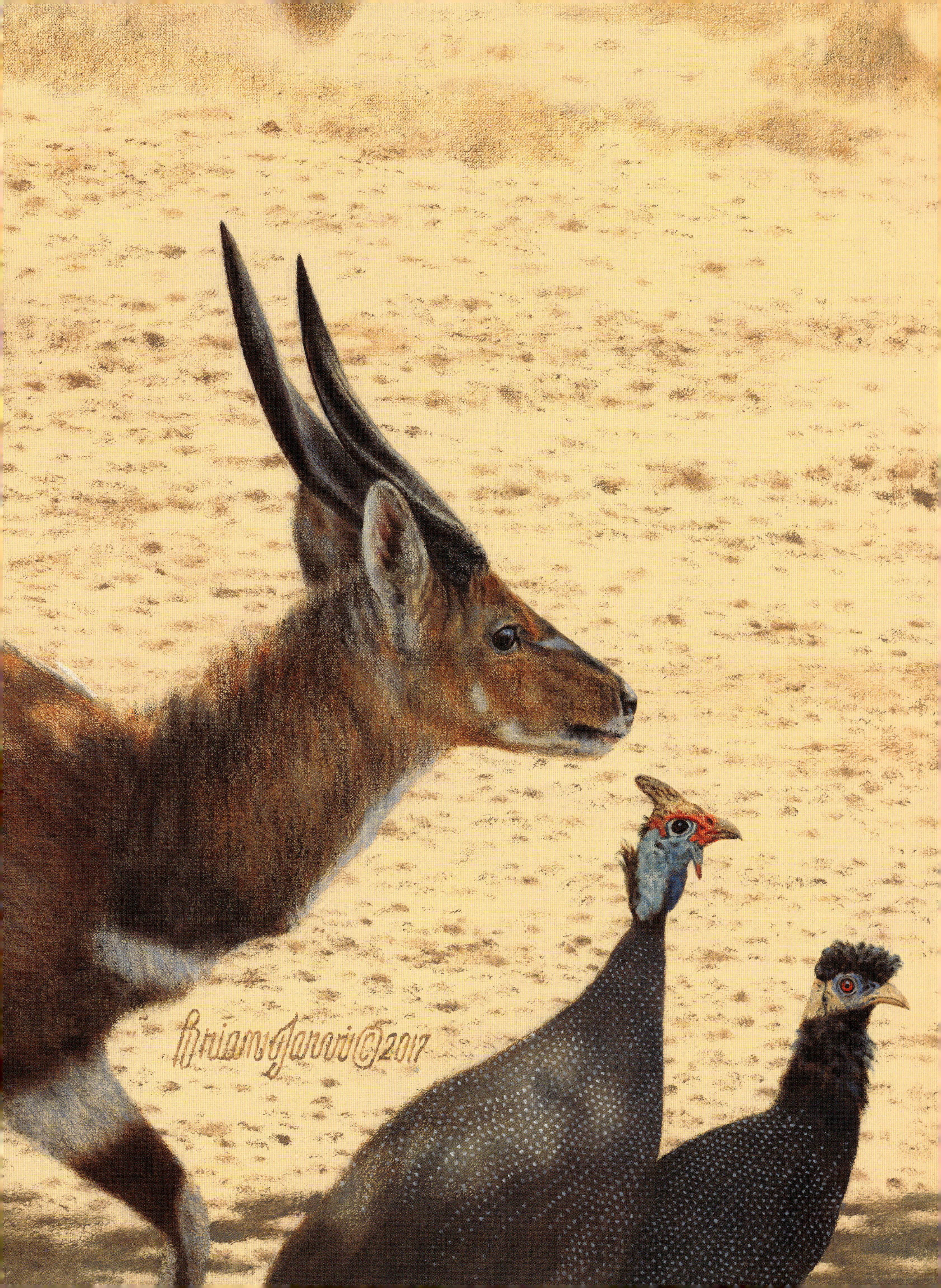
Brian Jarvi ©2017

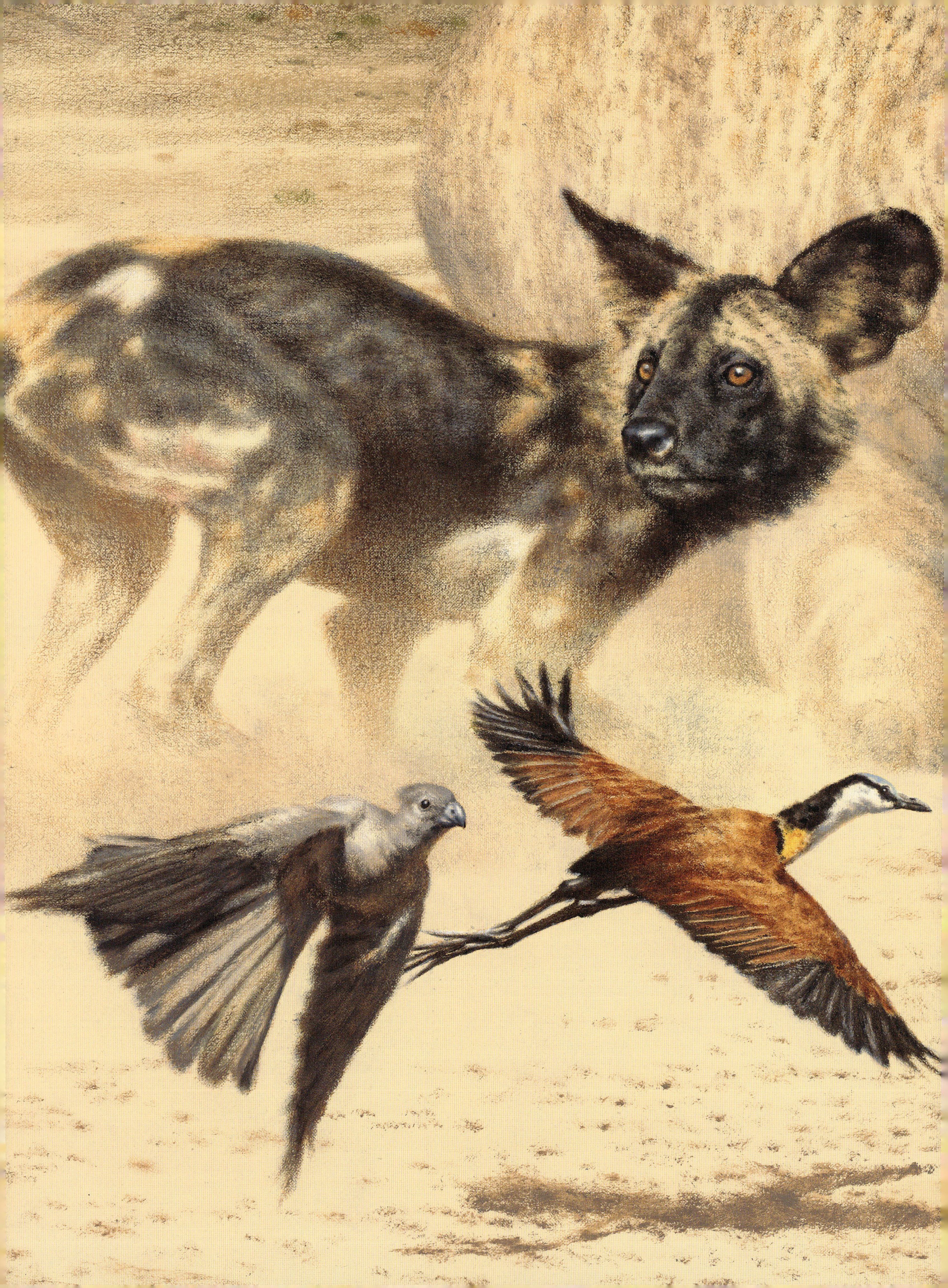

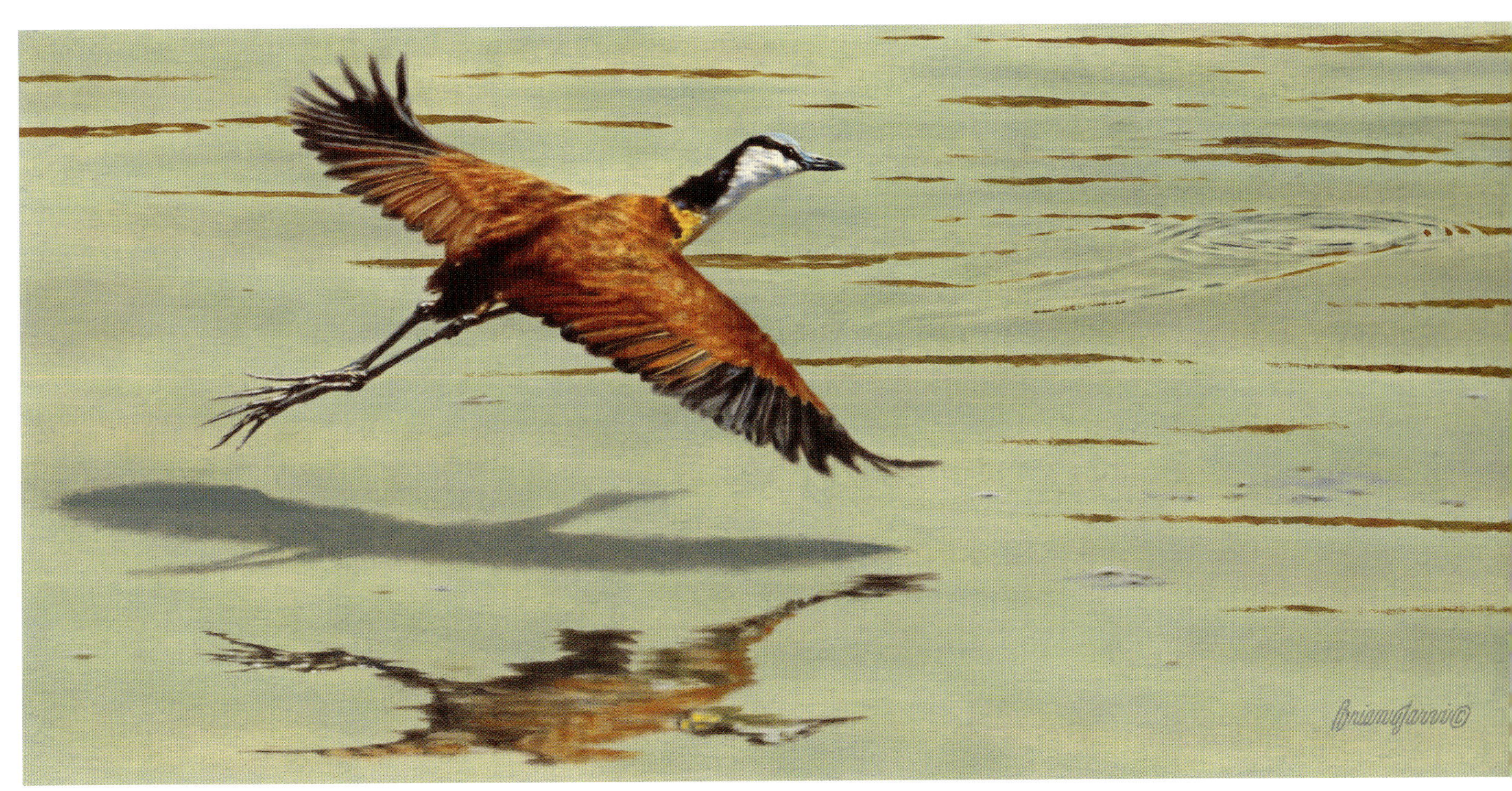

OPPOSITE
African Menagerie, Panel 2
Detail (Middle Right)
African Wild Dog, African Jacana, Grey Lourie

ABOVE
Jacana Study, 2016
African Jacana
Oil on Belgian Linen
12 x 24"
Collection of Steve and Vickie Wilcox

African Menagerie, Panel 2
Detail (Top)
Pink-Backed Pelican, Leopard, Little Egret

Like a painting within a painting, Jarvi's portrayal of a leopard in *African Menagerie* demonstrates the elegance with which he approaches individual subjects. He reveres the leopard both for its iconic status and its ability to adapt and survive.

OVERLEAF, LEFT
Cheetah Study, 2016
Cheetah
Pencil
11 x 13"
Collection of Lee and Mary Jo Jess

OVERLEAF, RIGHT
Panthera Pardus, 2014
Leopard
Oil on Belgian Linen
36 x 24"
Collection of Annette Bishop

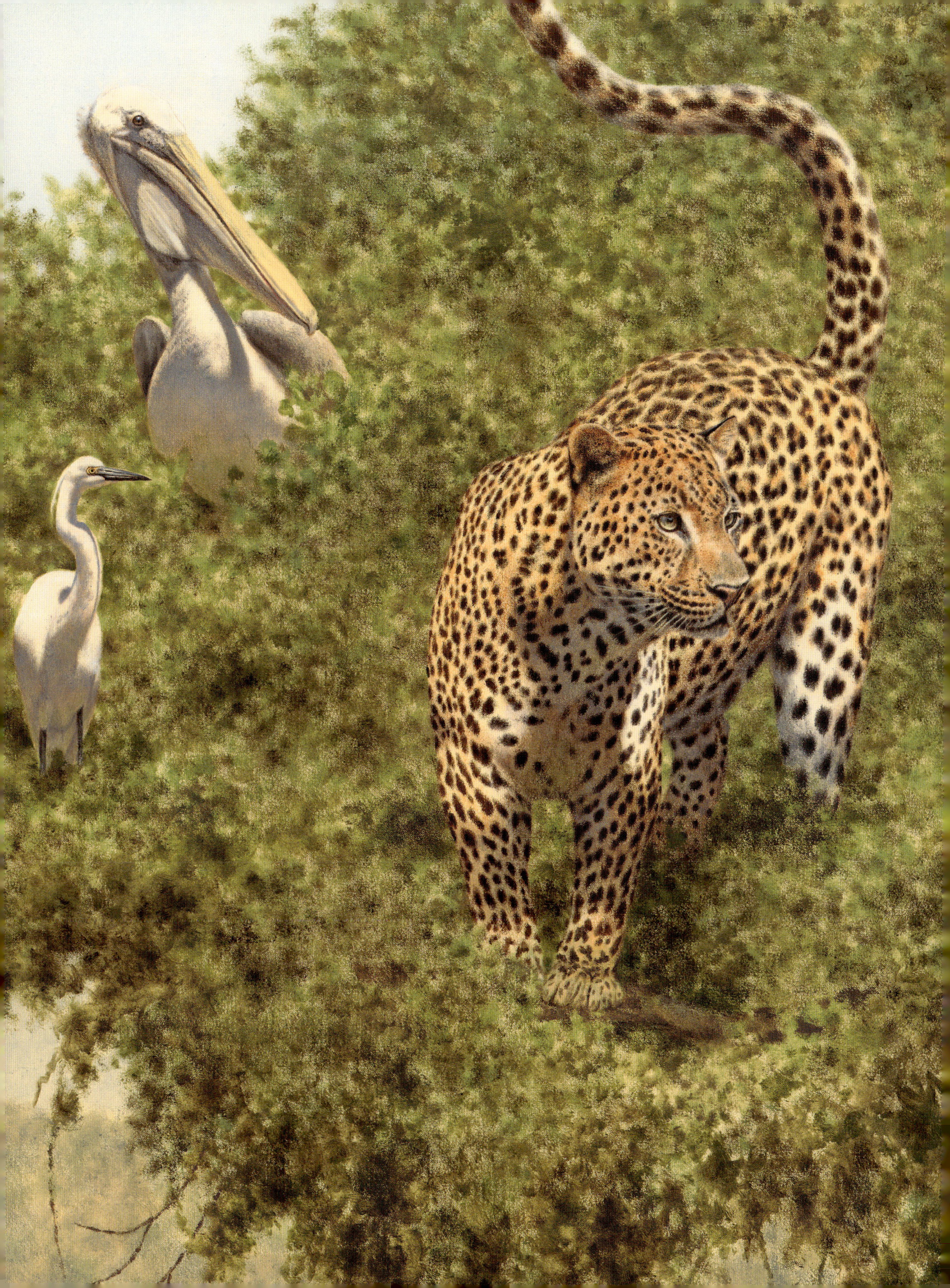

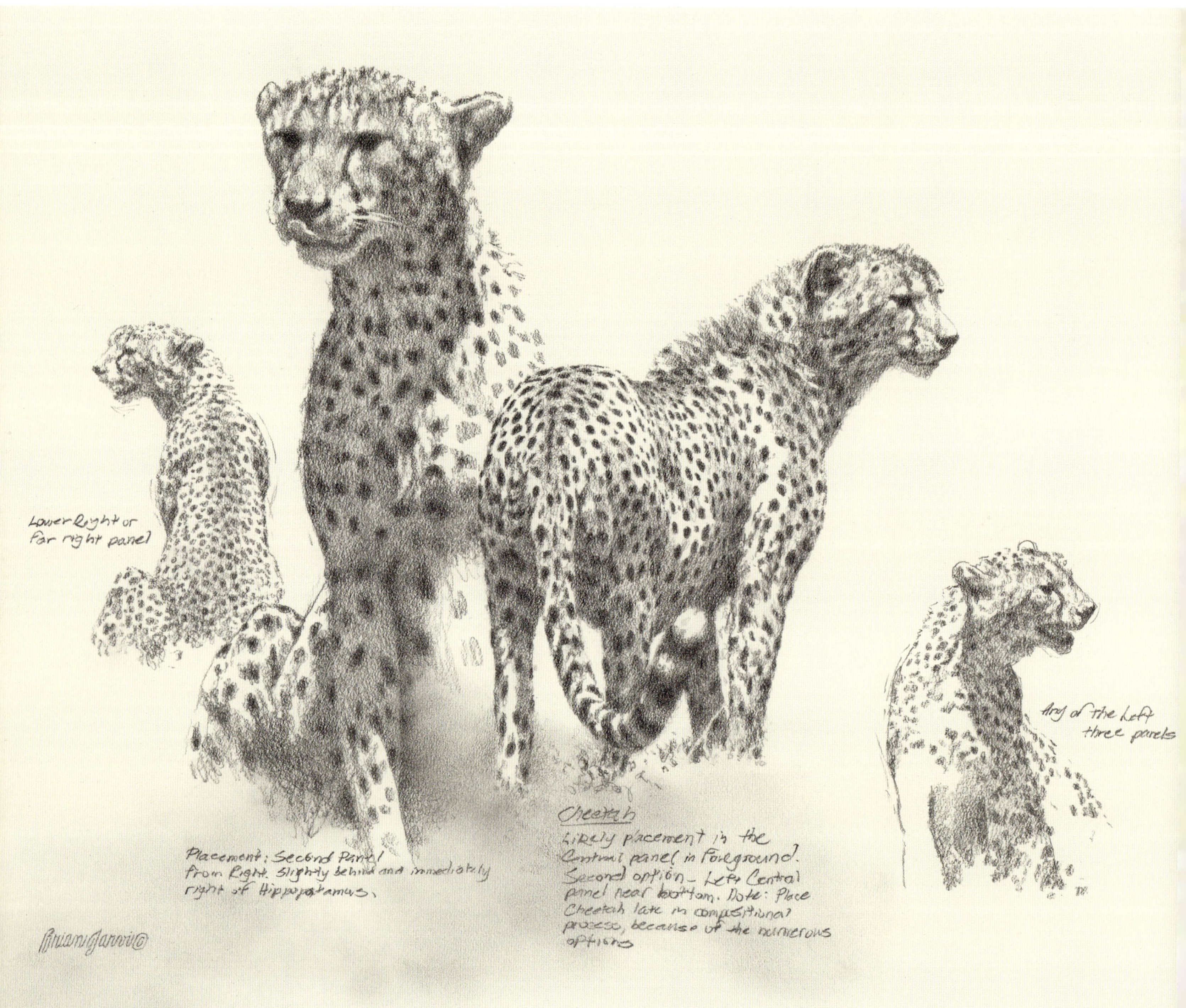
Lower Right or
Far right panel
Any of the Left
three panels
Placement: Second Panel
from Right. Slightly behind and immediately
right of Hippopotamus.
Cheetah
Likely placement in the
Central panel in Foreground.
Second option – Left Central
panel near bottom. Note: Place
Cheetah late in compositional
process, because of the numerous
options
Brian Jarvi ©

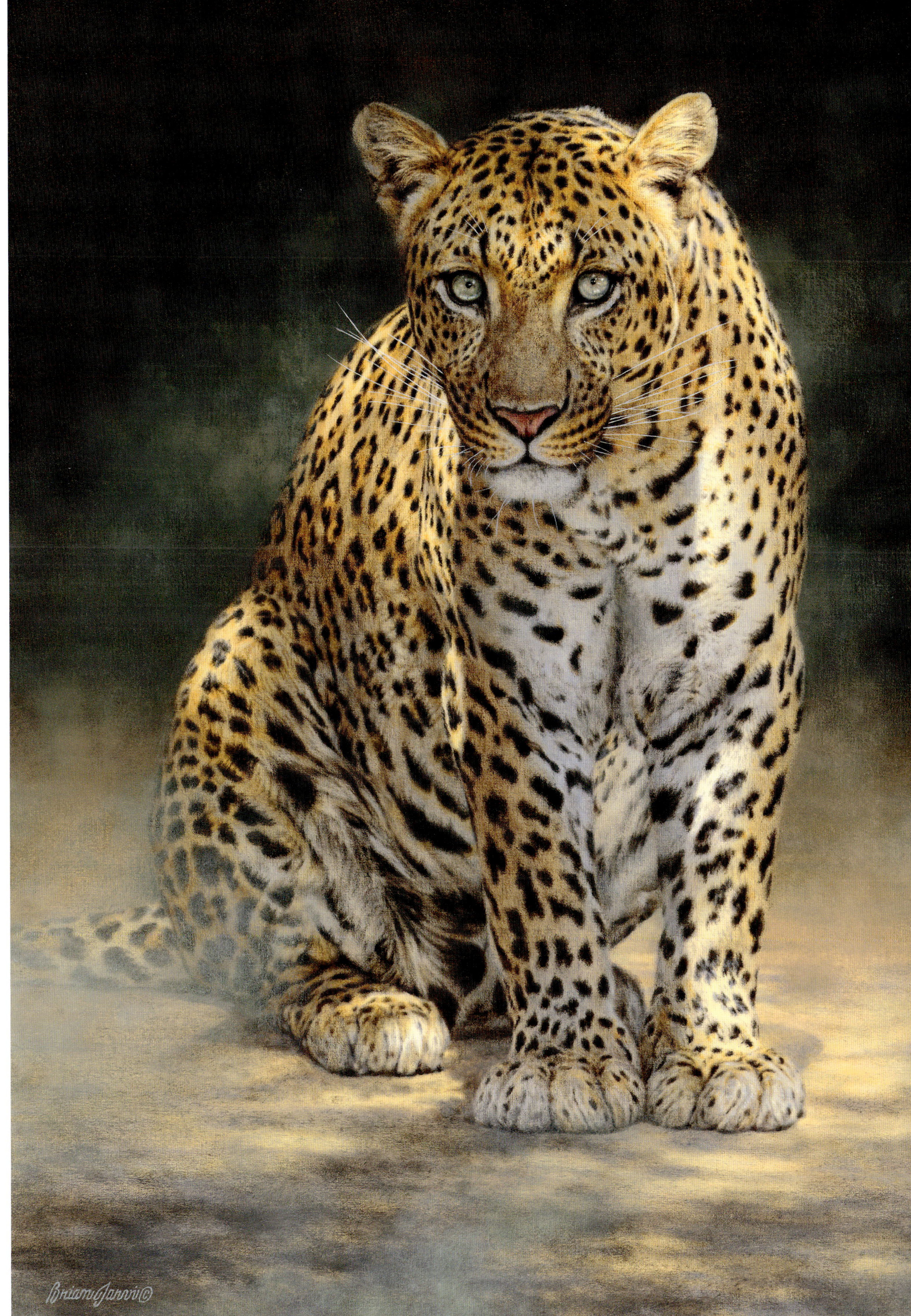
Brian Jarvi©

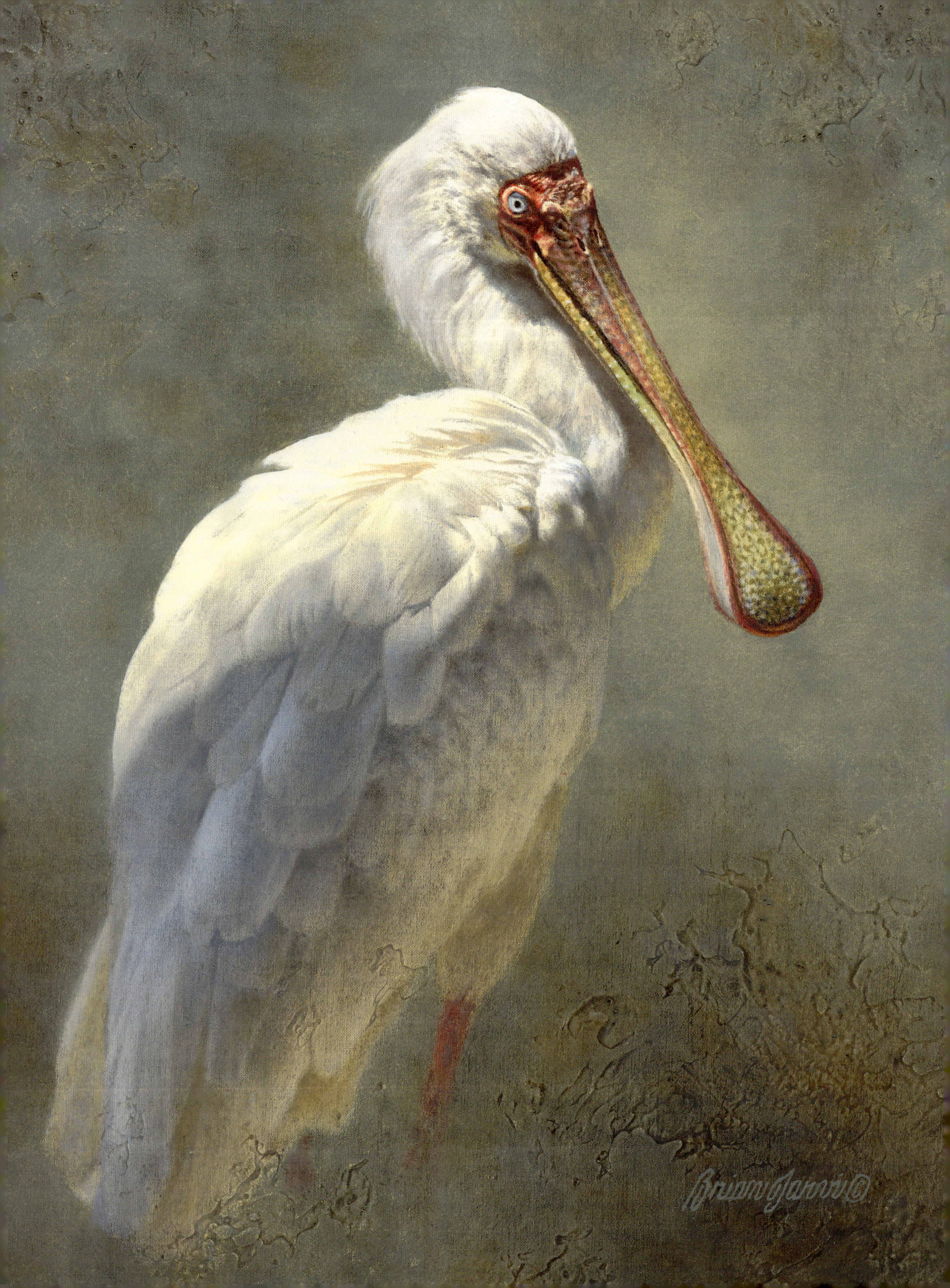
Brian Jarvi©

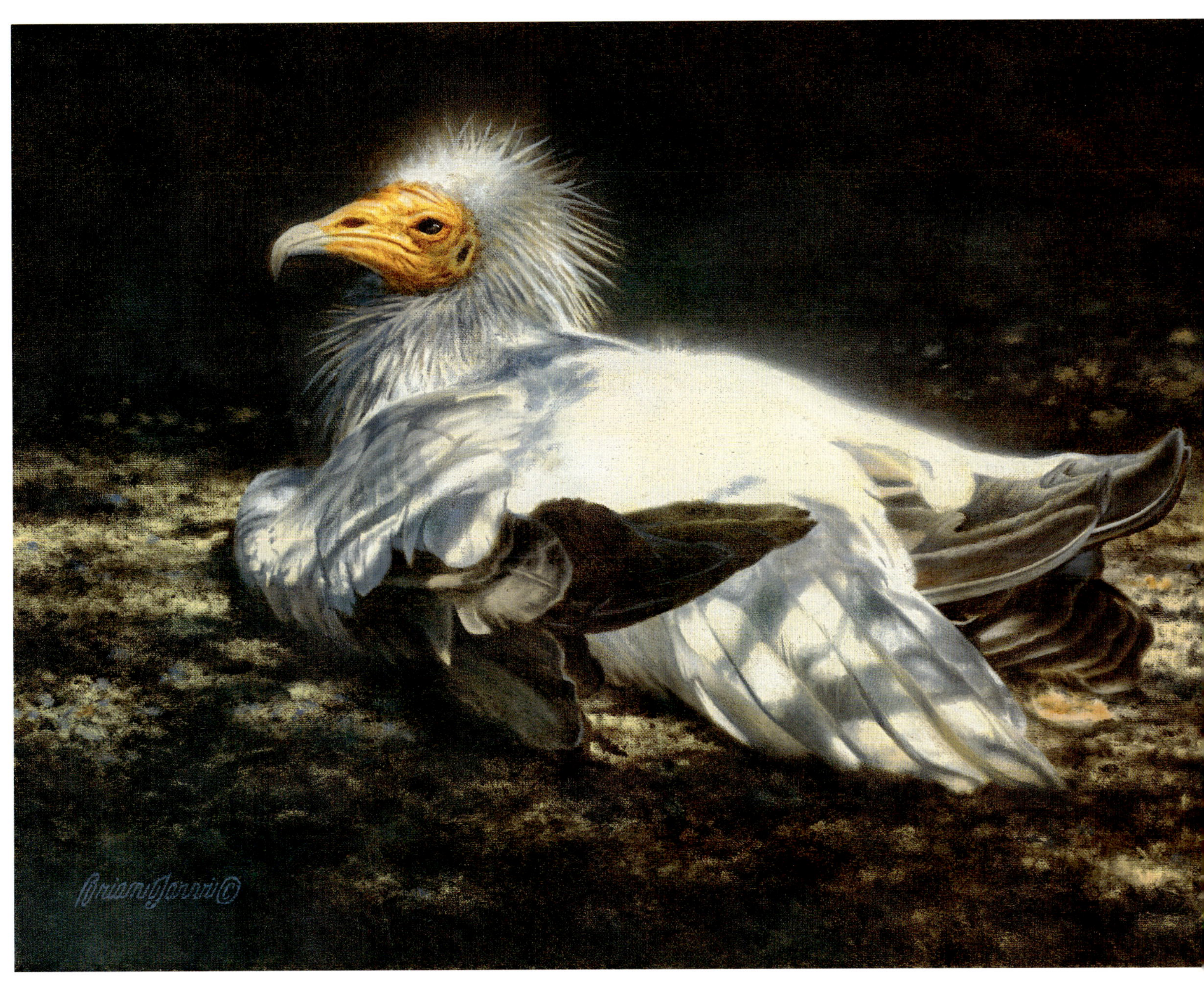

OPPOSITE
African Spoonbill Pre-Study, 2013
African Spoonbill
Oil on Belgian Linen
21 x 16"
Collection of Leigh Yawkey Woodson Art Museum

ABOVE
The Pharaoh's Deity, 2016
Egyptian Vulture
Oil on Belgian Linen
11 x 14"

In 2013, this spoonbill study was juried into the prestigious Leigh Yawkey Woodson Art Museum's annual *Birds in Art* exhibition, the premier venue for avian art in the world. Jarvi loved experimenting with spare simplicity and varied edges, making it seem like a gossamer from another time.

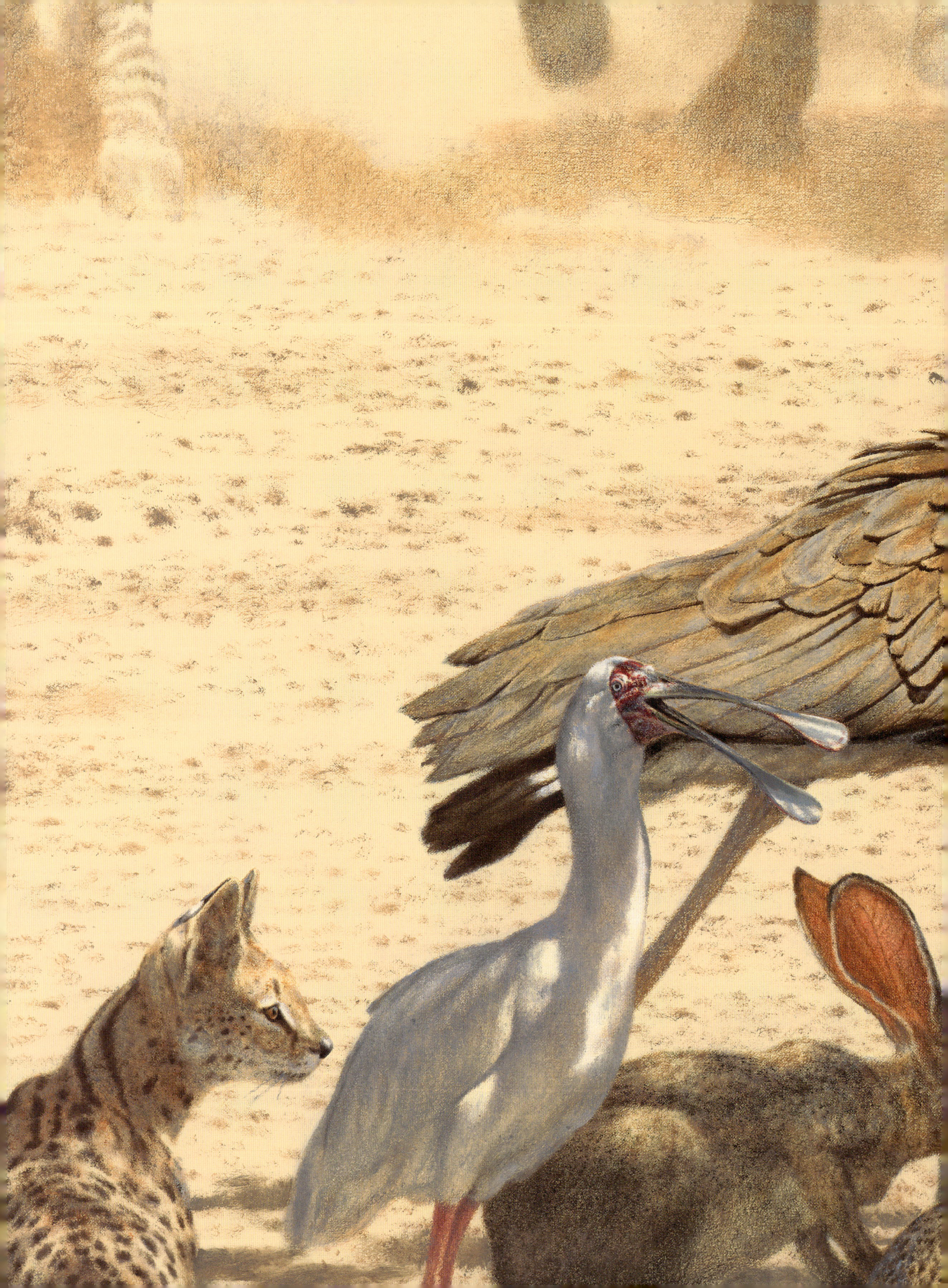

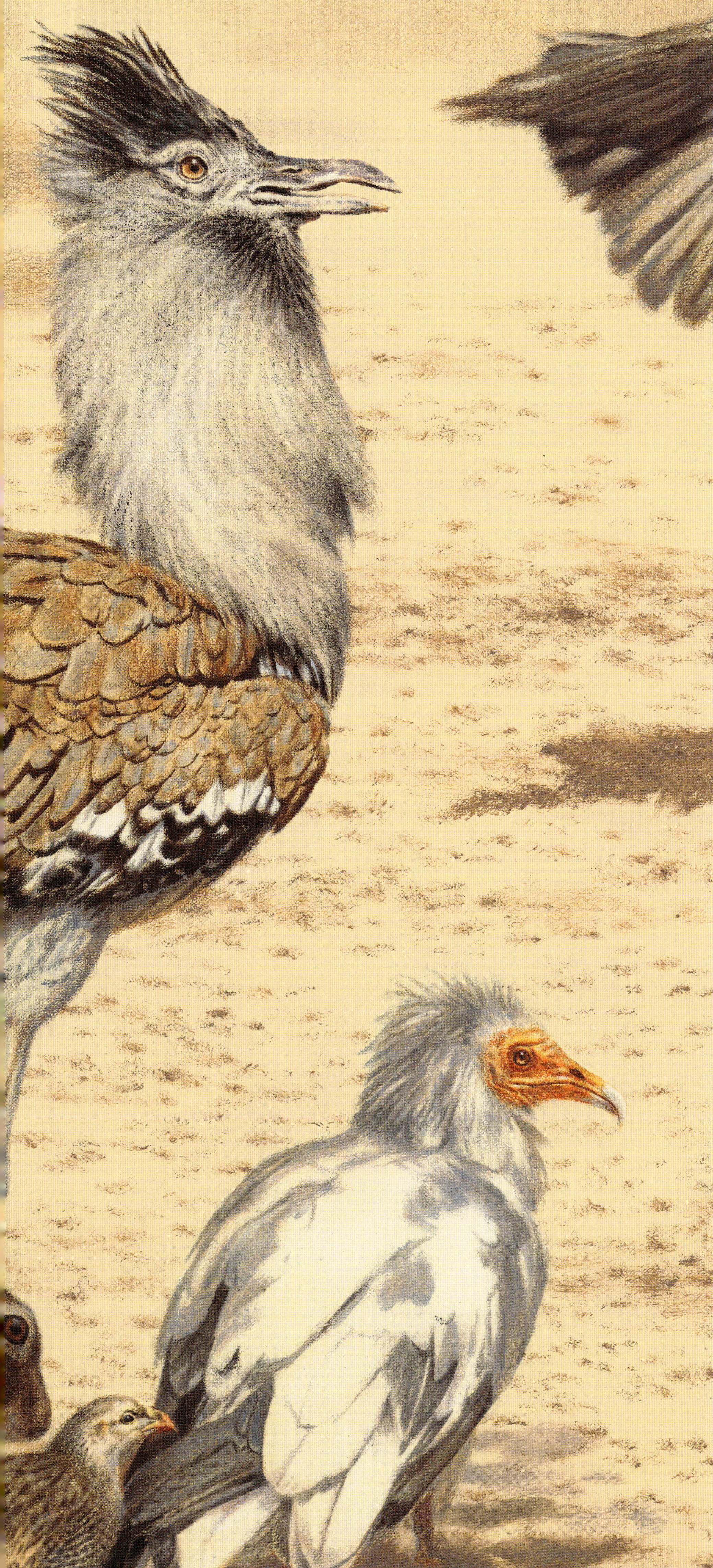

African Menagerie, Panel 2
Detail (Bottom)
Kori Bustard, African Spoonbill, Egyptian Vulture, Serval, African Hare, Natal Francolin

OVERLEAF
Serval Study, 2016
Serval
Oil on Belgian Linen
11 x 14"
Collection of Michael and Sarina Nordmarken

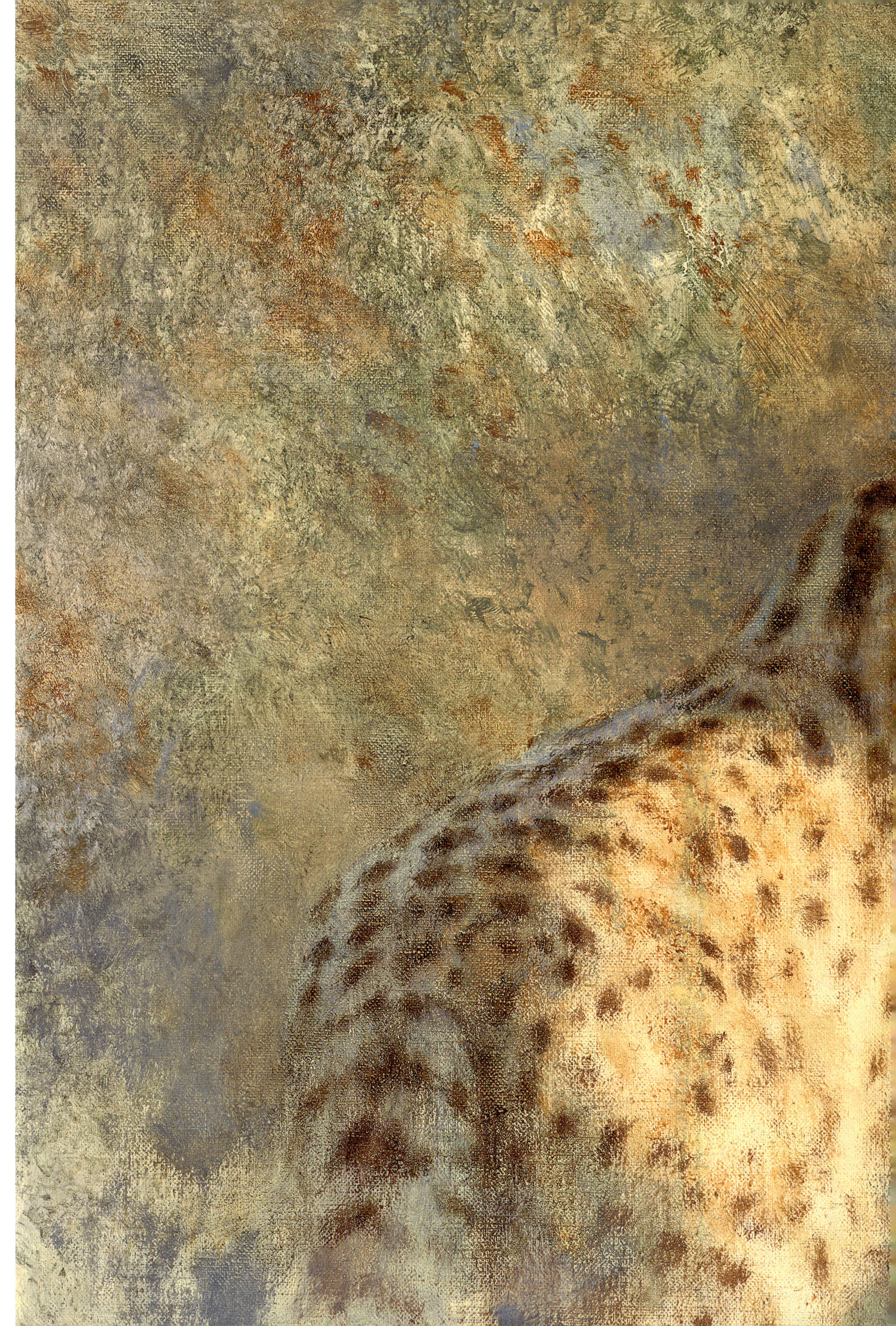

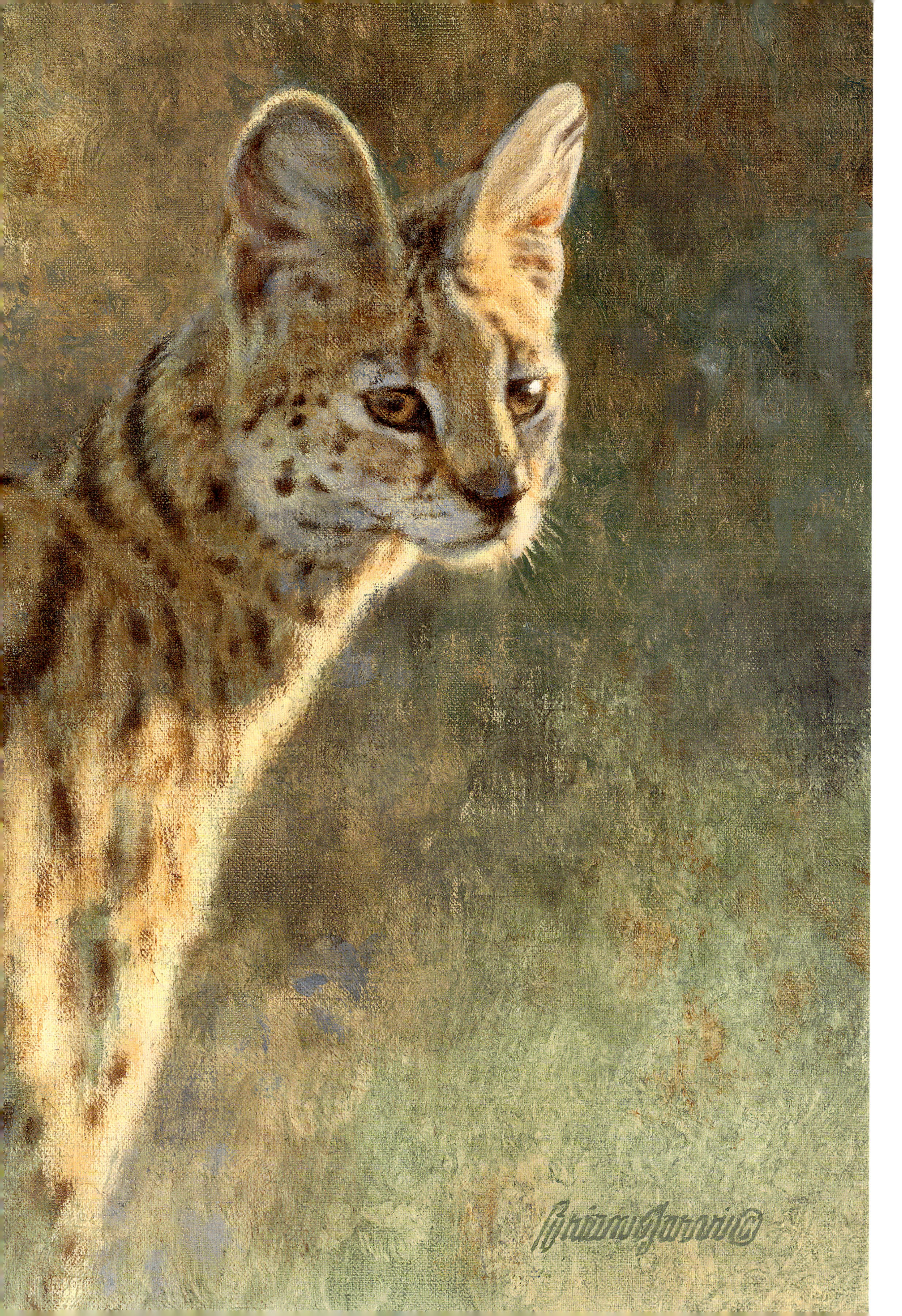

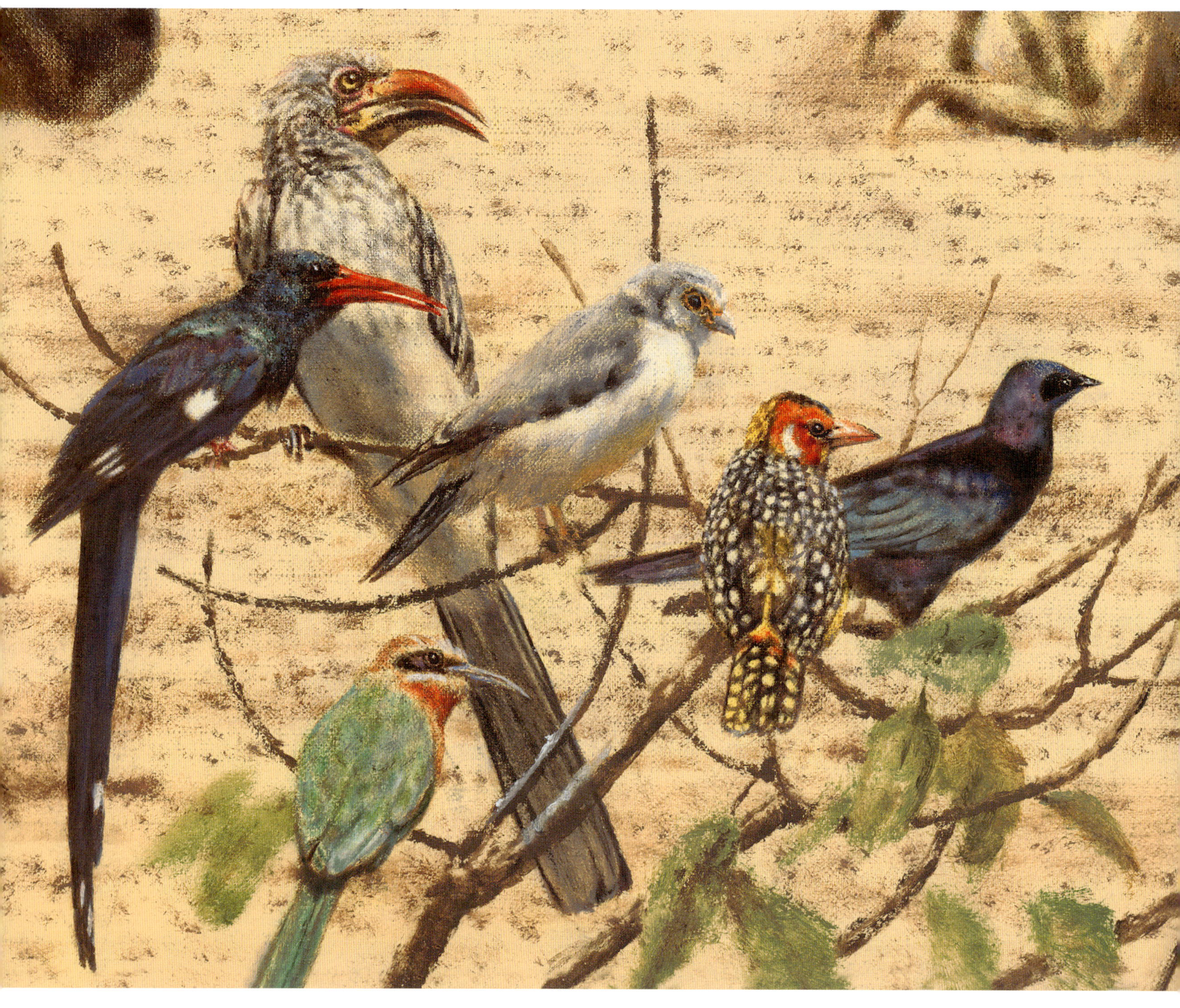

ABOVE

African Menagerie, Panel 3
Detail (Bottom Left)
Red-Billed Hornbill, Red-Billed Wood Hoopoe, African Pygmy Falcon, Burchell's Glossy Starling, Red-and-Yellow Barbet, White-Fronted Bee-Eater

OPPOSITE

Blue-Grey Flycatcher, 2014
Blue-Grey Flycatcher
Pencil
12½ x 9"
Collection of David and Kathleen Chesness

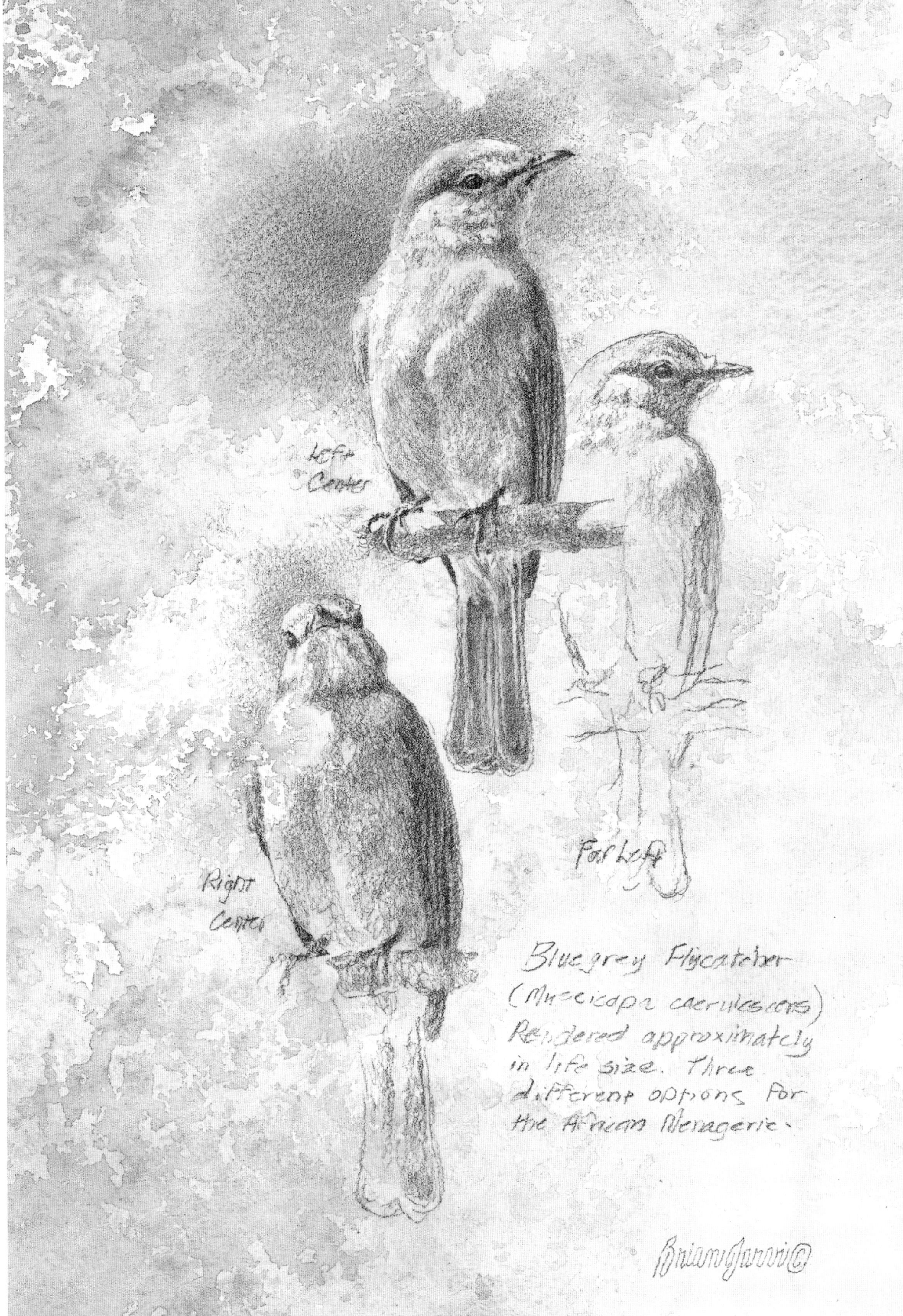
Left Center
Far Left
Right Center
Bluegrey Flycatcher
(Muscicapa caerulescens)
Rendered approximately
in life size. Three
different options for
the African Menagerie.
Brian Jarvi ©

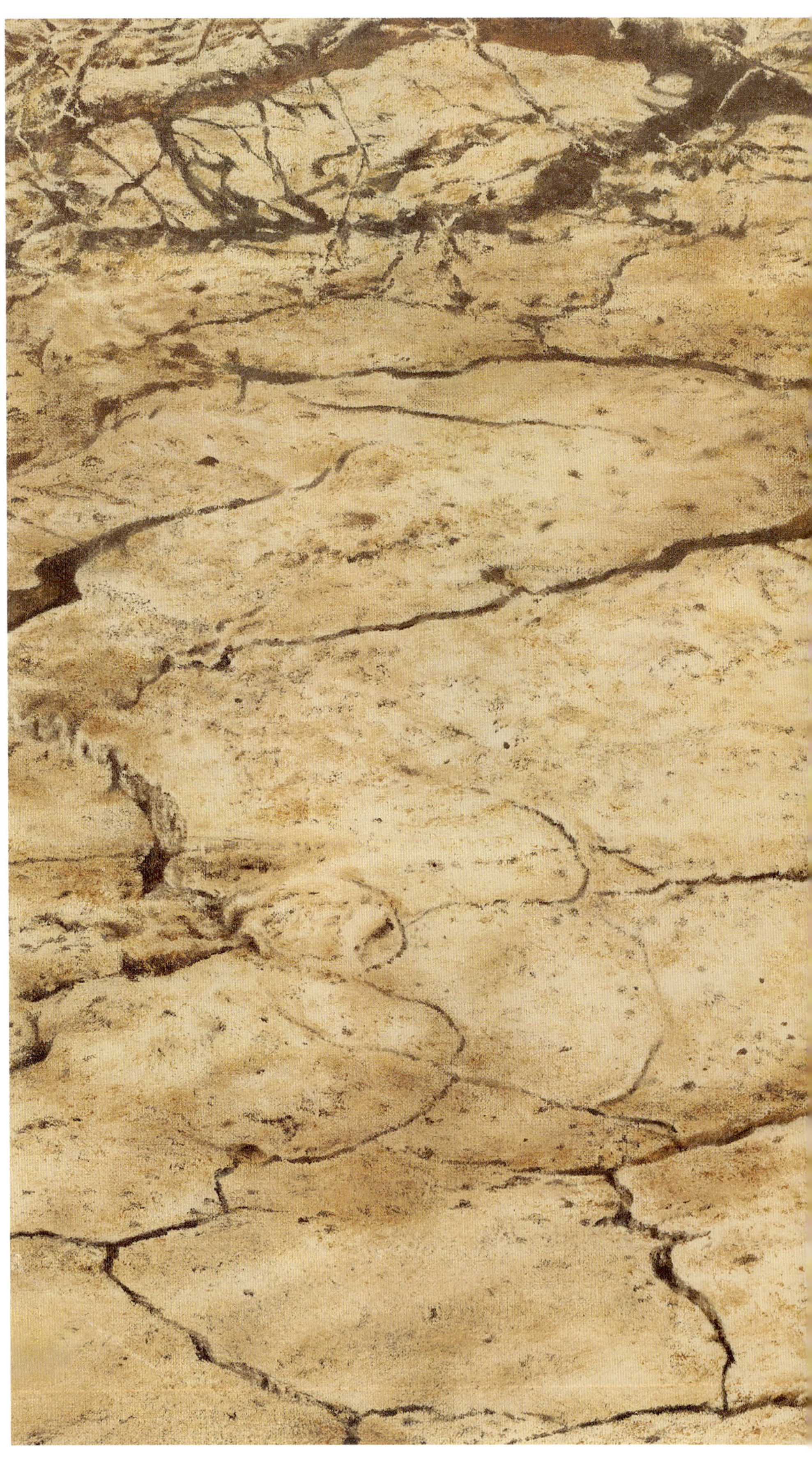

And the Meek, 2016
Red-and-Yellow Barbet
Oil on Belgian Linen
14 x 24"
Collection of Dan and Cynthia Margo

OVERLEAF, TOP LEFT
African Penguin Study, 2016
African Penguin
Pencil
14 x 21"

Few people realize there are penguins in Africa; they gather in colonies along the southern coast.

OVERLEAF, BOTTOM LEFT
Nile Monitor Study, 2014
Nile Monitor Lizard
Charcoal and Pencil
11 x 14"

OVERLEAF, RIGHT
African Menagerie, Panel 3
Detail (Bottom Center)
African Penguin, Leopard Tortoise, Nile Monitor Lizard, African Pygmy Falcon, Burchell's Glossy Starling, Red-and-Yellow Barbet, Blue-Grey Flycatcher

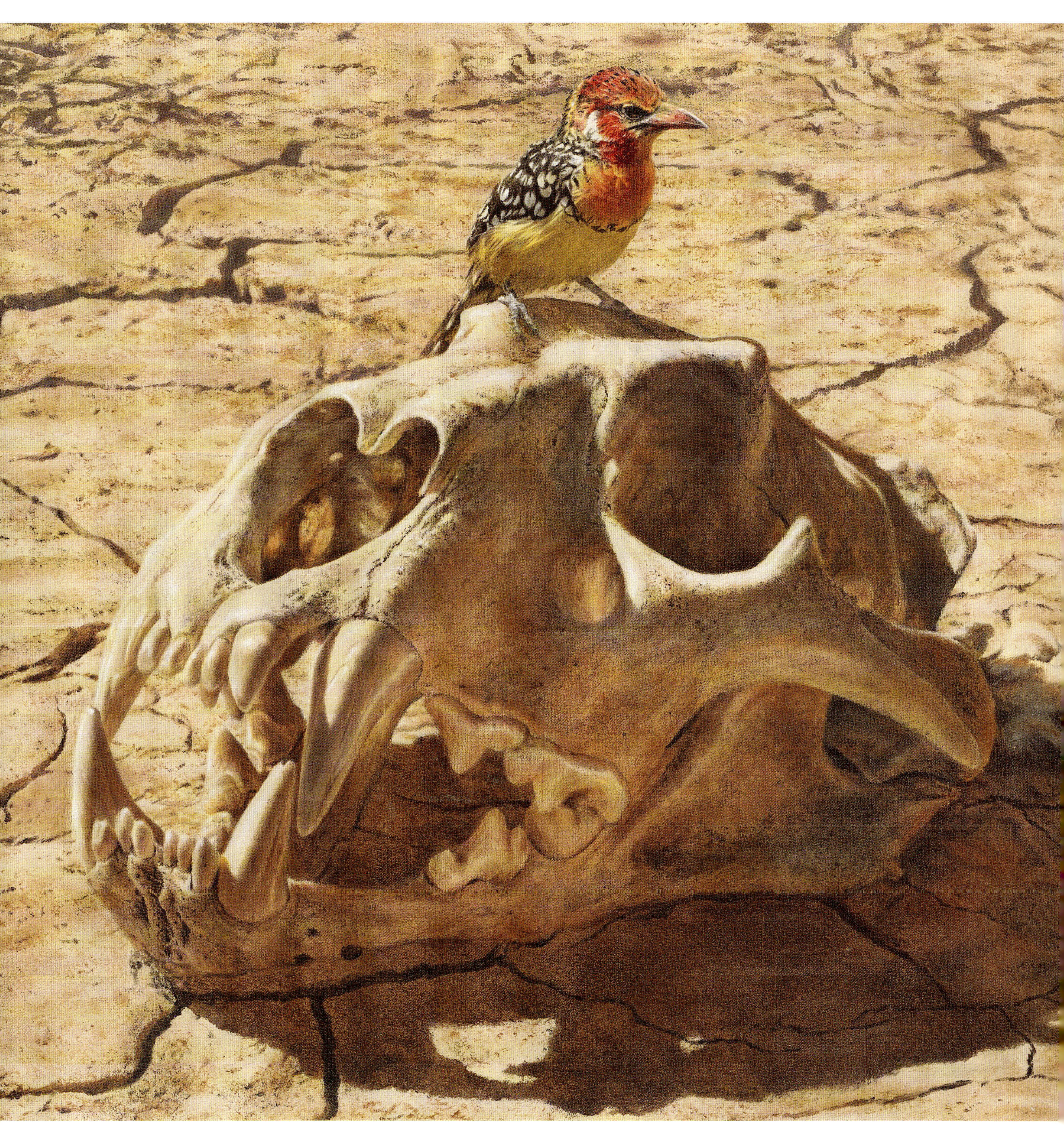

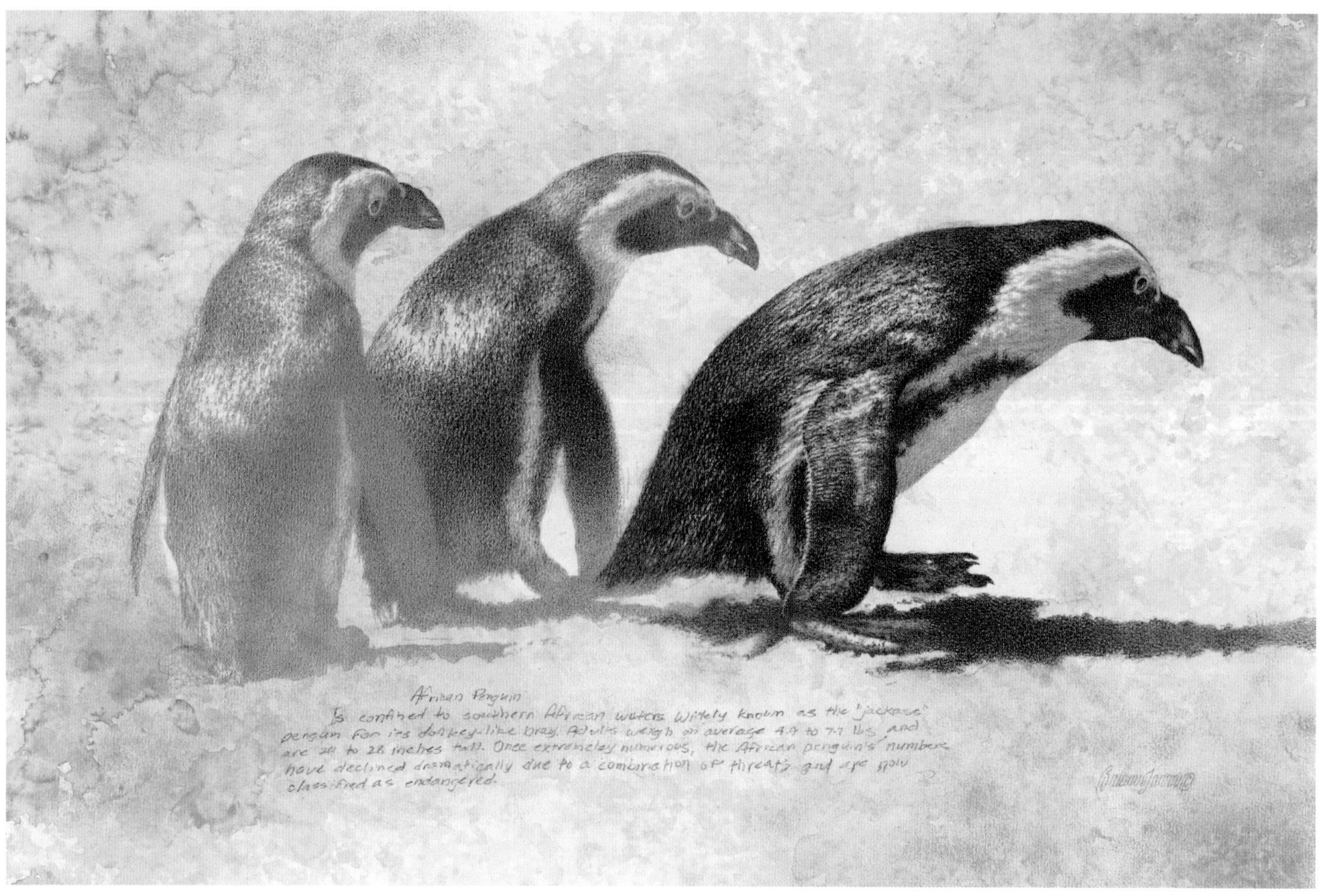
African Penguin
Is confined to southern African waters. Widely known as the "jackass"
penguin for its donkey-like bray. Adults weigh on average 4.9 to 7.7 lbs, and
are 24 to 28 inches tall. Once extremeley numerous, the African penguin's numbers
have declined dramatically due to a combination of threats and are now
classified as endangered.

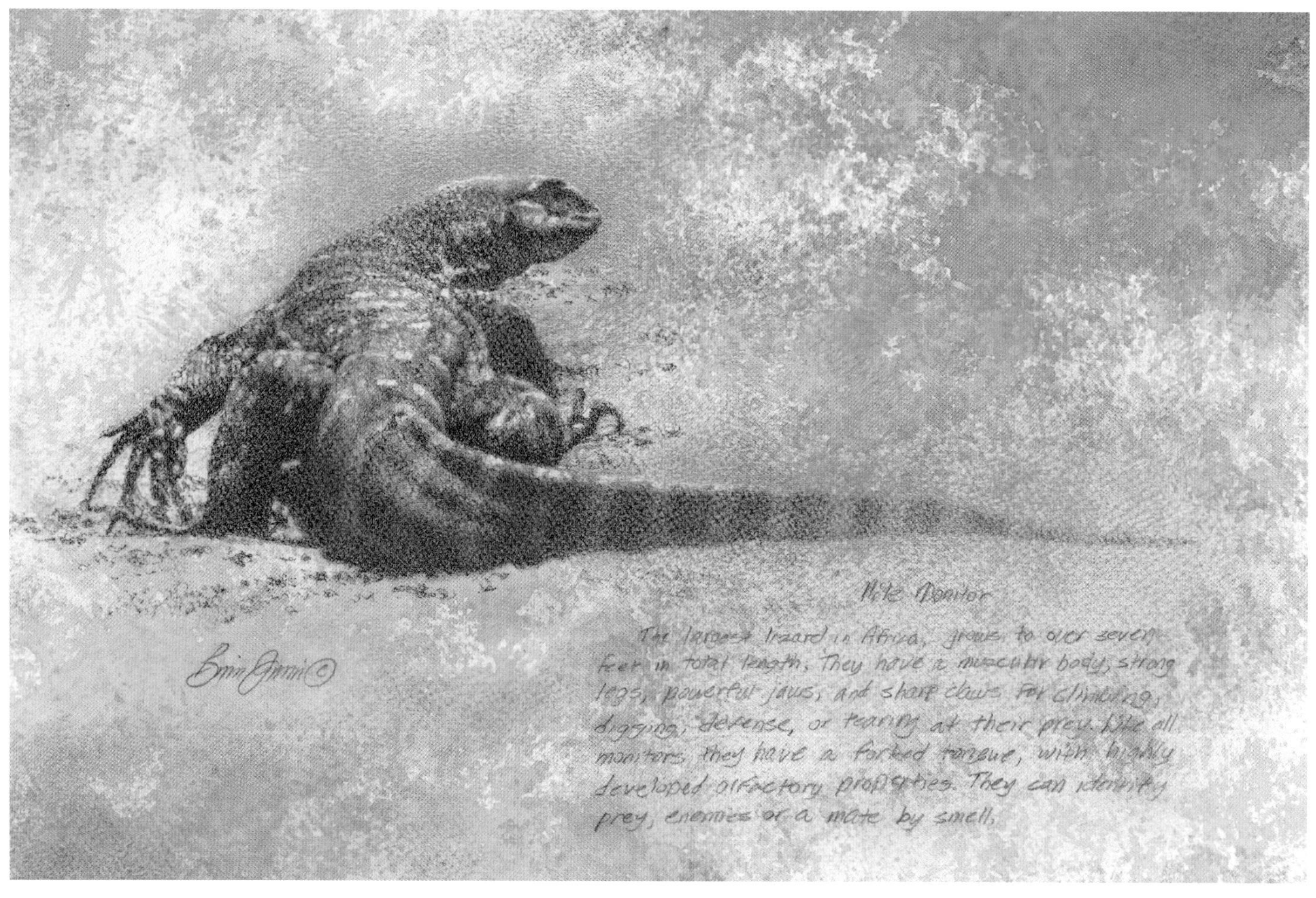
Nile Monitor
The largest lizard in Africa, grows to over seven
feet in total length. They have a muscular body, strong
legs, powerful jaws, and sharp claws for climbing,
digging, defense, or tearing at their prey. Like all
monitors they have a forked tongue, with highly
developed olfactory properties. They can identify
prey, enemies or a mate by smell.

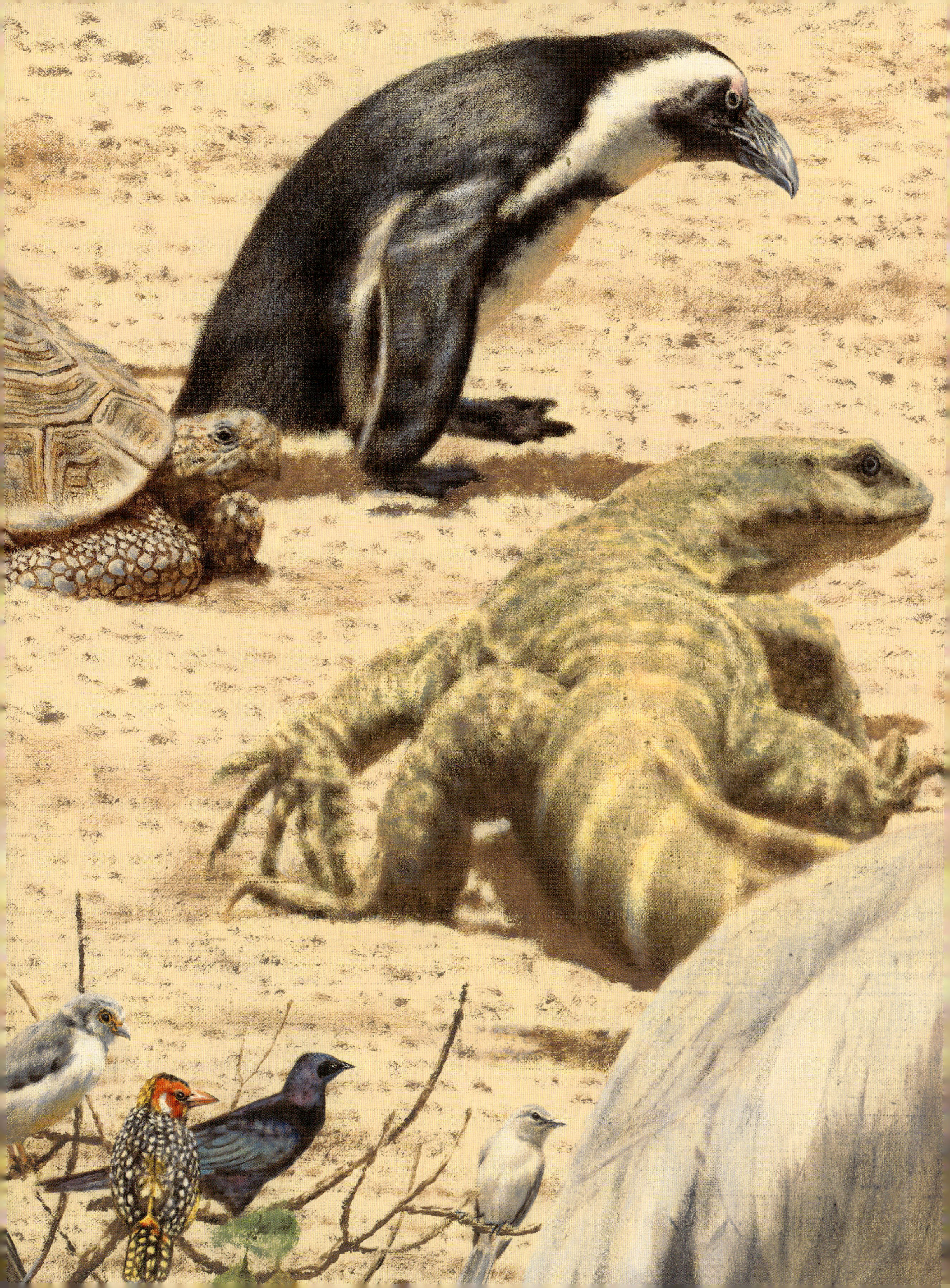

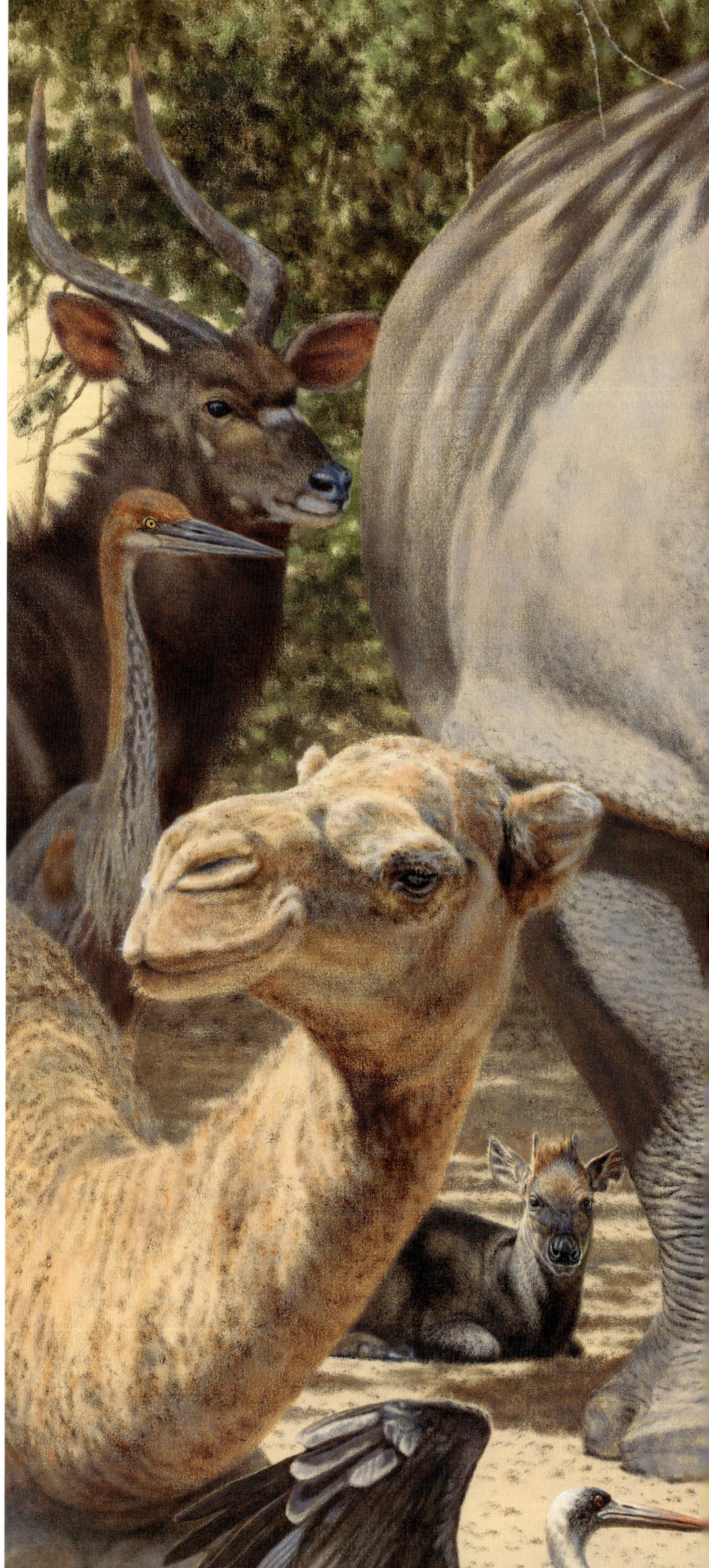

African Menagerie, Panel 3
Detail (Center)
Kurrichane Thrush, Black-Backed Puffback, Nyala, Somali Wild Ass, Goliath Heron, White Rhinoceros, Dromedary Camel, African Forest Buffalo, Black Duiker, Woolly-Necked Stork

The white rhino, like the black rhino, is a charismatic icon for reflection on the importance of human stewardship. Recently, the last male white rhino in the northern herd population died. Without augmentation by rhinos from more southern reaches, the subgroup of animals that has existed there for countless millennia will vanish. The combination of habitat loss and black-market demand for rhino horns has devastated several subpopulations of both species. According to Jarvi, this is one of the greatest challenges of our time—to ensure safeguards are in place before it's too late.

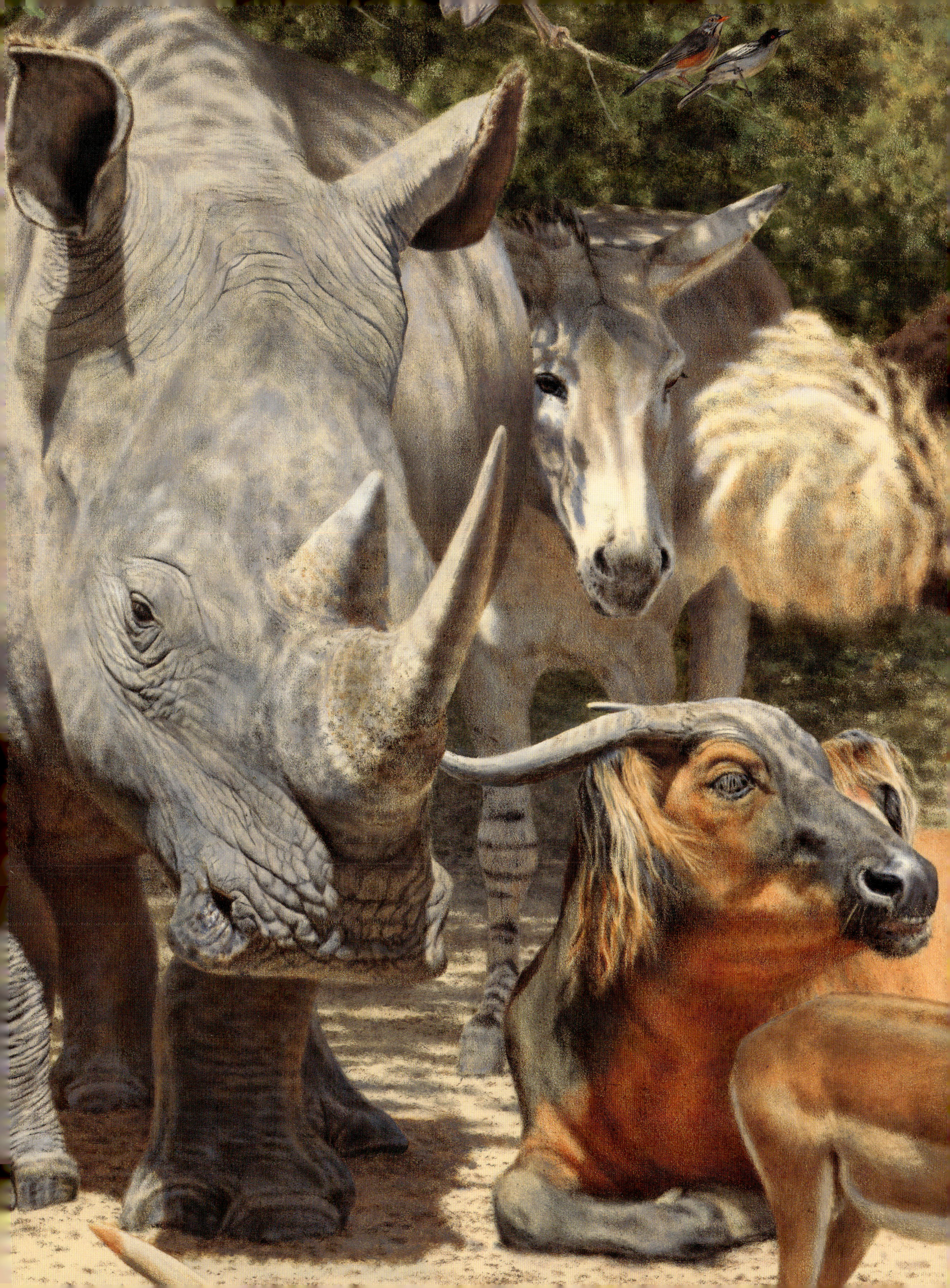

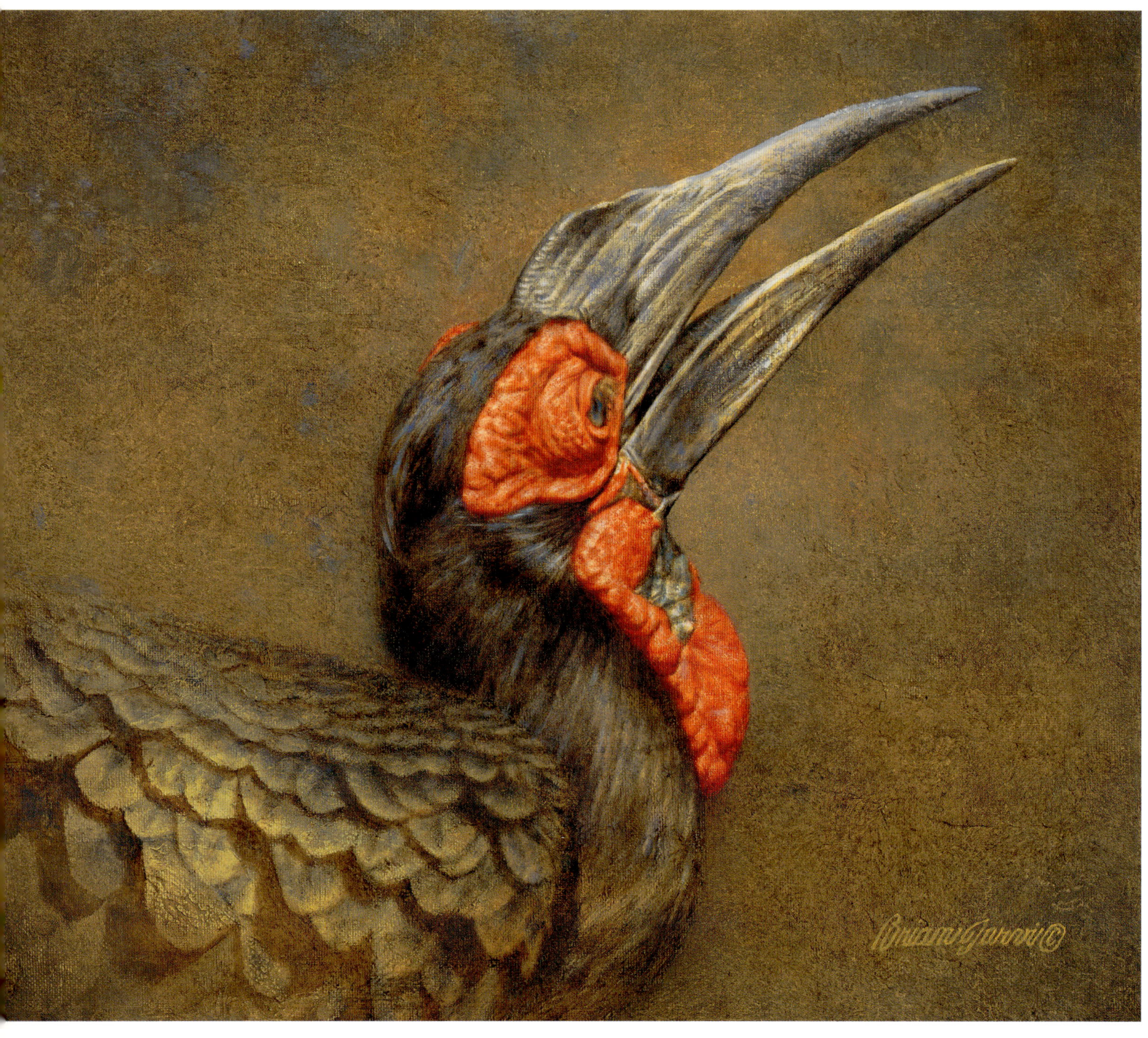

ABOVE
Ground Hornbill Study, 2016
Ground Hornbill
Oil on Canvas
14 x 16"

OPPOSITE
African Menagerie, Panel 3
Detail (Bottom Right)
Bongo, Wattled Crane, Ground Hornbill

During the early stages of designing and composing *African Menagerie*, Jarvi sought species in body shape, form of outline, color, and ecological interrelationships that might transition well from one panel to the next. The composition was—pun intended—an evolution. The bongo, which exudes a magnetic presence, is one of his favorites.

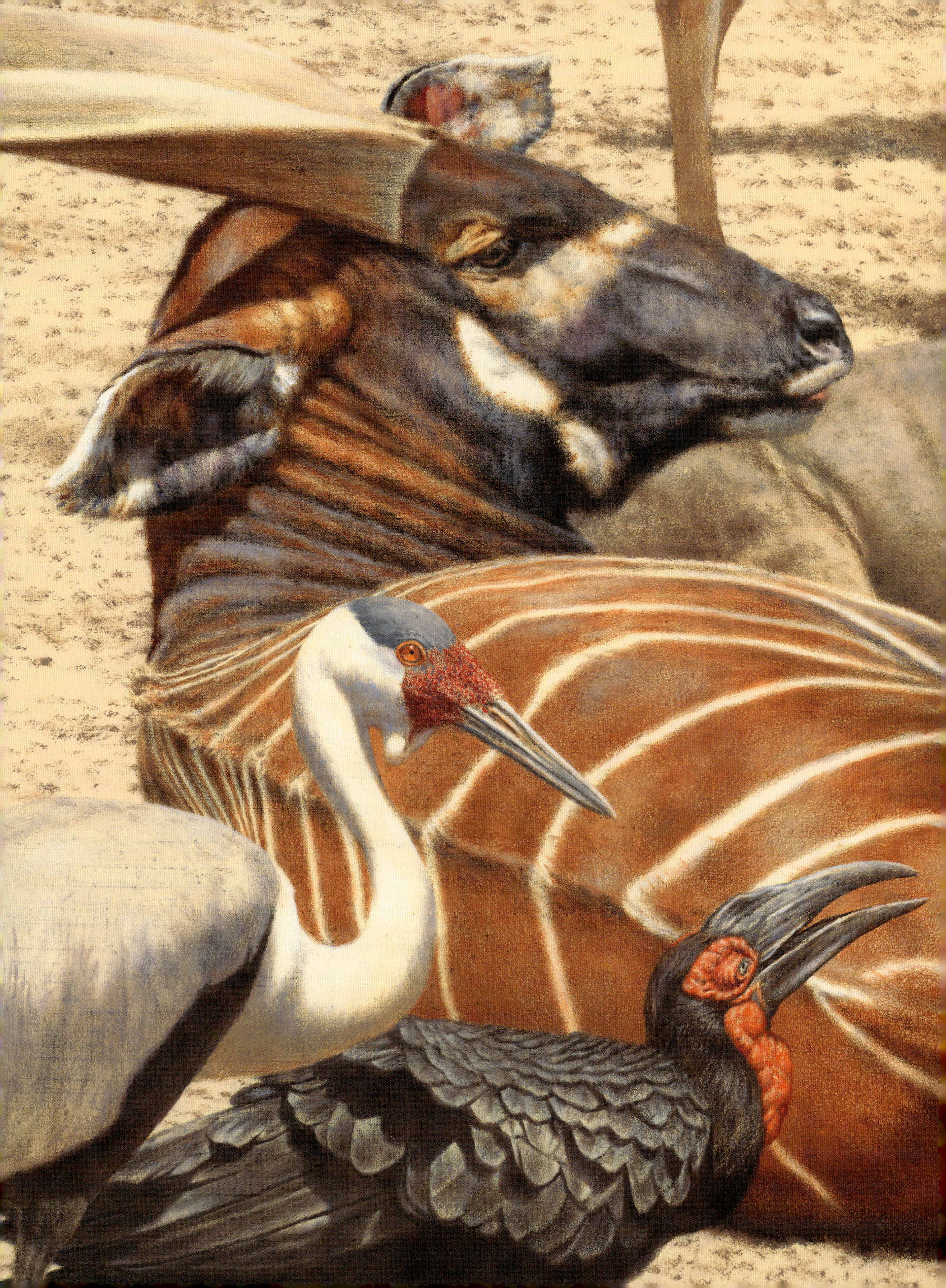

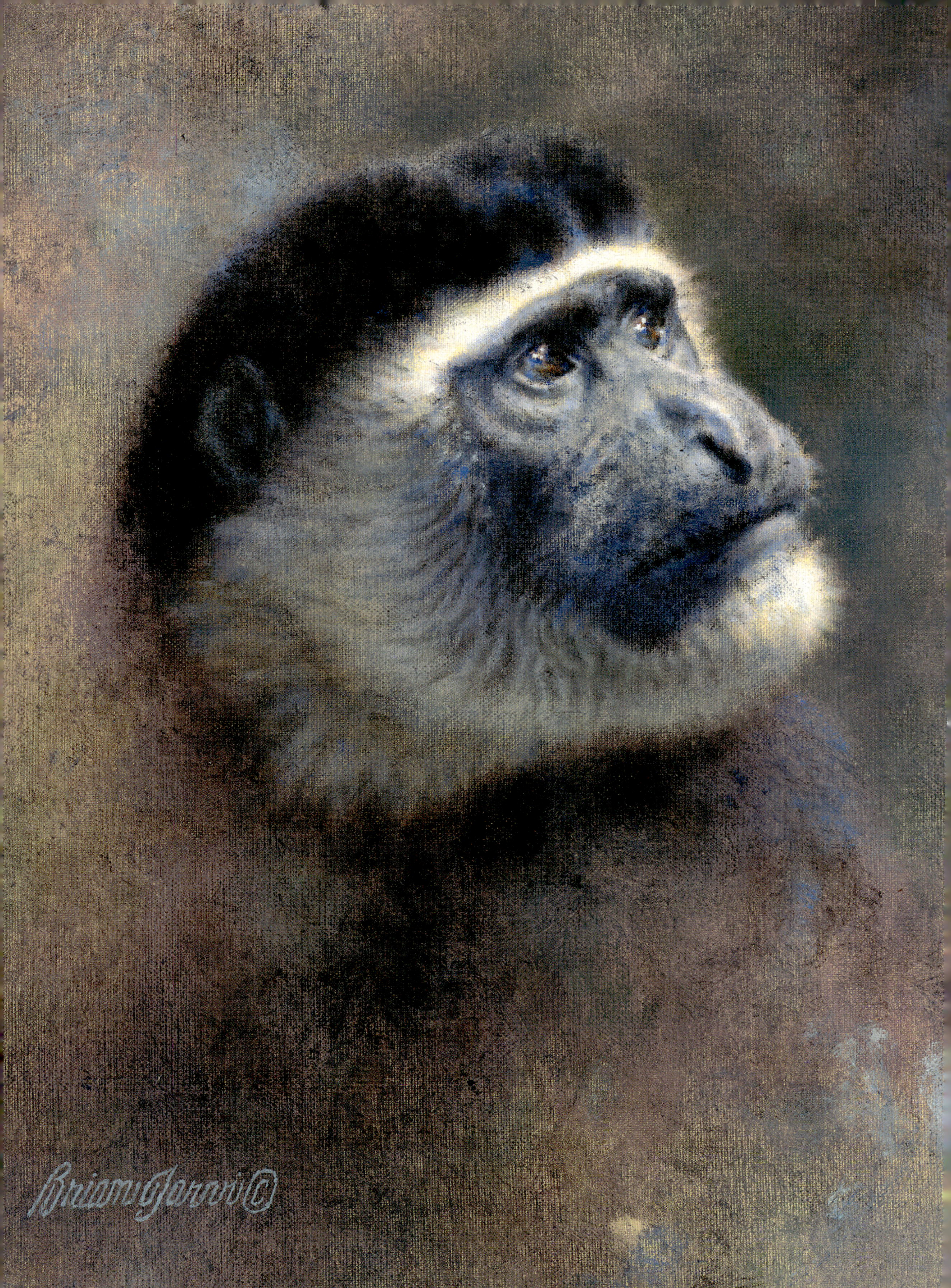
Brian Jarvi ©

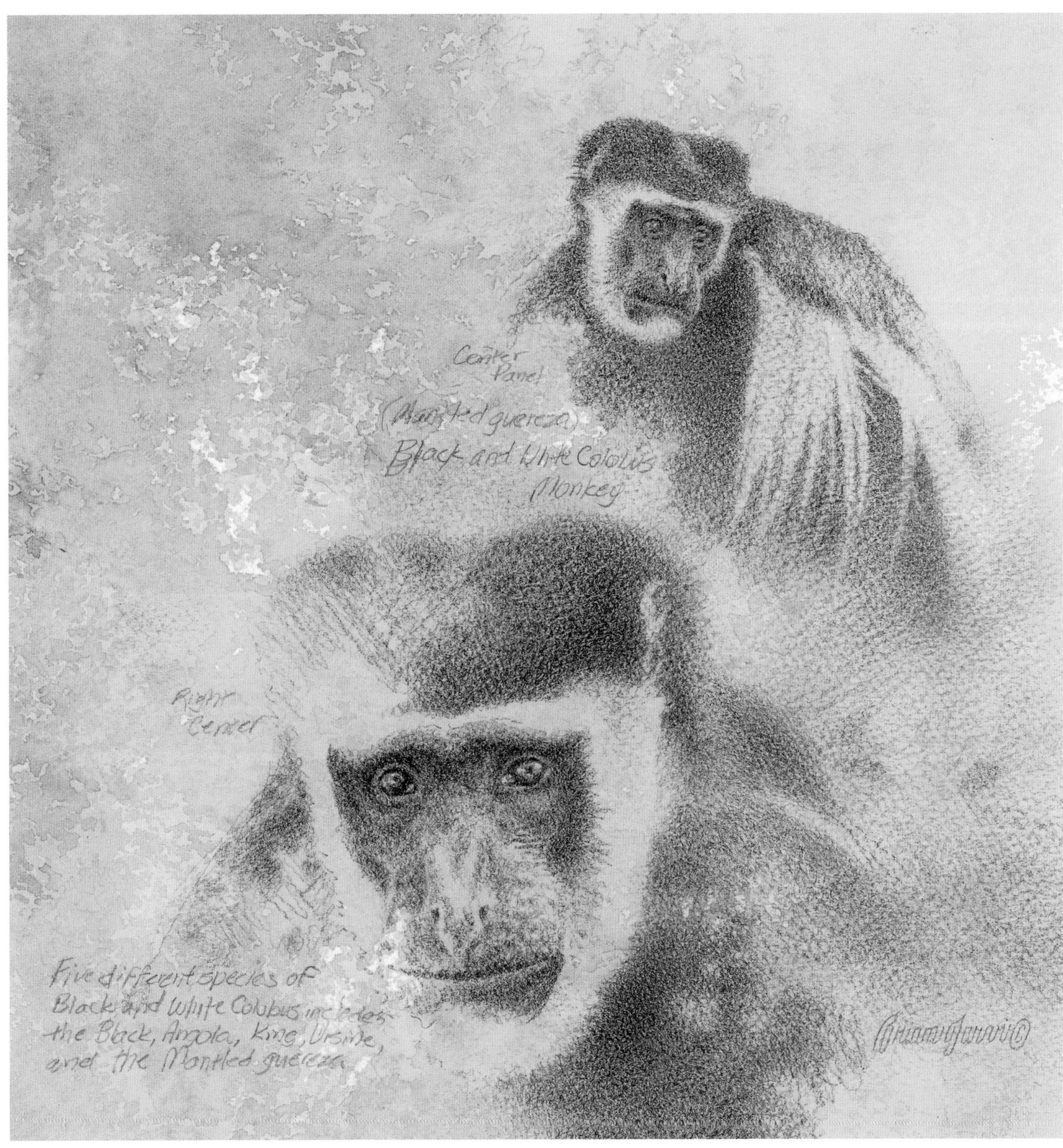

OPPOSITE
Colobus Head Study, 2013
Colobus Monkey
Oil on Belgian Linen
10 x 8"
Collection of Annette Bishop

ABOVE
Colobus Study II, 2016
Colobus Monkey
Pencil
9 x 9"
Collection of Stephanie Hunter

OVERLEAF
African Menagerie, Panel 3
Detail (Top Left)
De Brazza's Monkey, Colobus Monkey, Wolf's Mona Monkey, Great White Egret

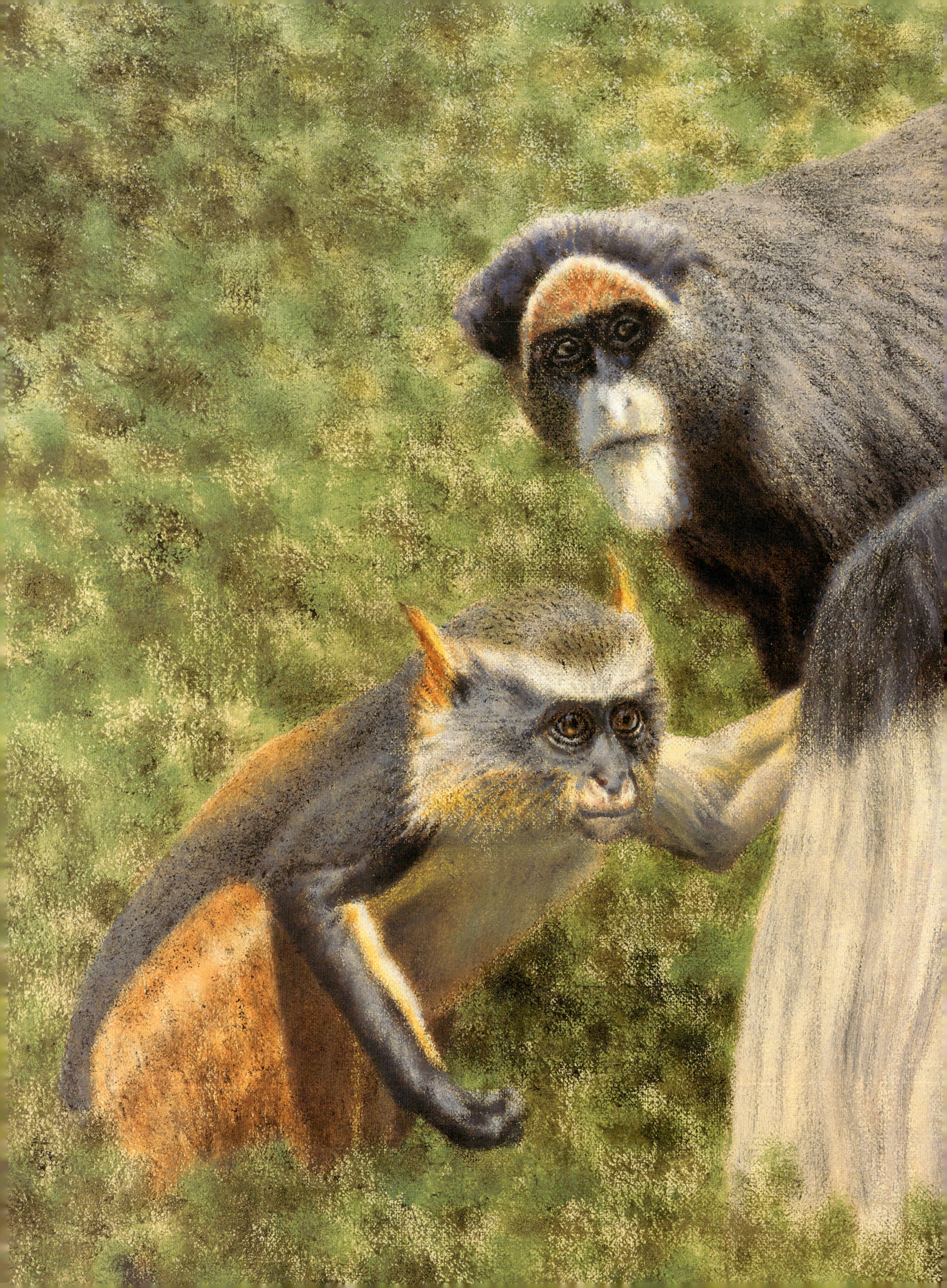

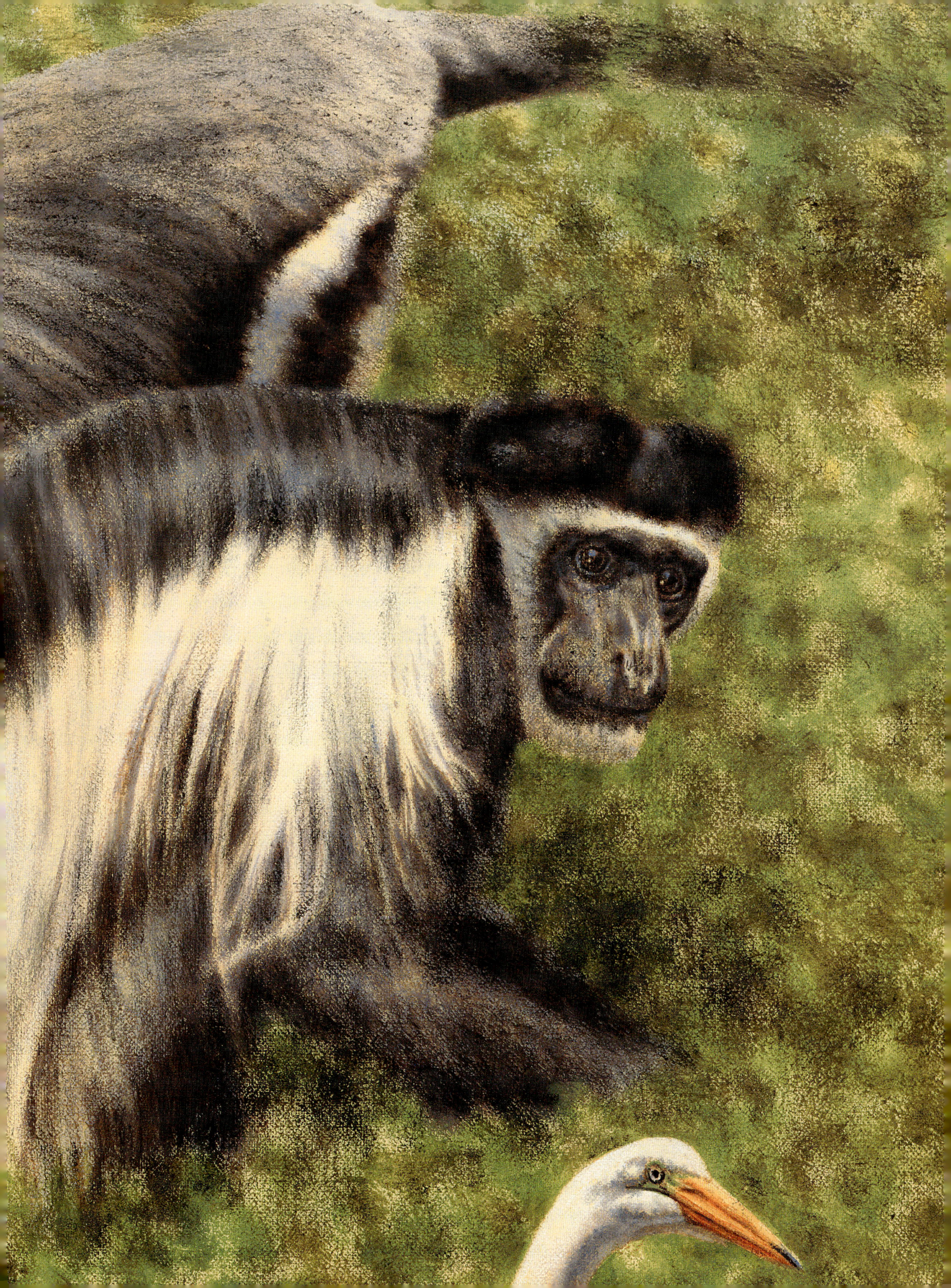

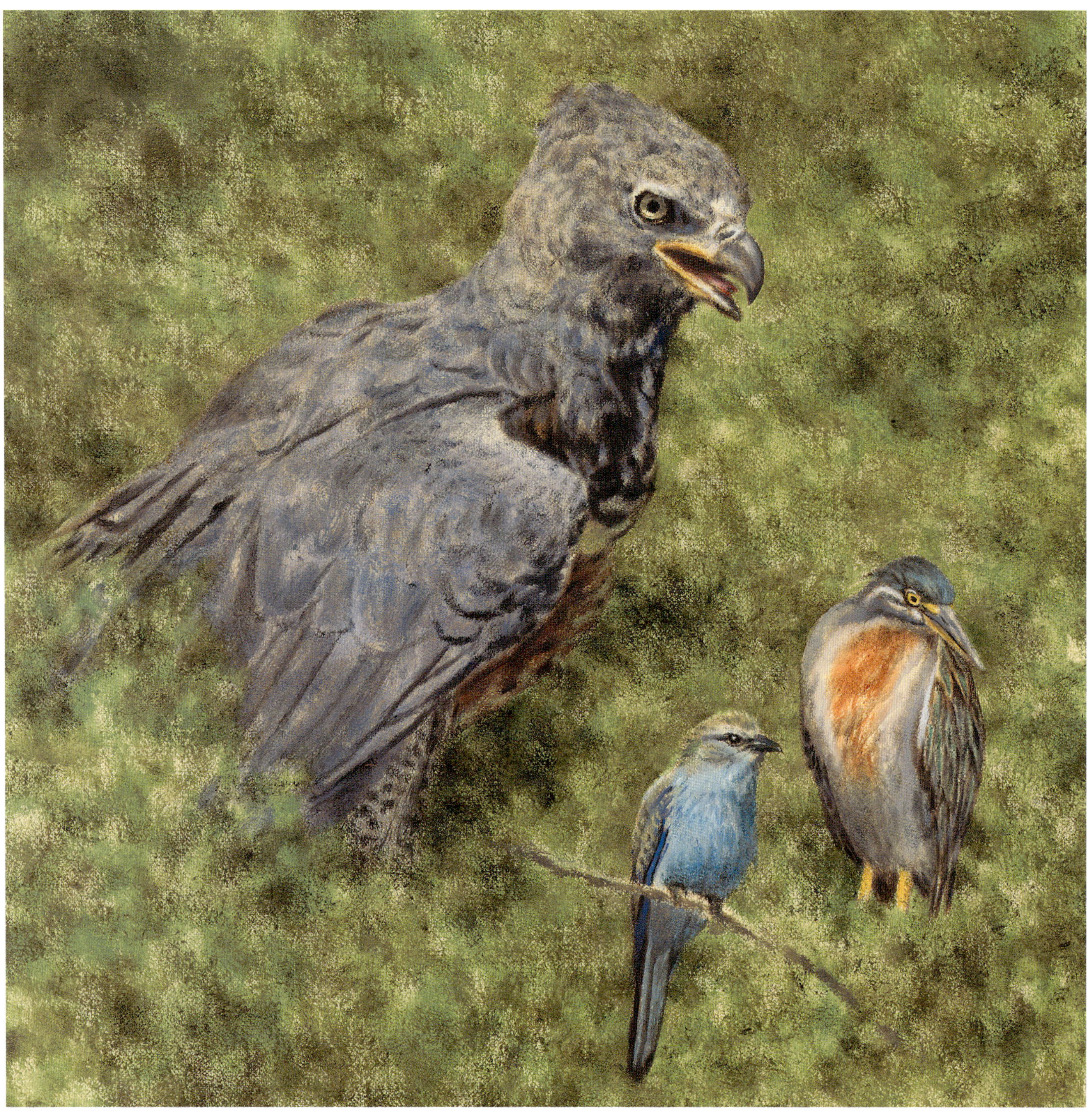

ABOVE
African Menagerie, Panel 3
Detail (Top Center)
Crowned Eagle, Green-Backed Heron, European Roller

OPPOSITE
Morgan's Mona, 2015
Wolf's Mona Monkey
Oil on Belgian Linen
12 x 9"
Collection of Annette Bishop

As this study began to take shape, the innocent demeanor of Wolf's mona monkeys reminded Jarvi of his young daughter, Morgan, admired for her astute and discerning curiosity. Upon completion, Jarvi named this study *Morgan's Mona*. Whether in studies or the larger *Menagerie*, many of Jarvi's subjects have hidden meanings.

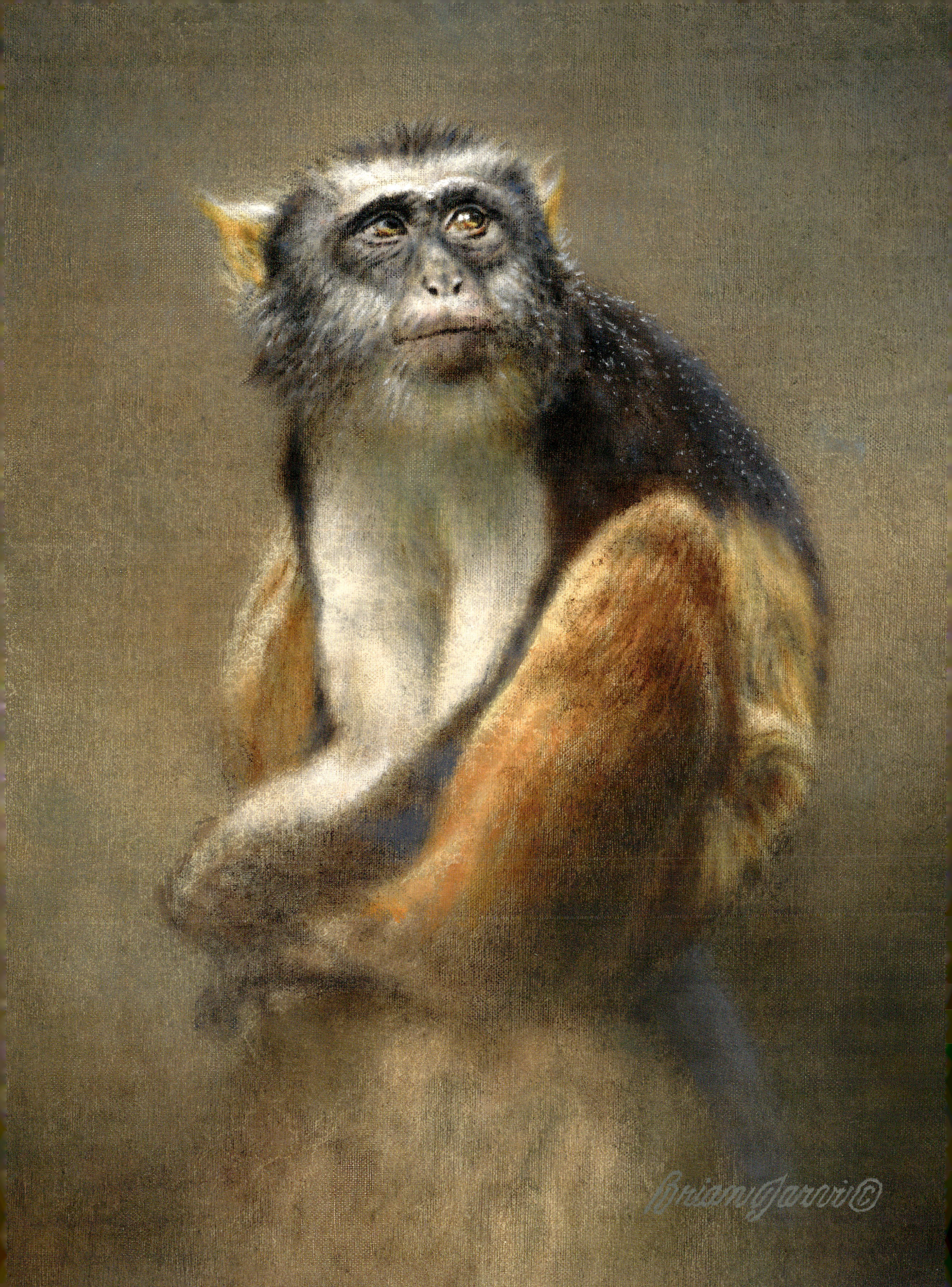
Brian Jarvi ©

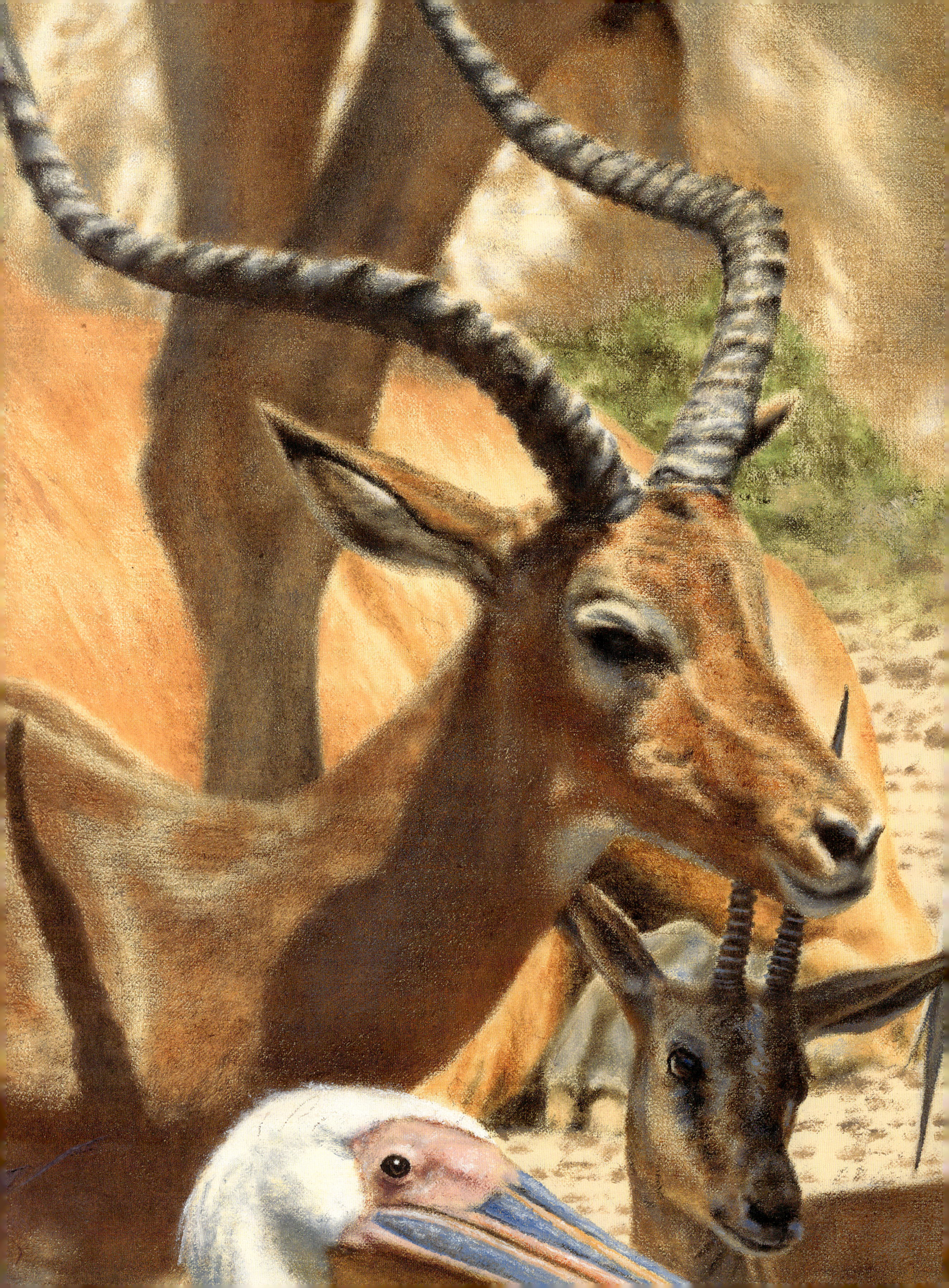

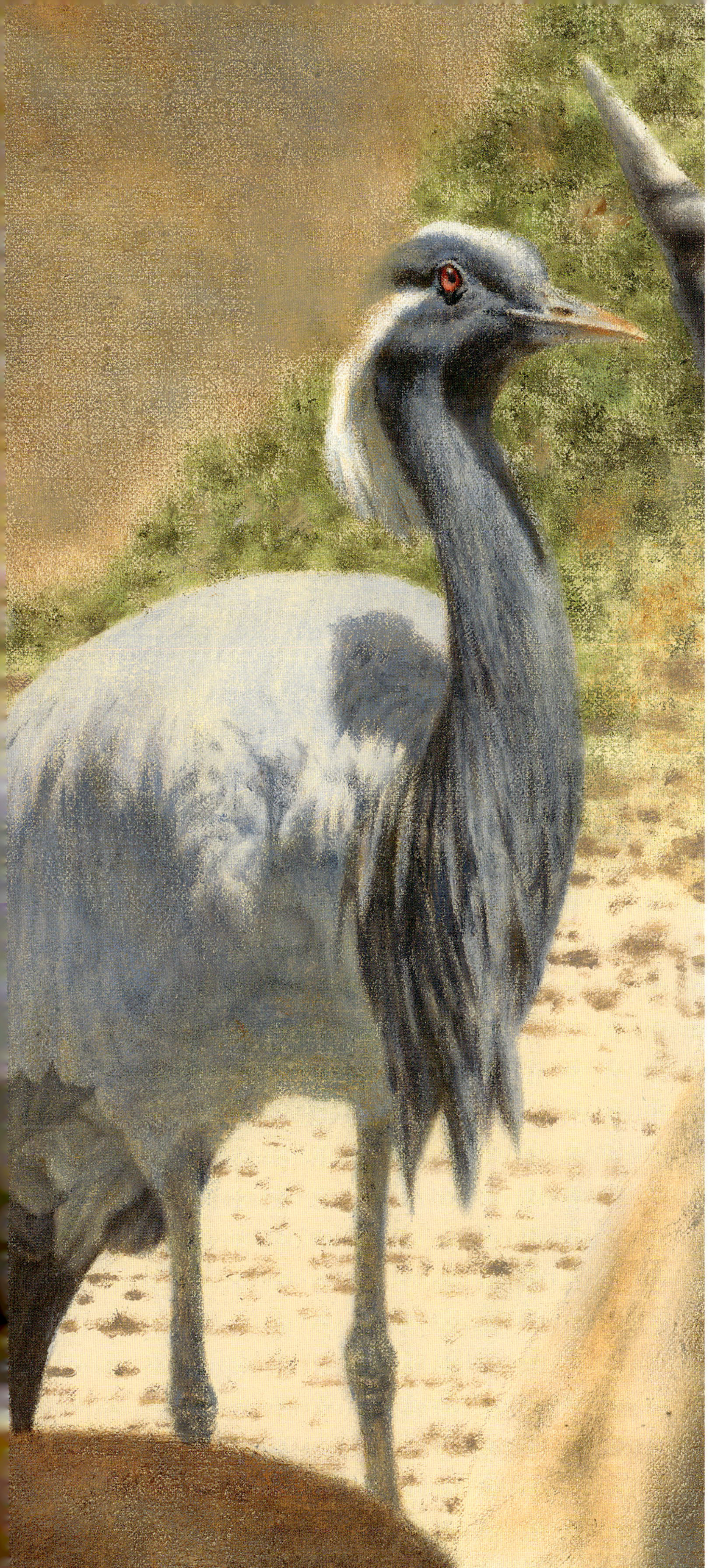

African Menagerie, Panel 4
Detail (Lower Middle Left)
Demoiselle Crane, Impala, Cuvier's Gazelle, Great White Pelican

The large umbrella-like shadow cast by the ostrich, though filled with tiny brushstrokes of color, was choreographed to provide a design element in order to highlight the adjacent Cuvier's gazelle. Blended together, the creatures, palette choices, and meticulous arrangement of light and shadow were intended to provide mood and tonal harmony.

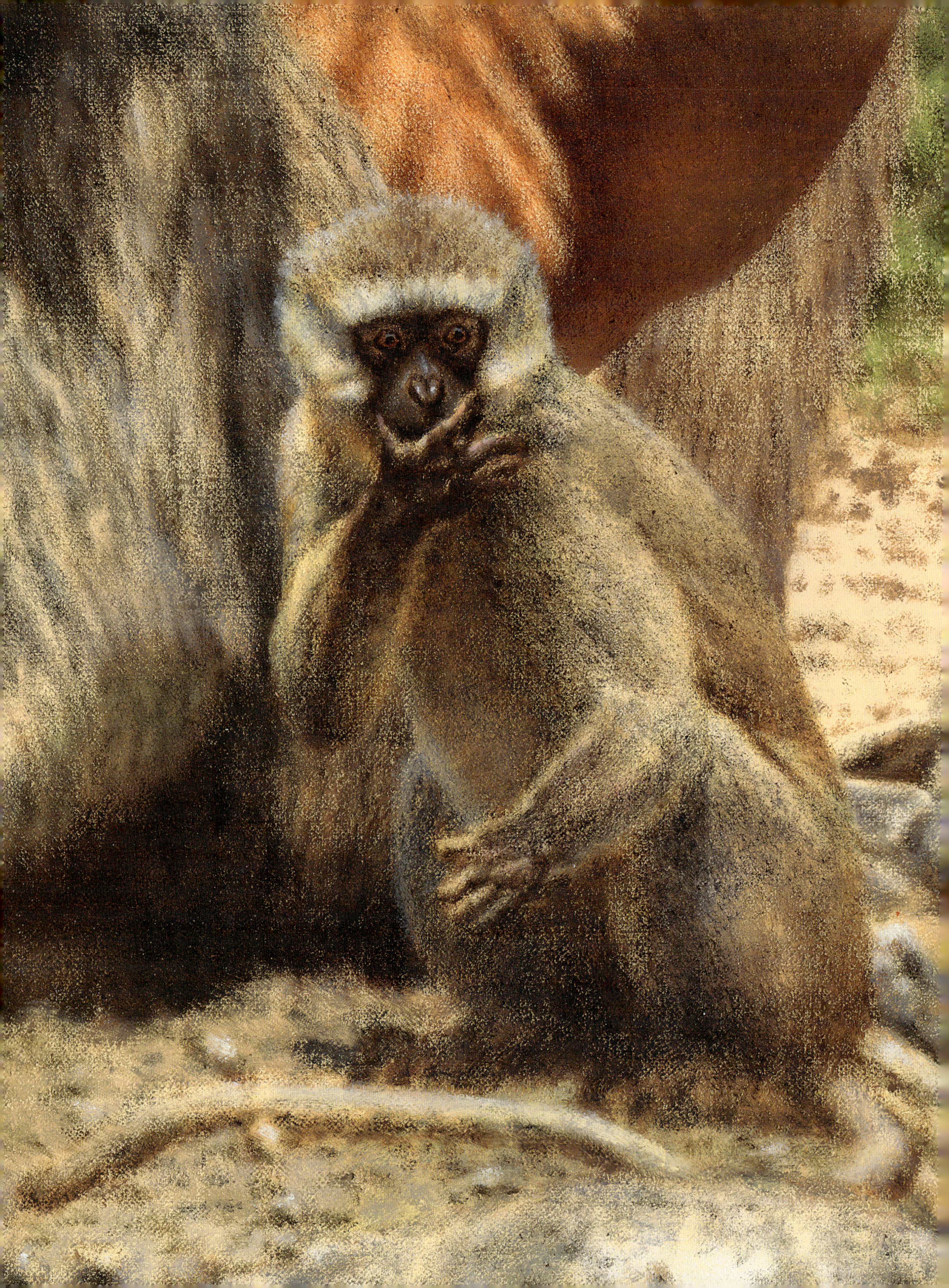

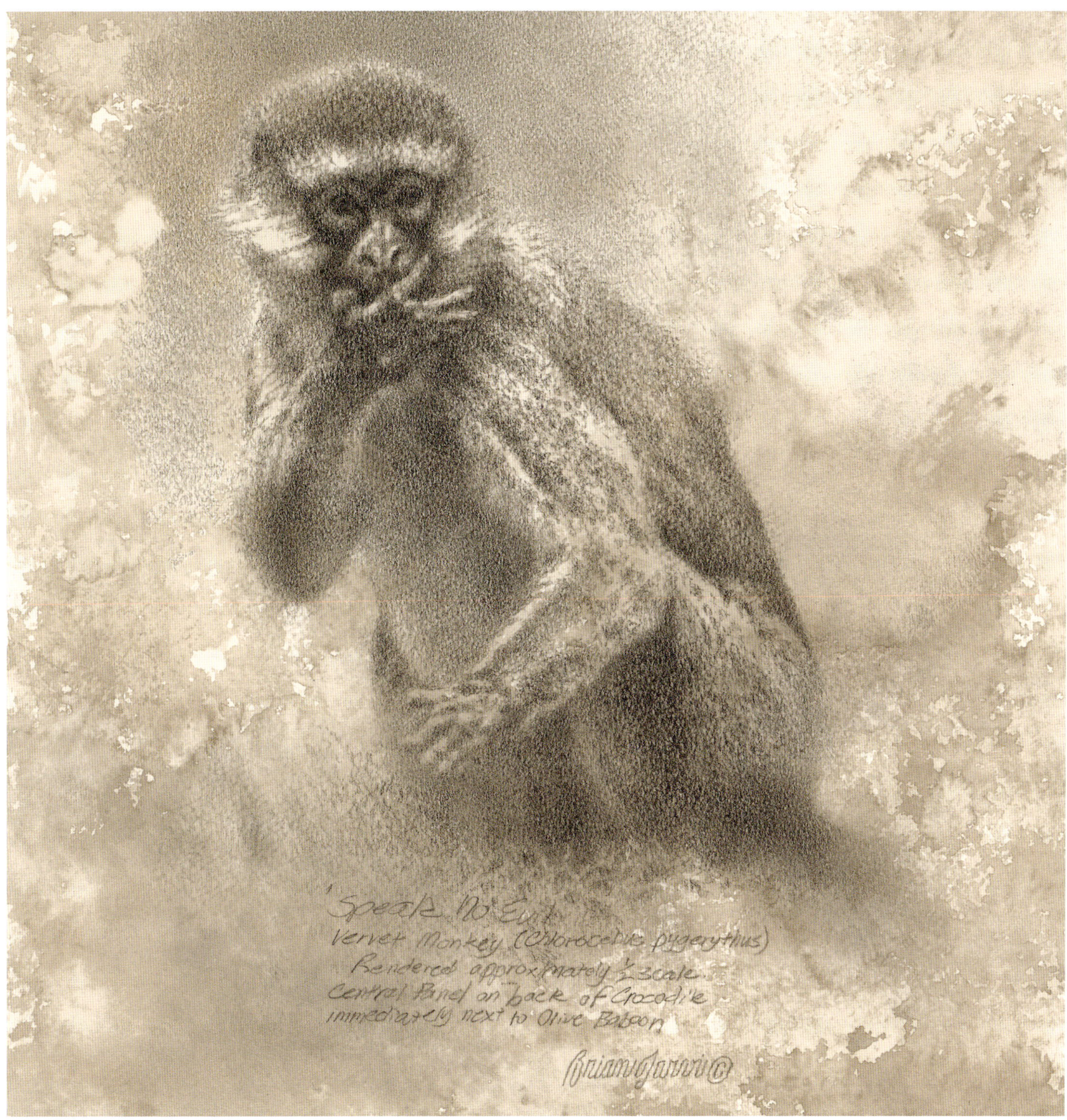

OPPOSITE
African Menagerie, Panel 4
Detail (Lower Middle Right)
Vervet Monkey

ABOVE
Speak No Evil, 2014
Vervet Monkey
Pencil
12 x 10"
Collection of Noah and Julie Wilcox

Many of the *African Menagerie* studies were done to familiarize Jarvi with physiology, musculature, and the natural histories of species. Sweet in their own right, the artful finished works—such as this rendering of a vervet monkey—are also coveted by collectors. It is Jarvi's hope that viewers of *Menagerie* will see these animals and then embark on their own discovery of what makes them unique.

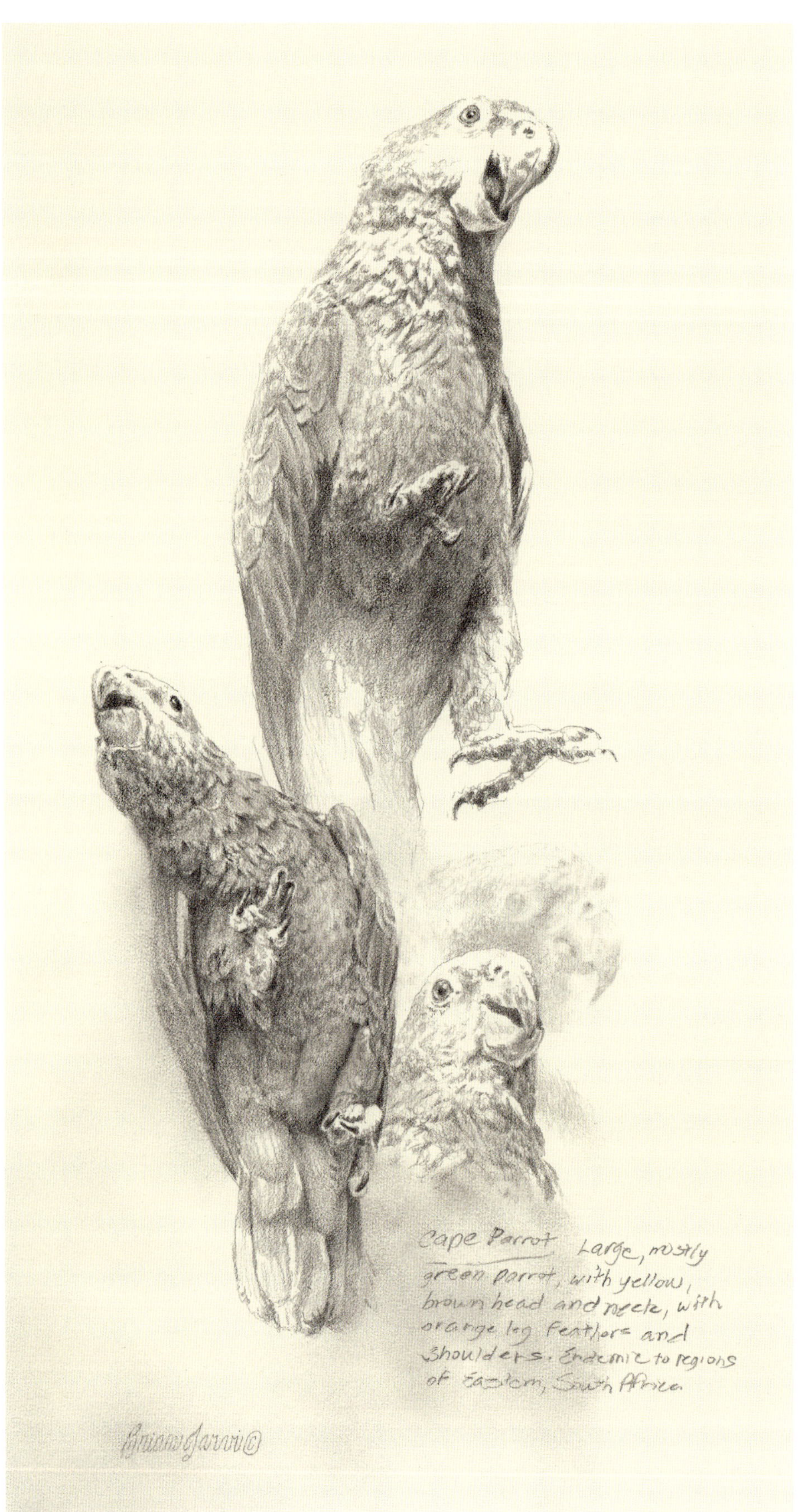

LEFT

Cape Parrot Study, 2014
Cape Parrot
Pencil
18 x 9"
Collection of Bill and Kathy Linder

OPPOSITE

African Menagerie, Panel 4
Detail (Middle Left)
Ostrich, Long-Crested Eagle,
Great Blue Turaco, Laughing Dove

OVERLEAF, LEFT

Grant's Gazelle Study, 2014
Grant's Gazelle
Pencil
18 x 13"
Collection of David and
Kathleen Chesness

OVERLEAF, RIGHT

African Menagerie, Panel 4
Detail (Middle Right)
Grant's Gazelle

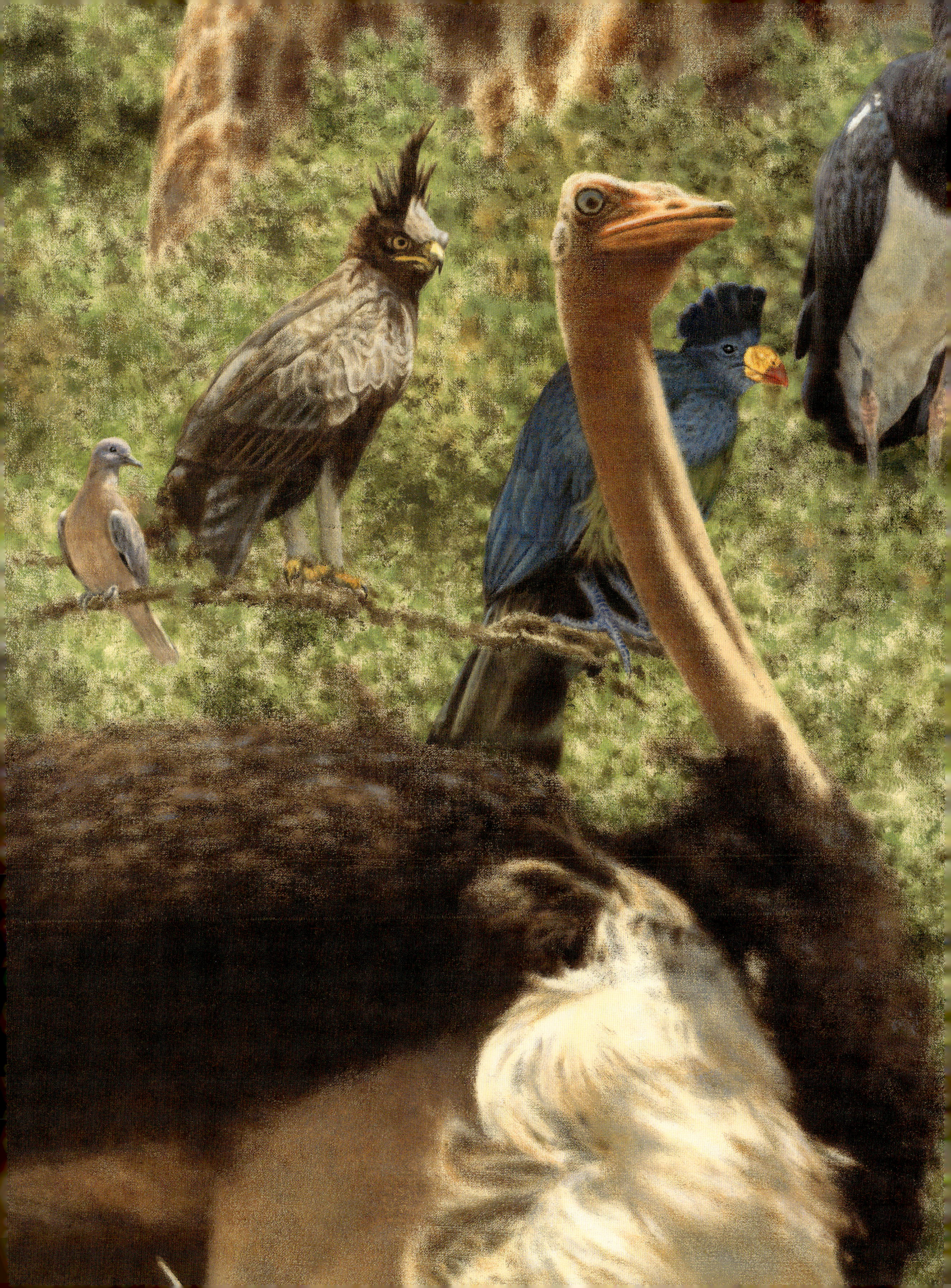

Grant's Gazelle
Various study postures. These
options allow for a range
of different locations in
the "Inquisition"

Right center panel

center panel

left middle panel
or left center

right middle

Bonnie Gurri ©

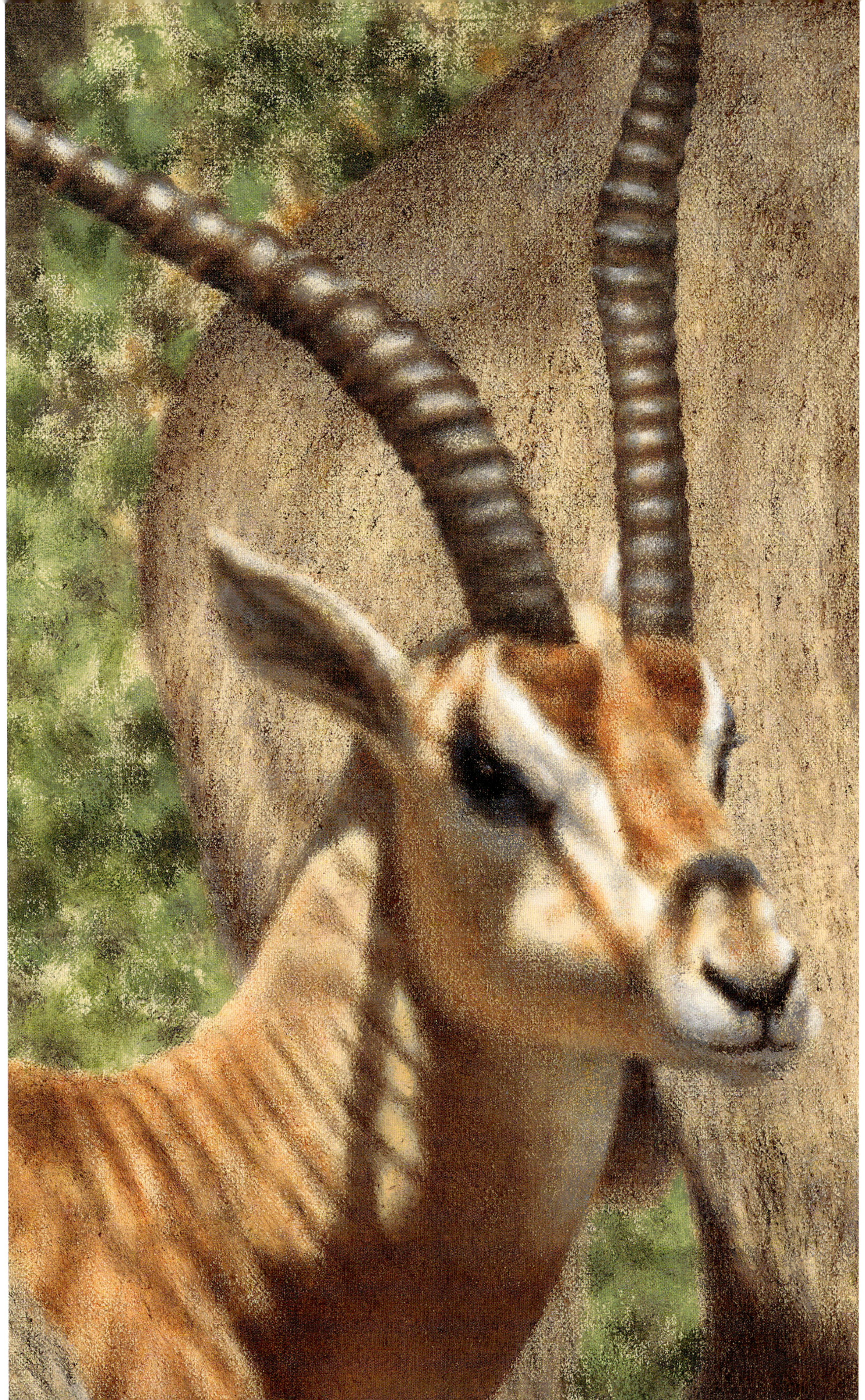

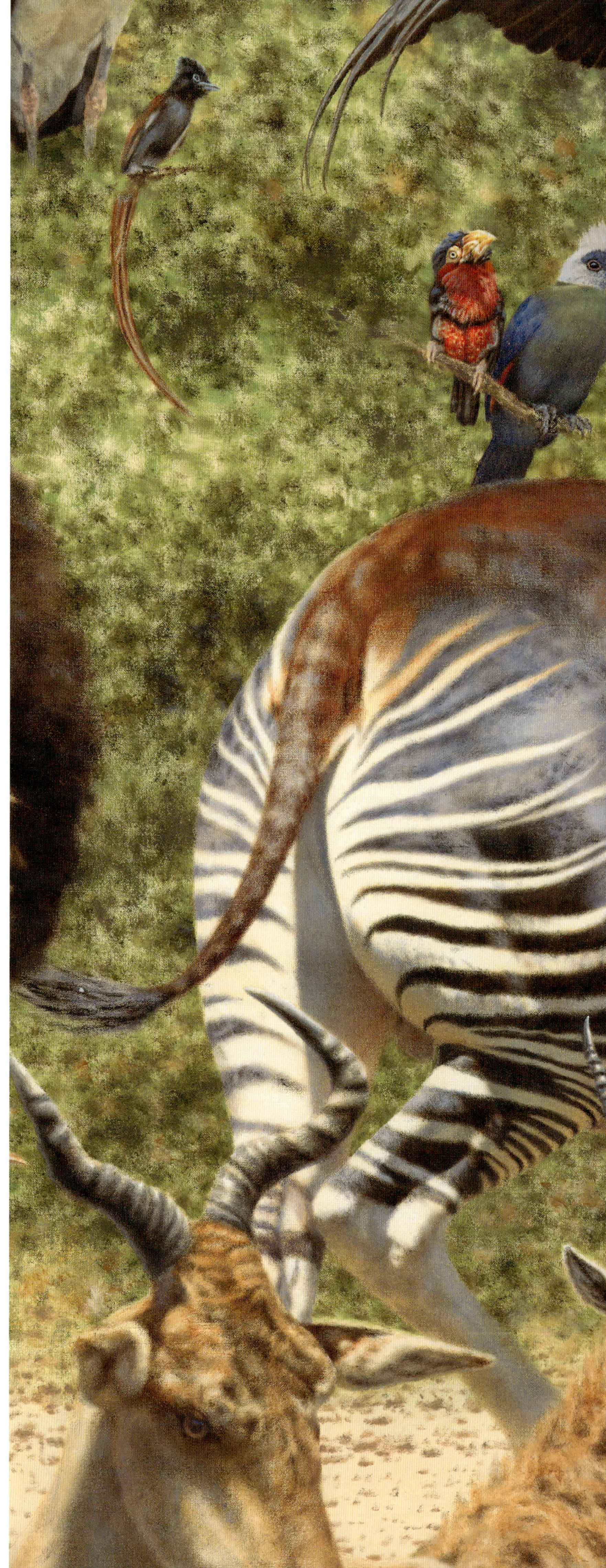

African Menagerie, Panel 4
Detail (Middle Right)
Yellow-Billed Stork, Paradise Flycatcher, Bearded Barbet, White-Crested Turaco, Okapi, Grant's Gazelle, Olive Baboon, Spotted Hyena, Vervet Monkey, Thomson's Gazelle, Hartebeest

Jarvi has a special fascination with more obscure subjects beyond the Big Five for which his work is most closely associated. In this depiction of the stunning okapi, his first goal was to showcase the fabulous markings on its hindquarters. This was the first species to be completed in full color, and therefore it holds a special spot in Jarvi's heart.

OVERLEAF, LEFT
Long-Crested Eagle Pre-Study, 2016
Long-Crested Eagle
Oil on Belgian Linen
18 x 14"

OVERLEAF, TOP RIGHT
Spotted Hyena Study, 2015
Spotted Hyena
Charcoal, Pencil, and Watercolor
12 x 23"
Collection of Brian and Jeanne Nicklason

OVERLEAF, BOTTOM RIGHT
Hartebeest Study, 2014
Hartebeest
Charcoal, Pencil, and Watercolor
18 x 23"

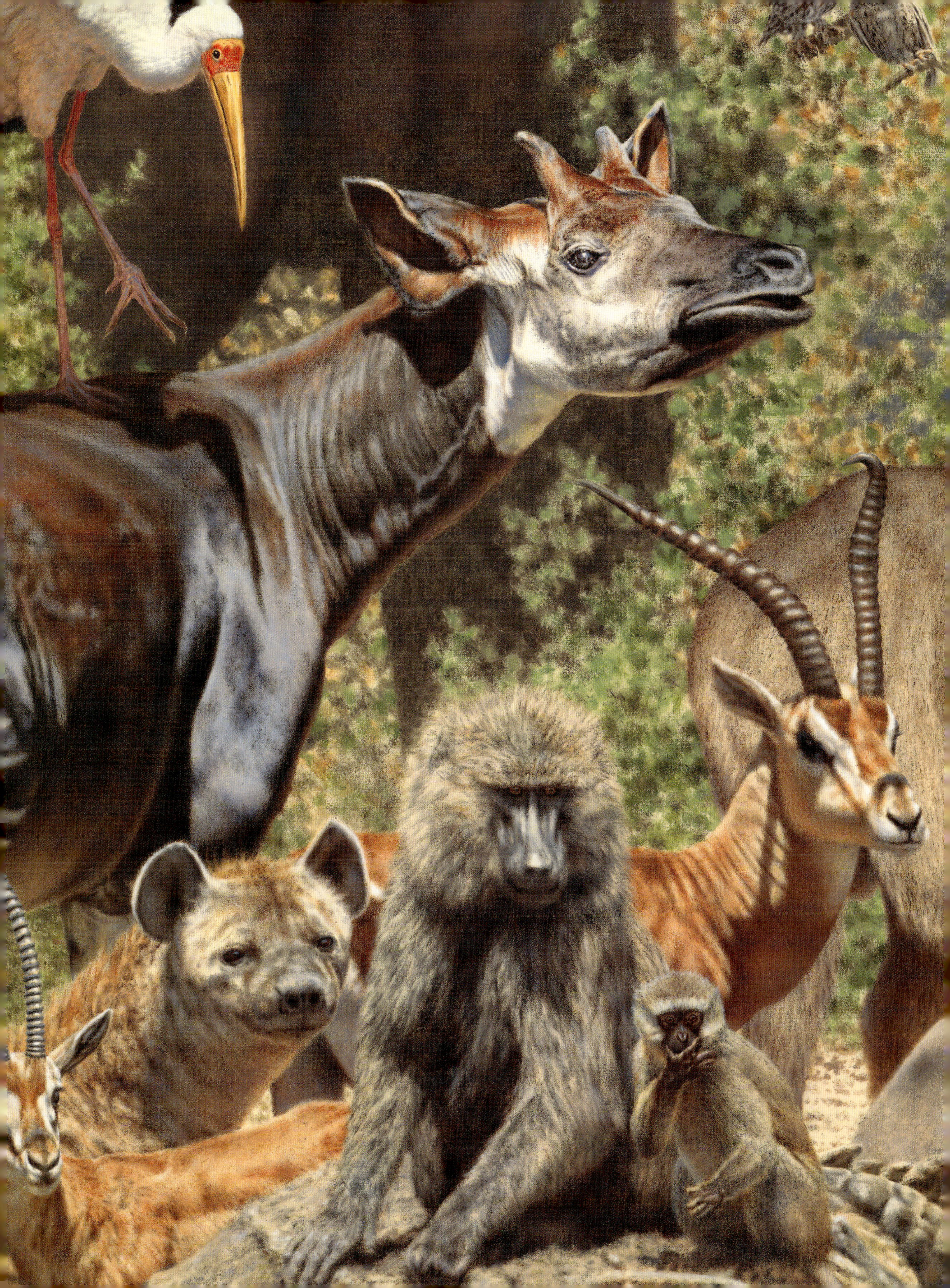

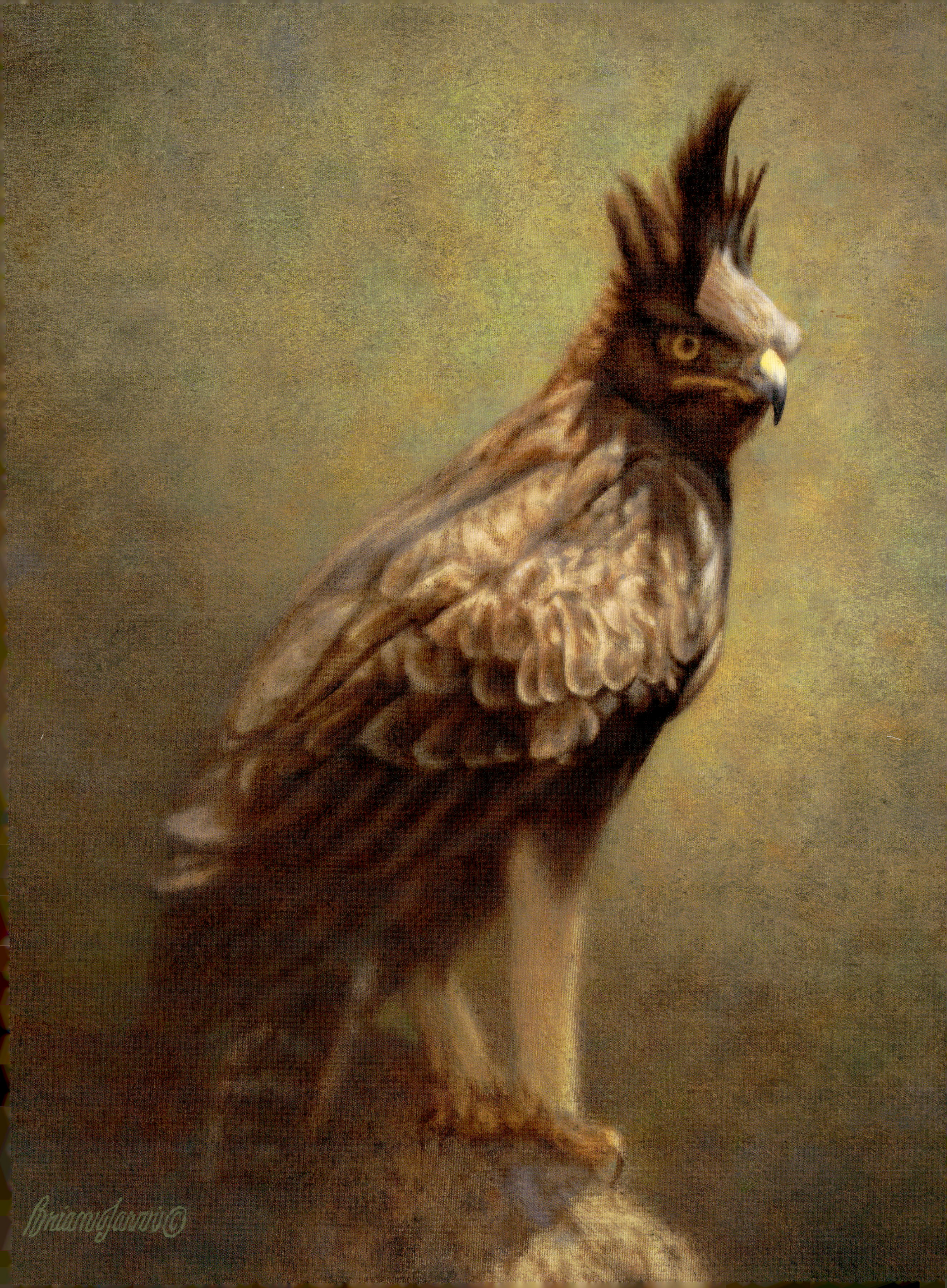

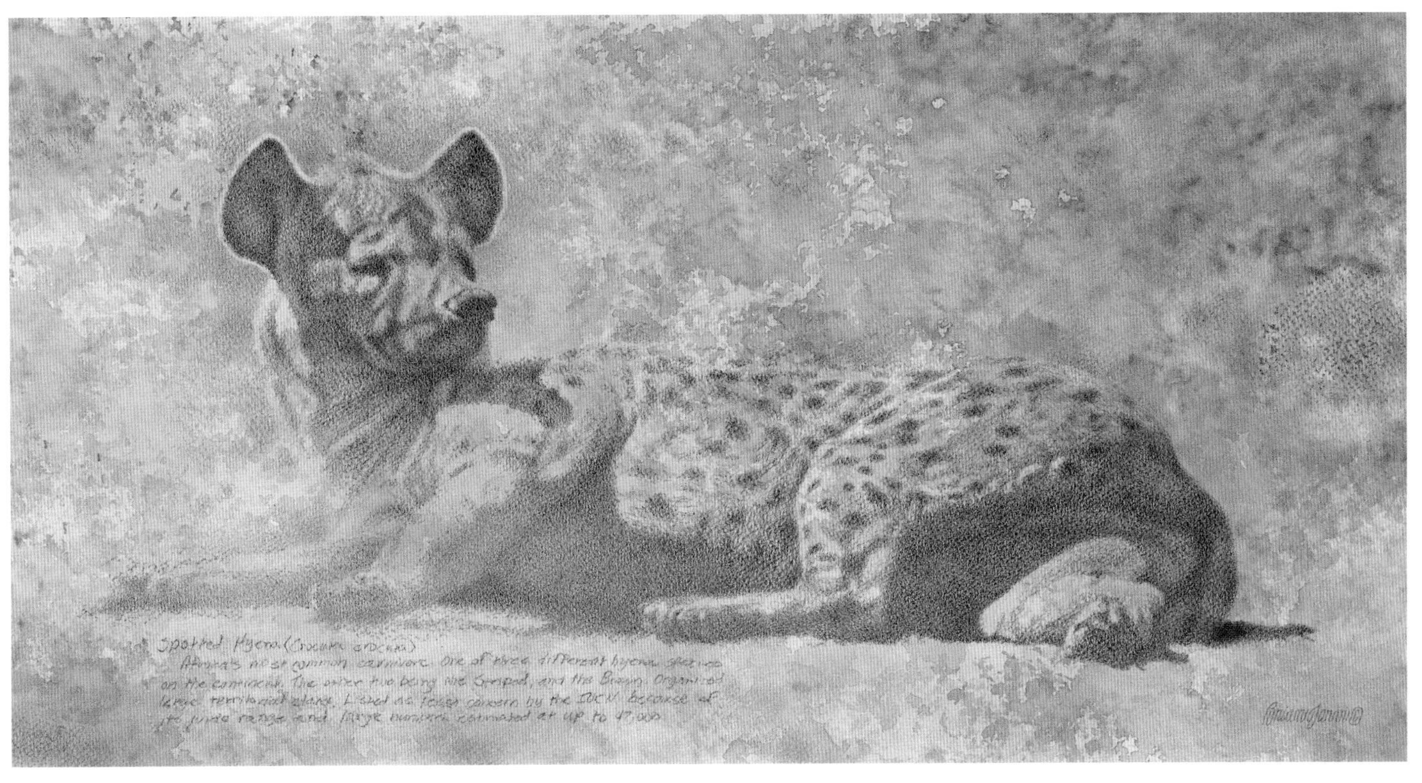

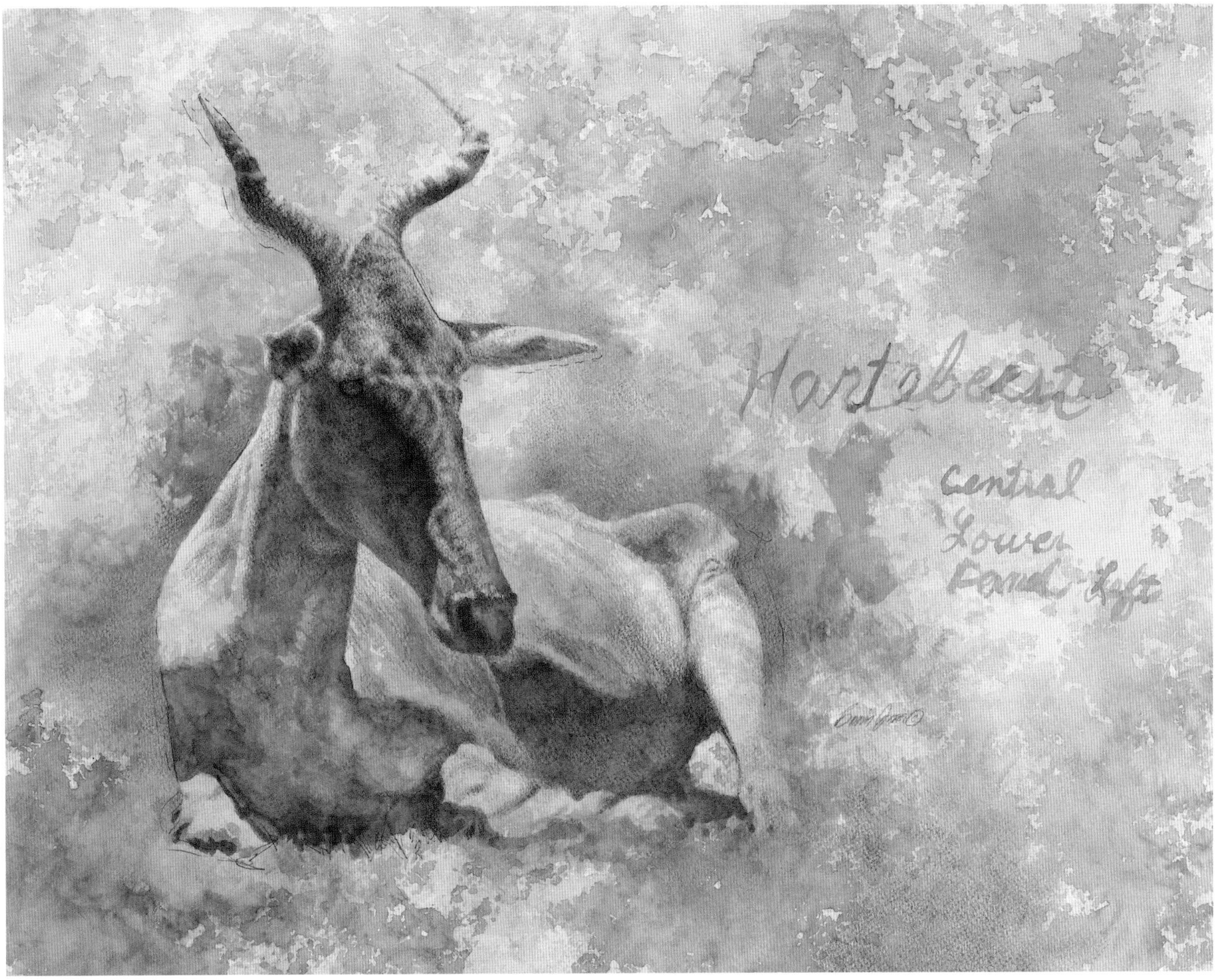

Hartebeest

Central
Lower
Panel-Left

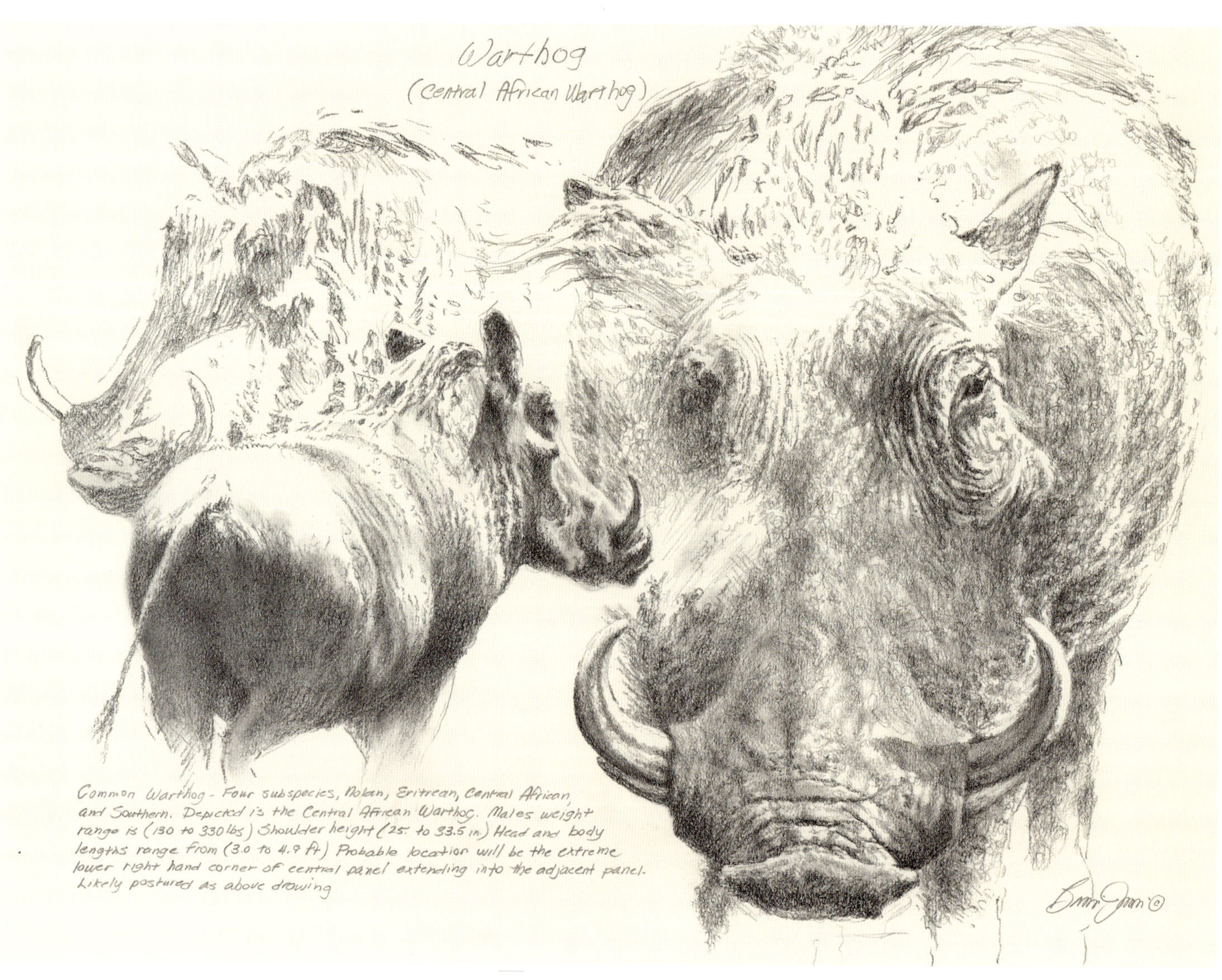

ABOVE
Warthog Study, 2015
Warthog
Pencil
18 x 23"
Collection of Kevin and Tuesdy Small

OPPOSITE
African Menagerie, Panel 4
Detail (Bottom Right)
Warthog

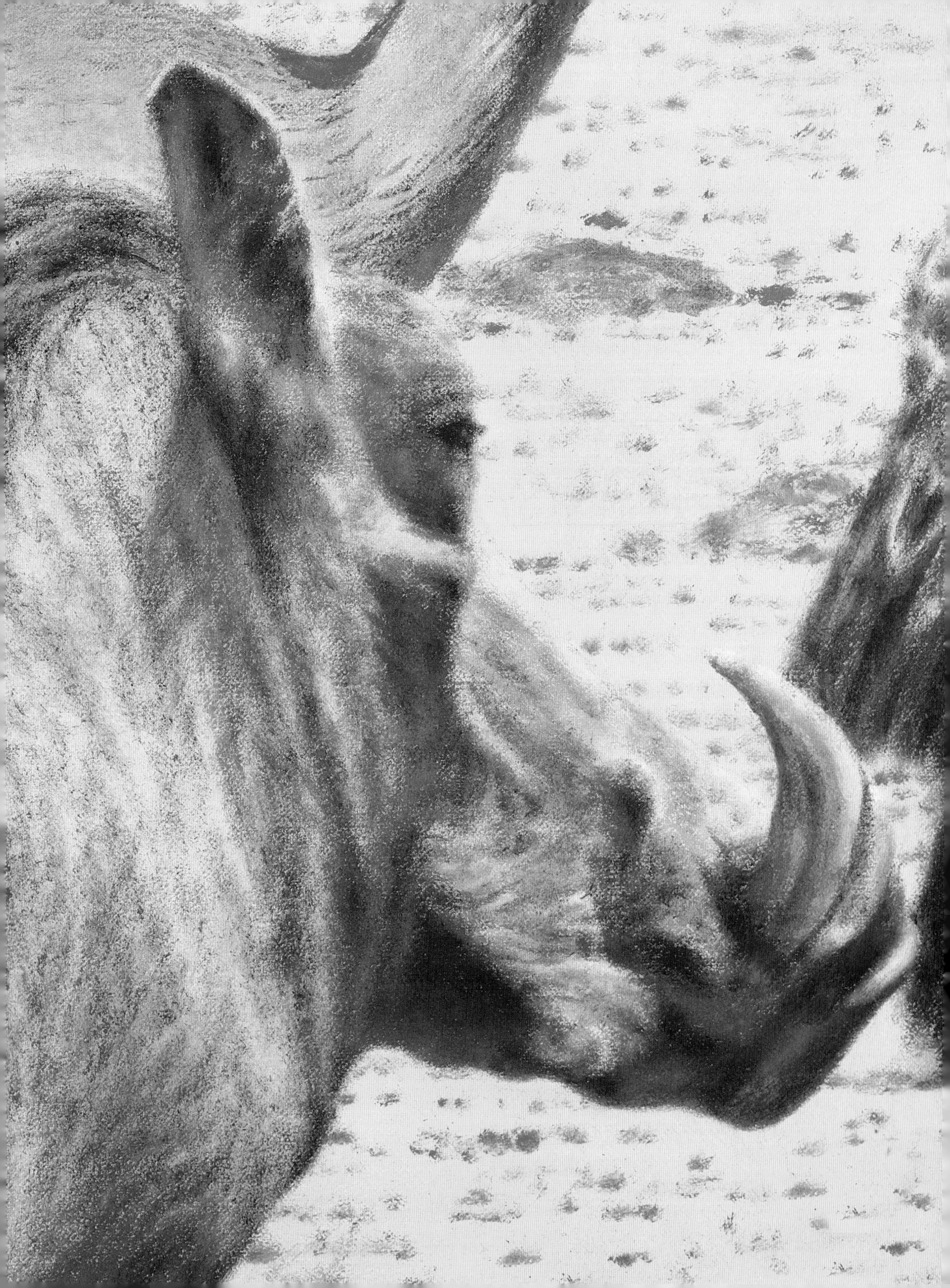

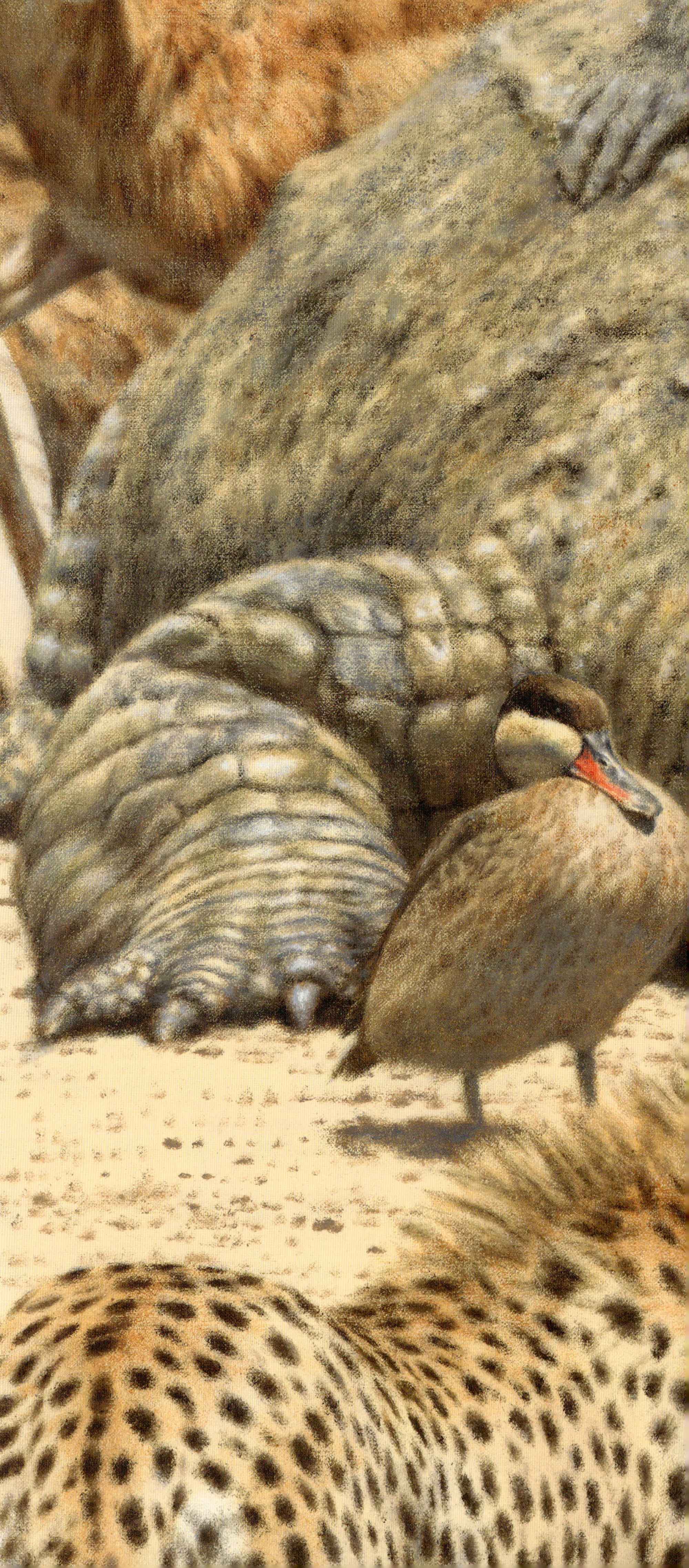

African Menagerie, Panel 4
Detail (Bottom Right)
Nile Crocodile, Red-Billed Teal, Cheetah, Bush Duiker

The Nile crocodile has a notorious reputation and is rooted in Egyptian lore. Few creatures better evoke raw wildness and sensations of danger.

OVERLEAF, LEFT
African Menagerie, Panel 4
Detail (Bottom Center)
Secretary Bird, Snouted Cobra, White Stork, Mwanza Flat-Headed Rock Agama, Rainbow Skink, Common Moorhen

Jarvi placed the snouted cobra and secretary bird side by side as an insinuation of irony. The bird is an incredibly adept predator of the poisonous snake.

OVERLEAF, RIGHT
Secretariat, 2015
Secretary Bird
Oil on Belgian Linen
23 x 12"

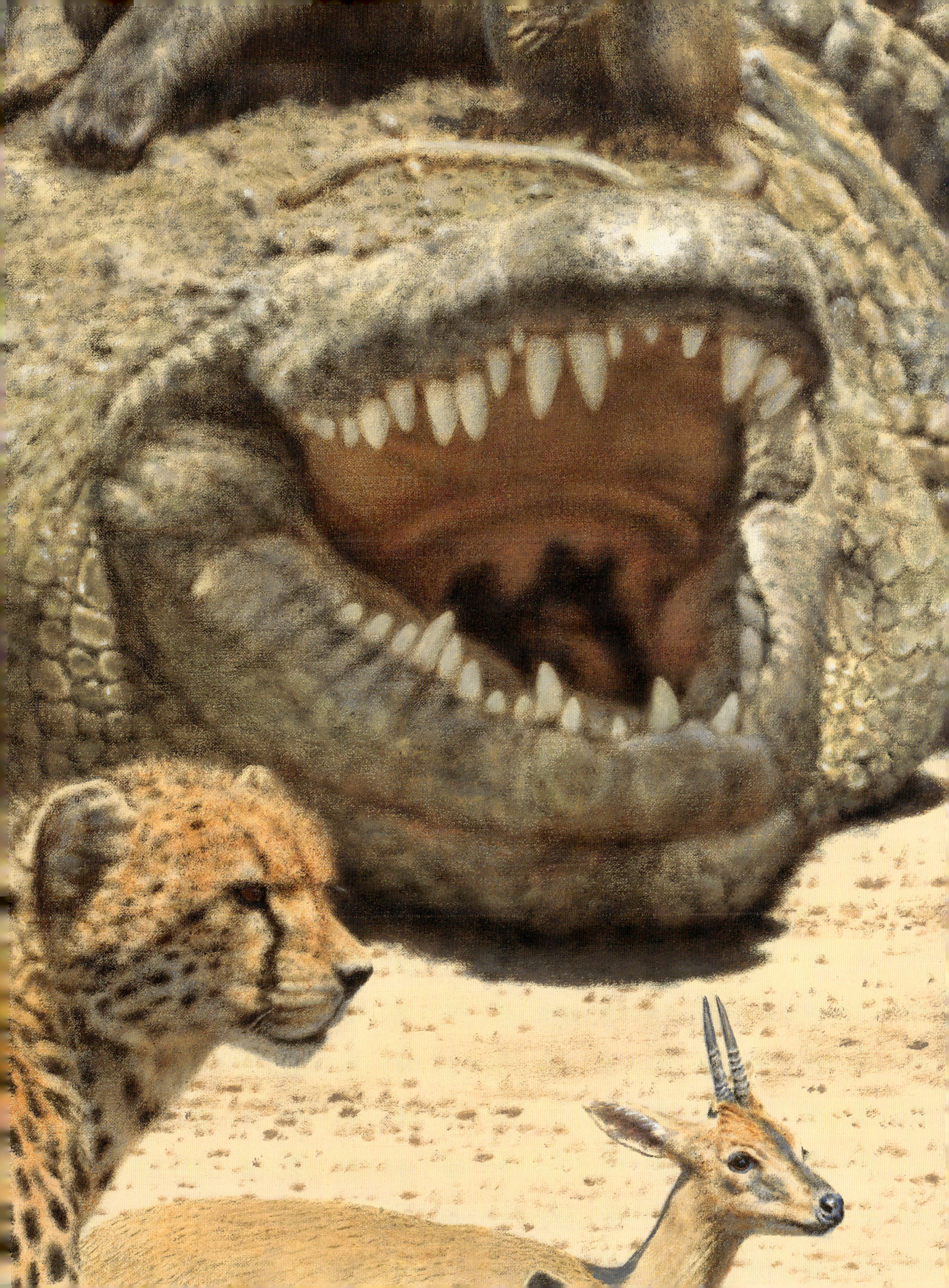

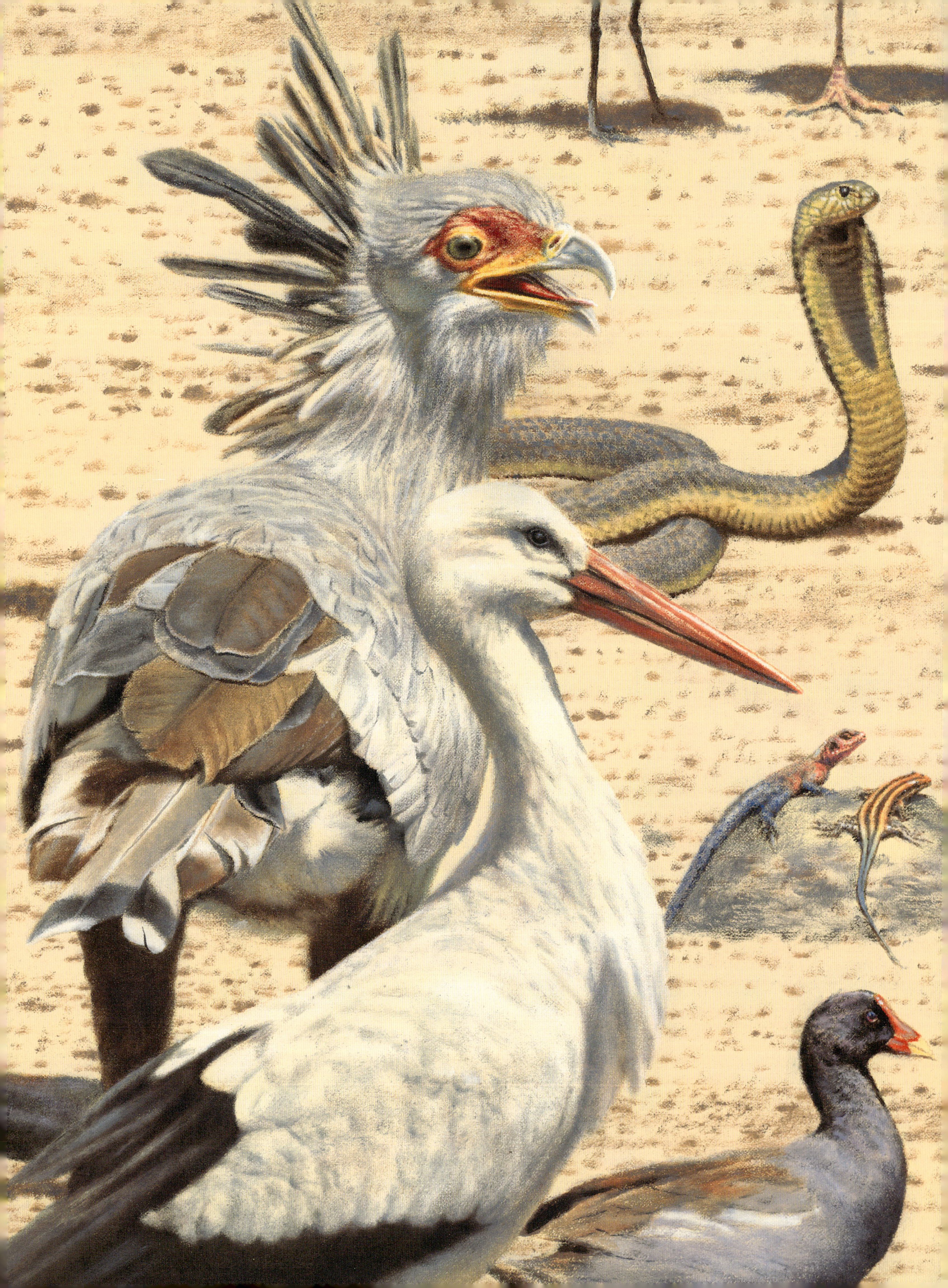

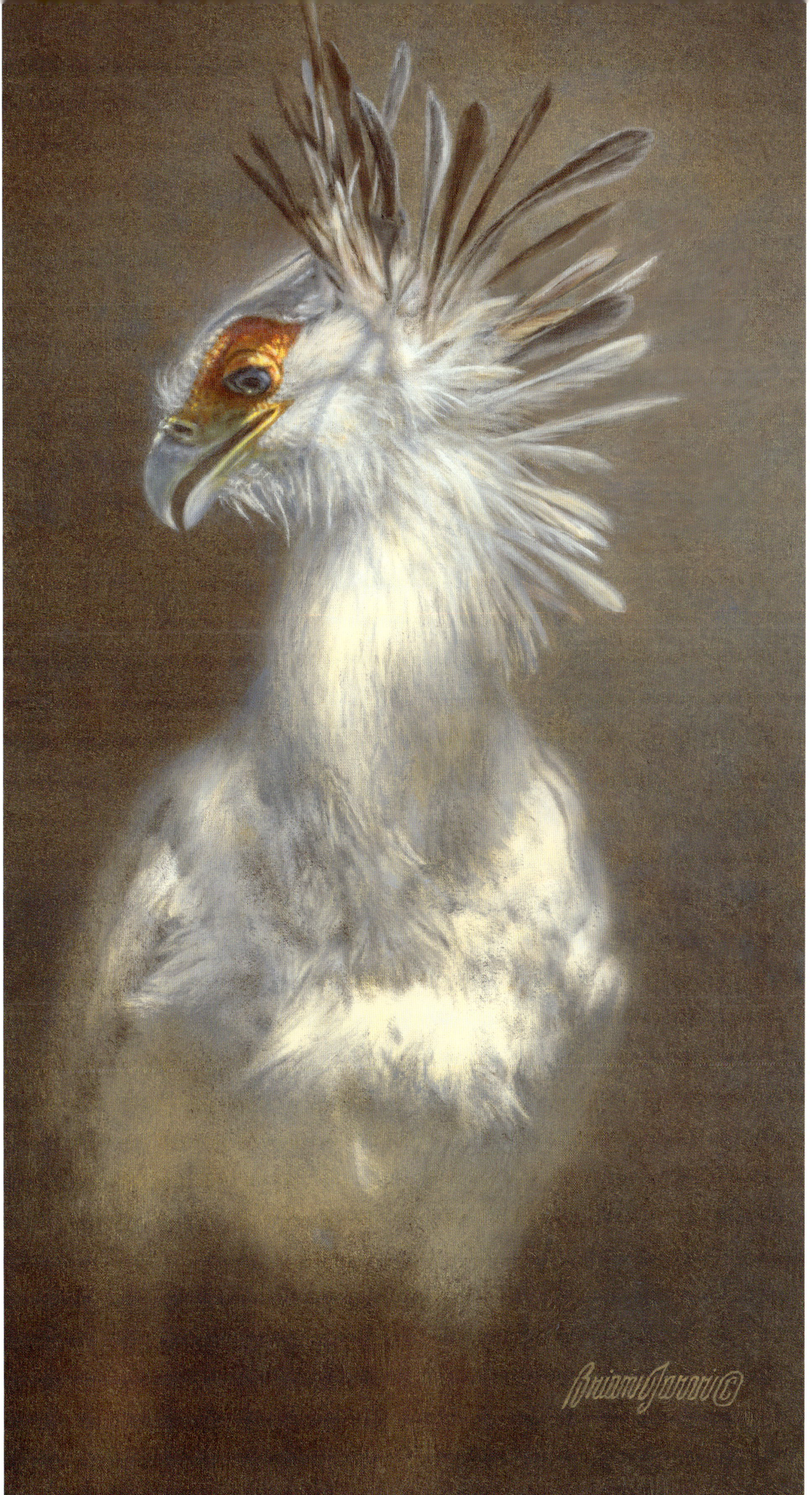

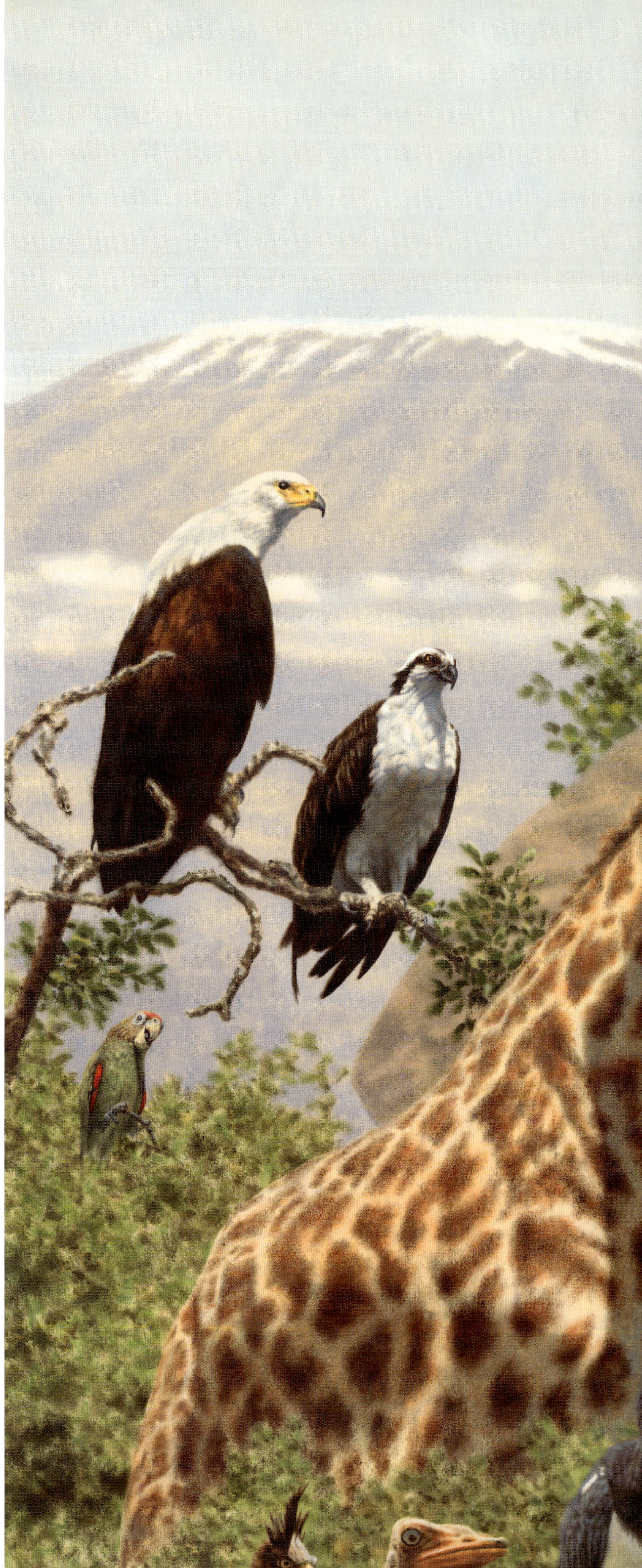

African Menagerie, Panel 4
Detail (Top)
Masai Giraffe, Bonobo, African Fish Eagle, Osprey, African Elephant, Cape Parrot, Abdim's Stork, Ostrich, Long-Crested Eagle

The added height of the central panel was implemented to accommodate the giraffe, elephant, and also the profile of Mount Kilimanjaro, which is rapidly losing its snowpack. The idea of symbolically placing the bonobo in this crown area flashed in Jarvi's mind years earlier. Bonobos, only a slight genetic variation away from humans, are known for their higher intelligence. Will we supply *our* superior wherewithal to deal with climate change?

OVERLEAF, LEFT
Bonobo Pre-Study, 2014
Bonobo
Charcoal
14 x 11"
Collection of Ed May

OVERLEAF, RIGHT
Giraffe Triptych, 2015
Giraffe
Oil on Belgian Linen
48 x 18"
Collection of Annette Bishop

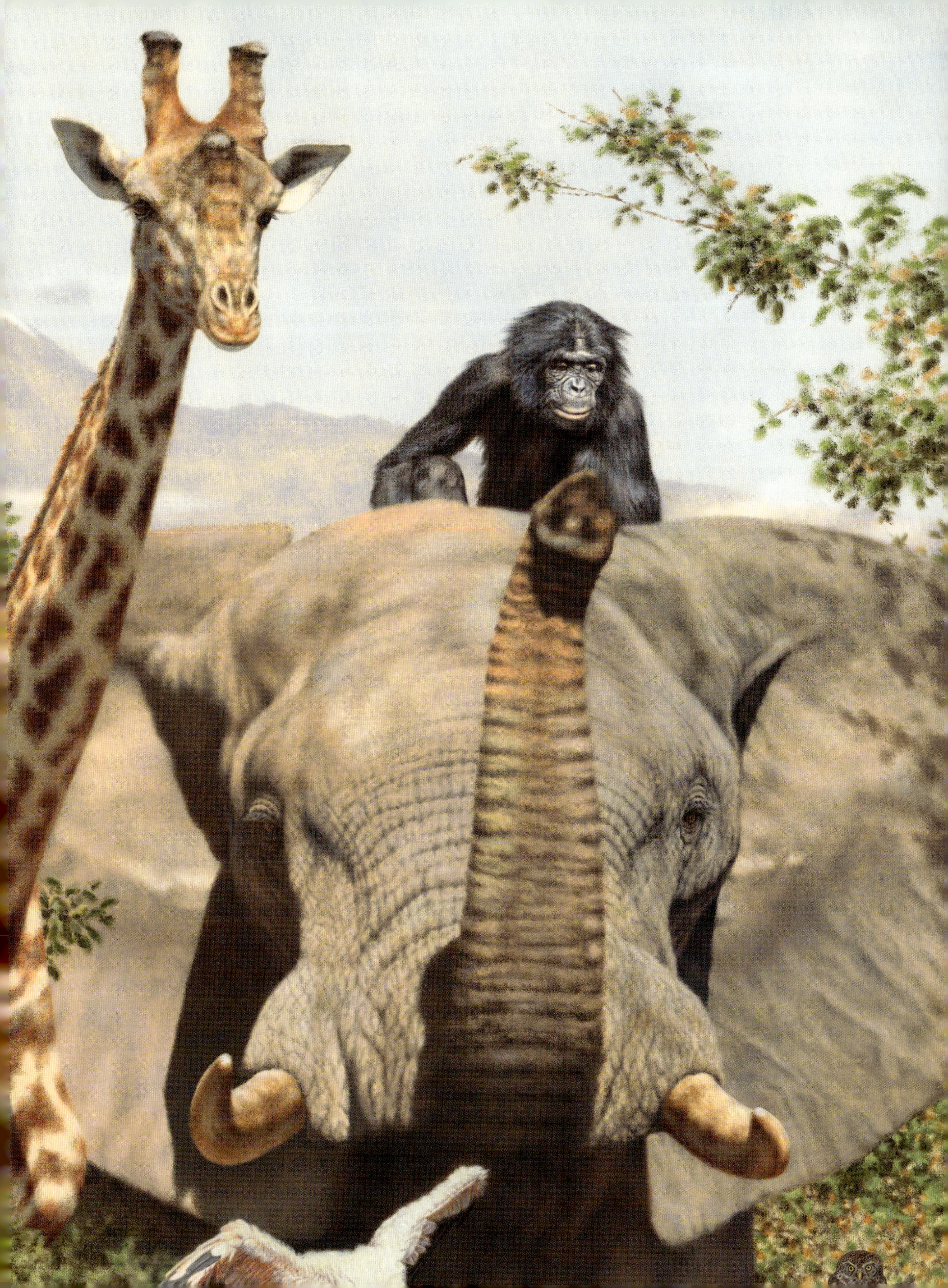

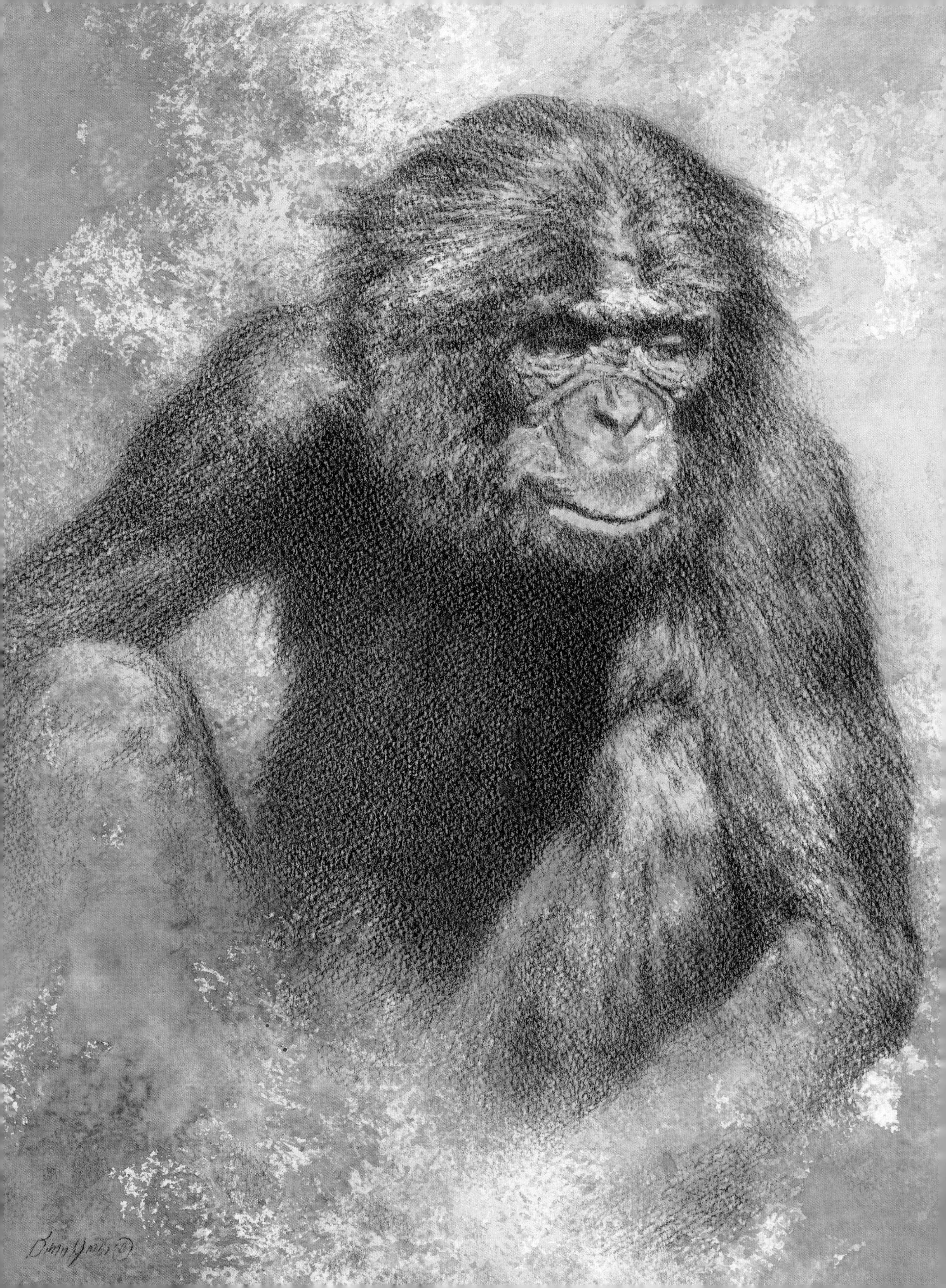

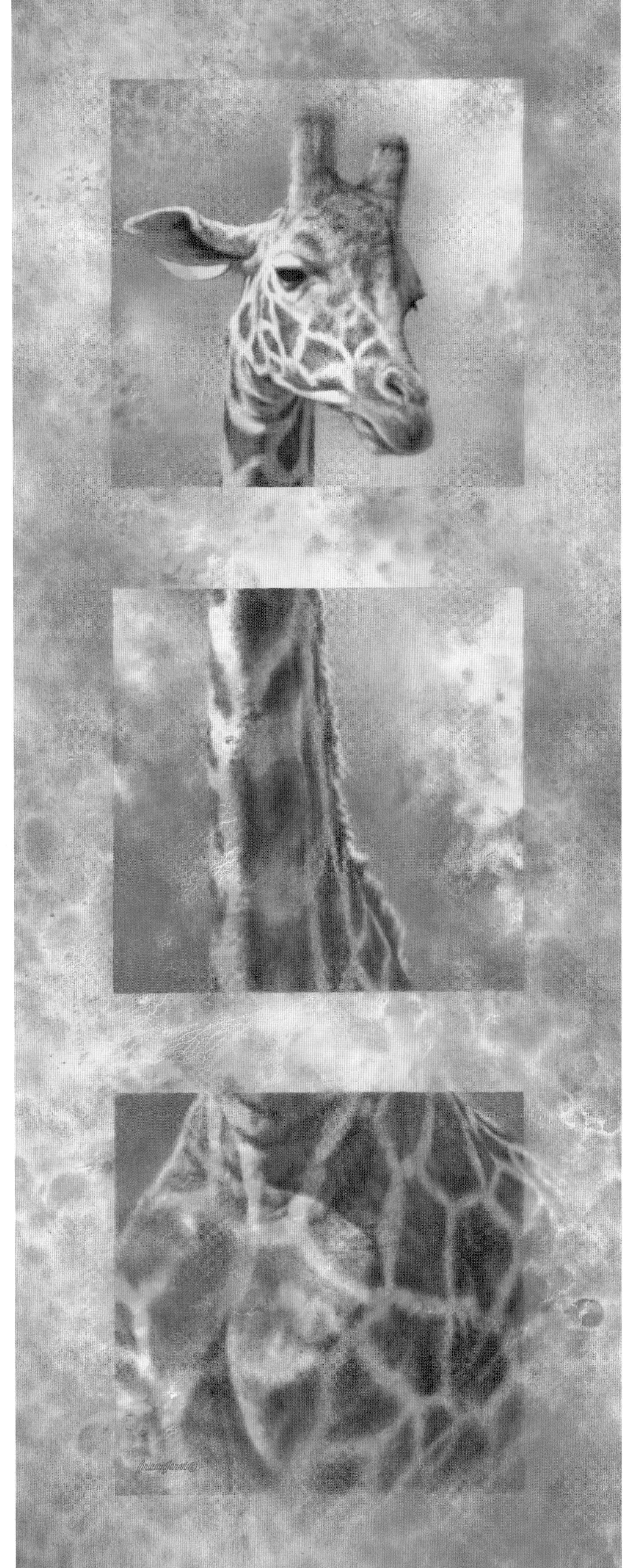

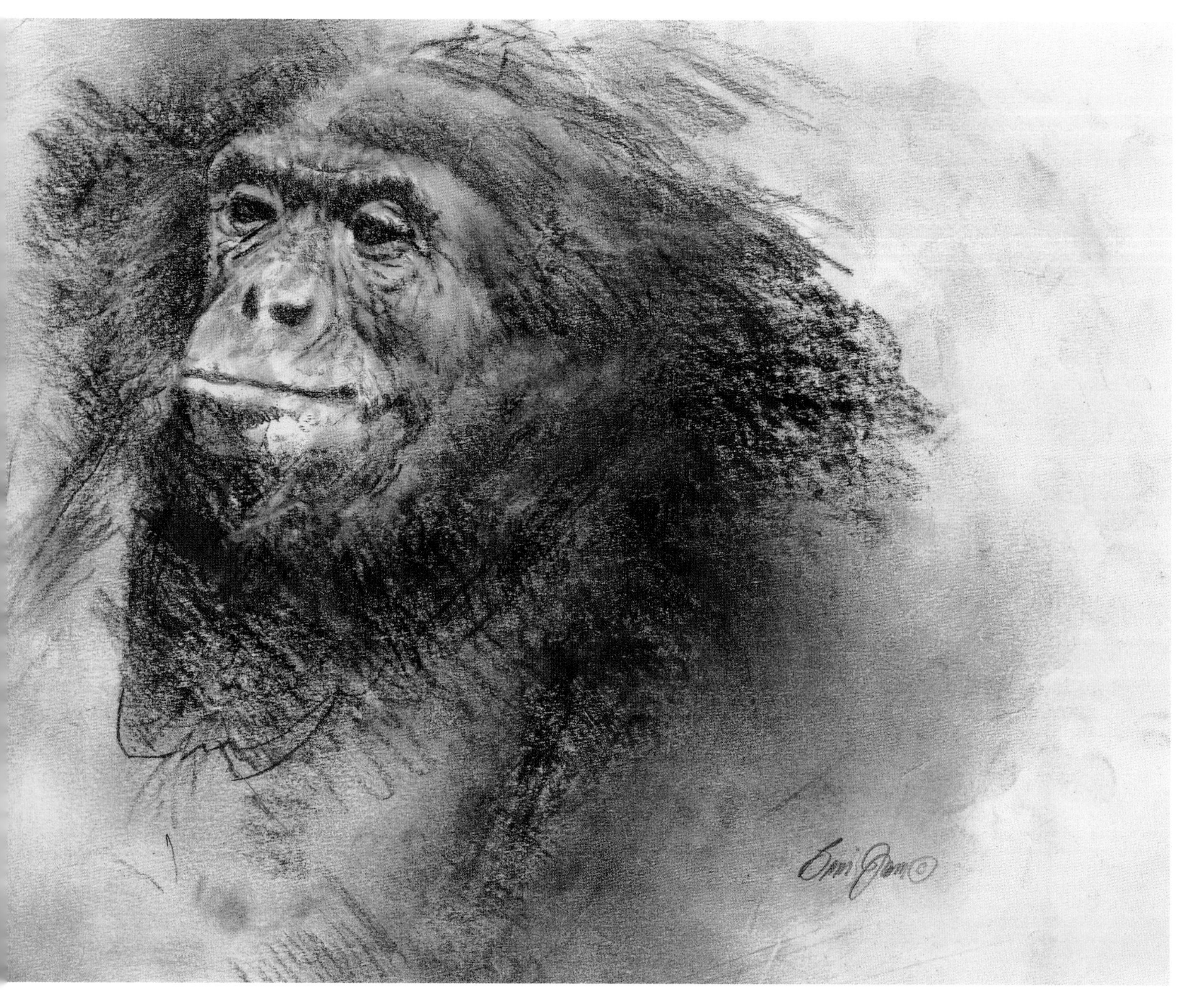

ABOVE
Bonobo Quick-Study, 2014
Bonobo
Charcoal
9 x 11"
Collection of Lee and Mary Jo Jess

OPPOSITE
Laughing Dove Pre-Study, 2016
Laughing Dove
Oil on Belgian Linen
12 x 9"

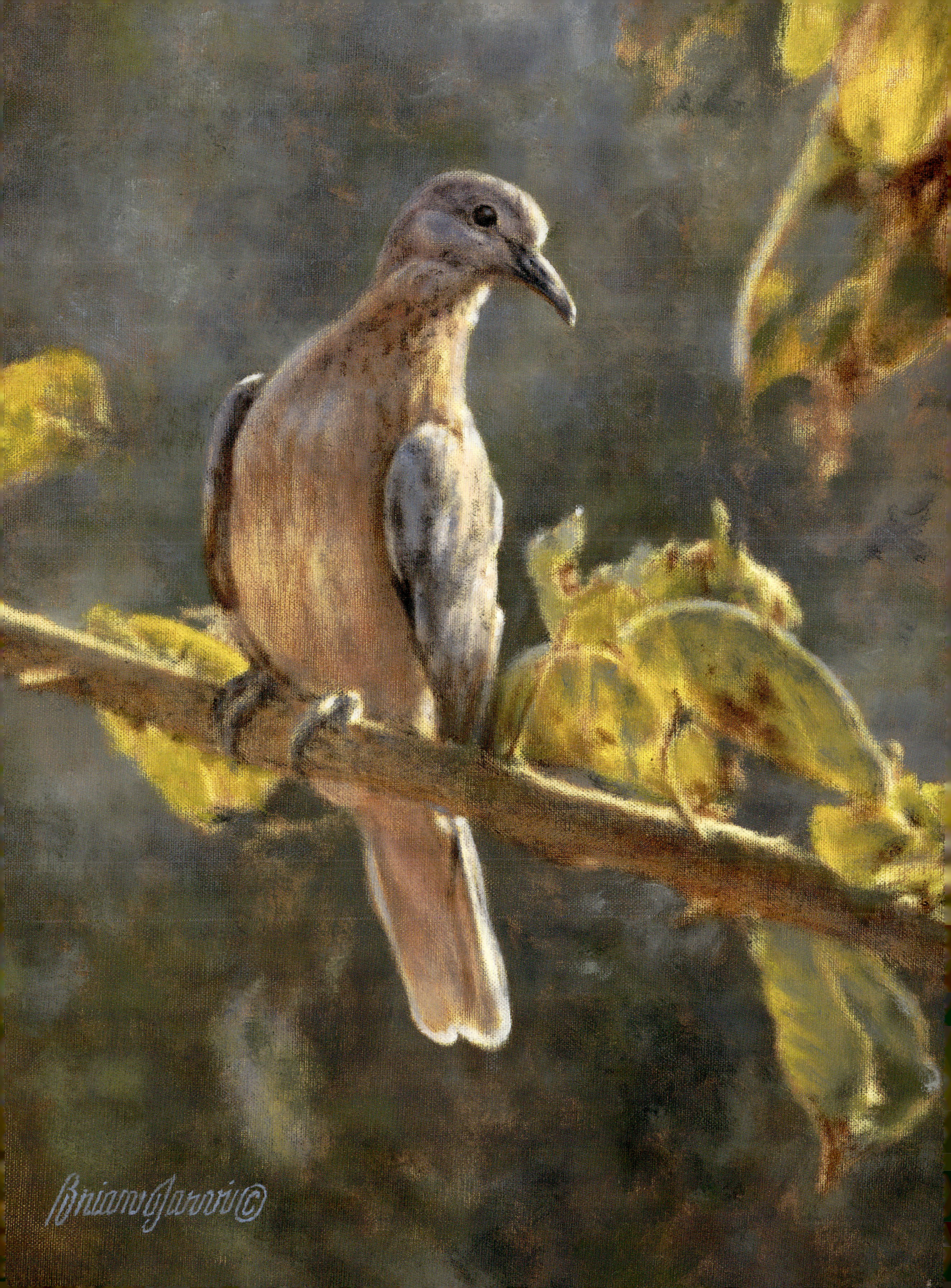
Brian Jarvi©

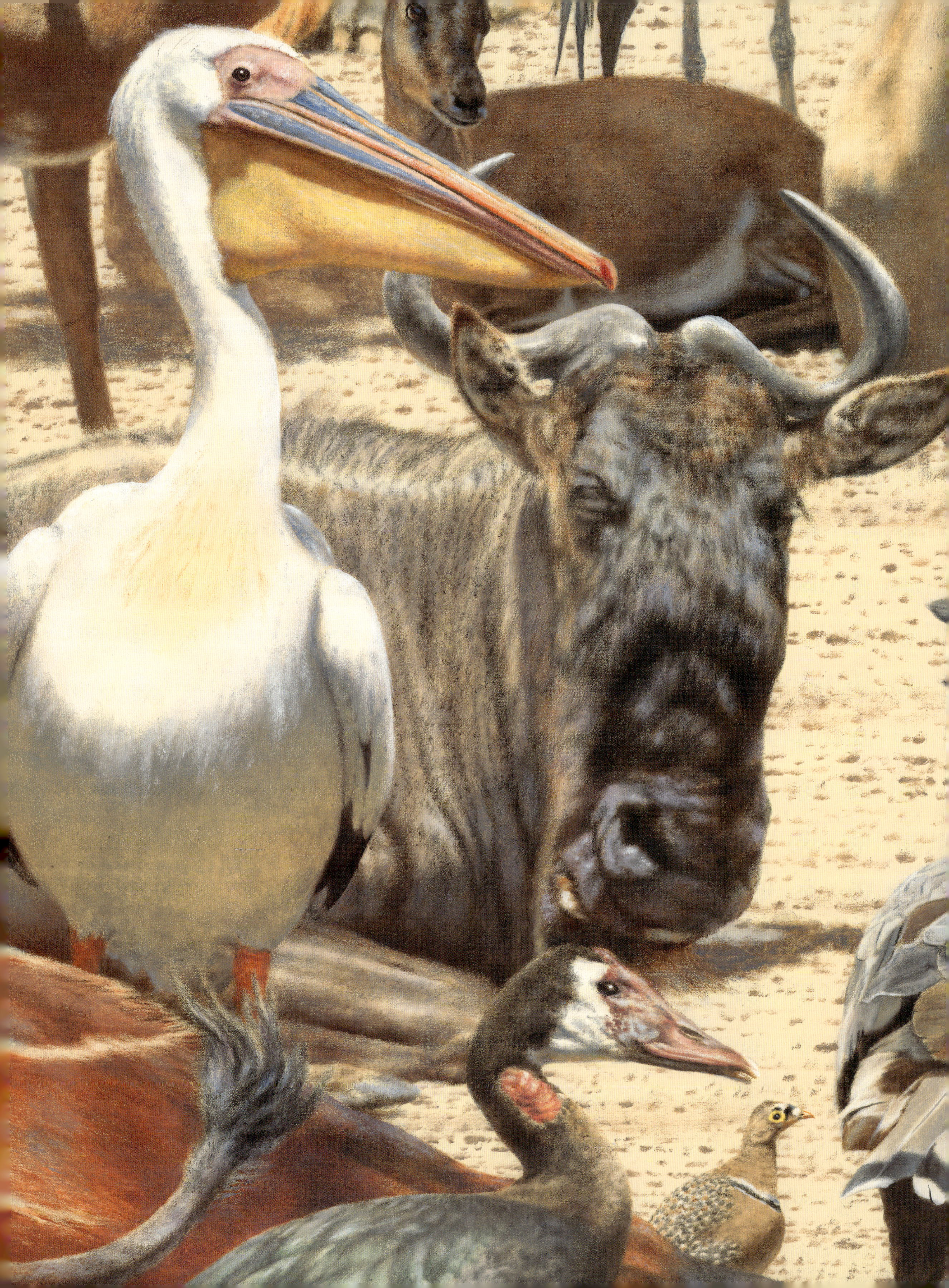

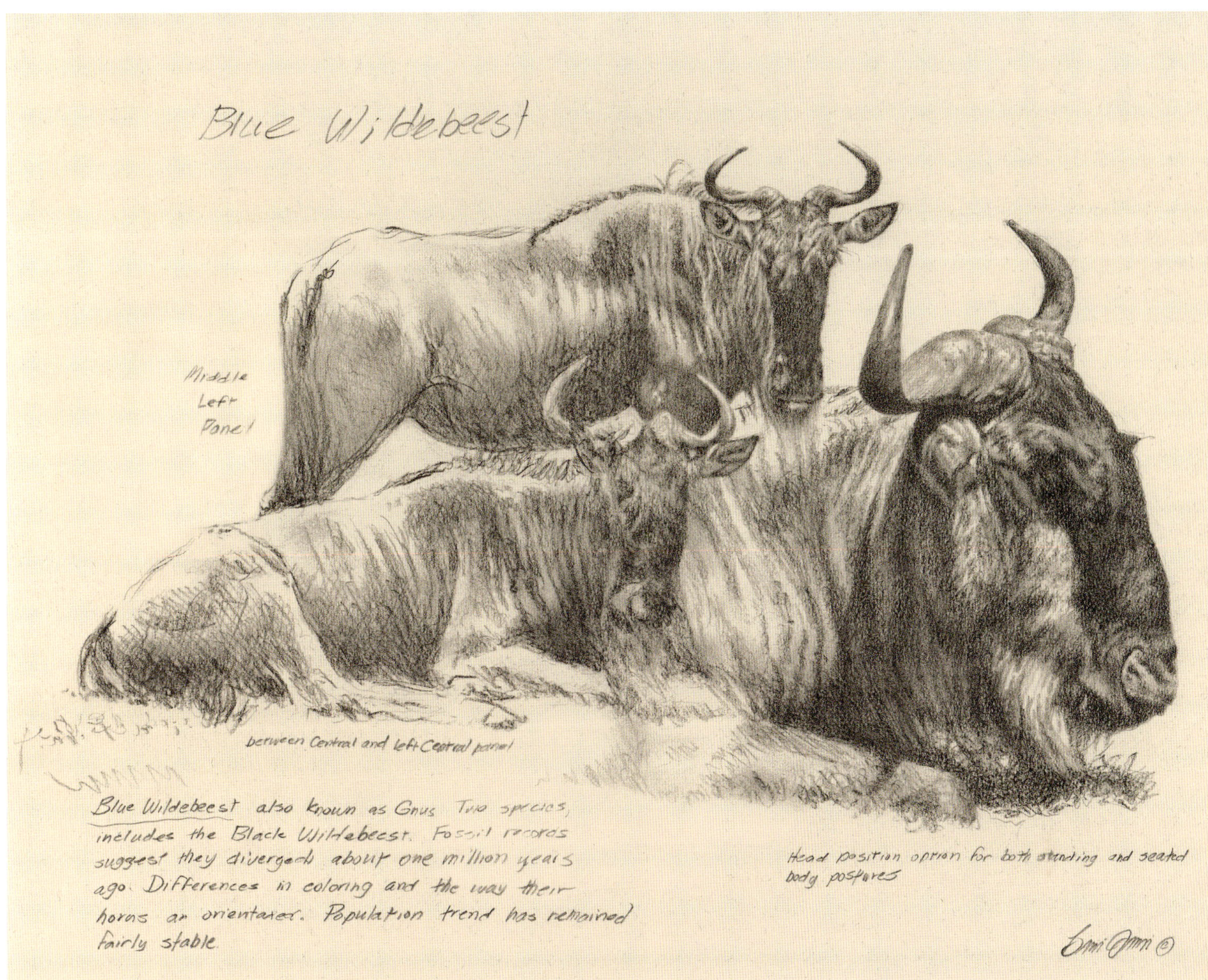

OPPOSITE
African Menagerie, Panel 4
Detail (Bottom Left)
Cuvier's Gazelle, Great White Pelican, Blue Wildebeest, Spur-Winged Goose, Double-Banded Sandgrouse

ABOVE
Wildebeest Study, 2014
Blue Wildebeest
Charcoal
19 x 24"

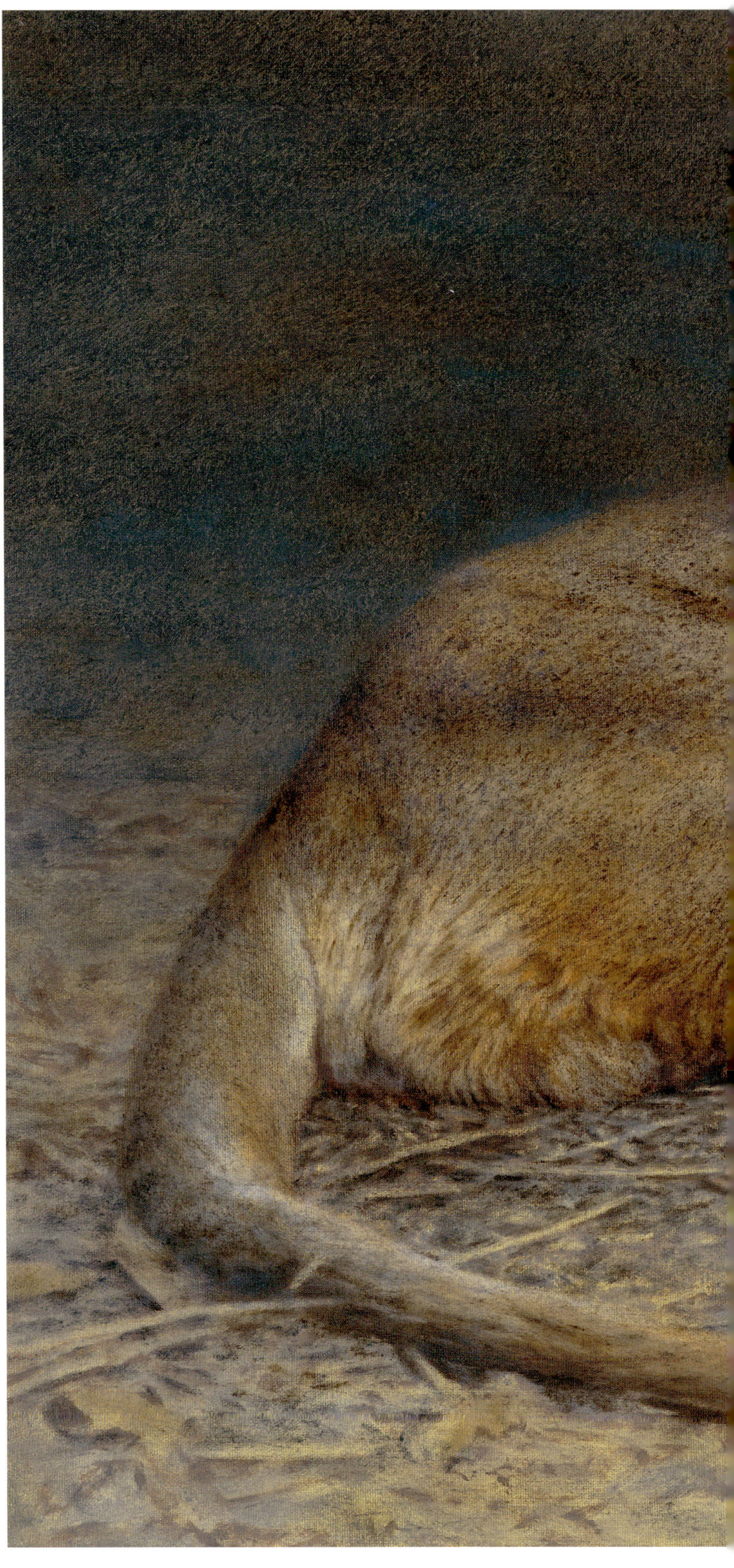

The Fossa Study, 2016
Fossa
Oil on Belgian Linen
15 x 24"

While the island of Madagascar is known for its lemurs, the fossa is an apex predator and a handsome muse. This piece was juried into the 2017 Society of Animal Artists' *Art and the Animal* exhibition.

OVERLEAF, LEFT
Ring-Tailed Lemur Study, 2015
Ring-Tailed Lemur
Pencil
18 x 10"
Collection of Julie Sandstrom

OVERLEAF, RIGHT
Windows to the Soul, 2016
Mandrill
Oil on Belgian Linen
33 x 26"
Collection of Todd and Shari Nelson

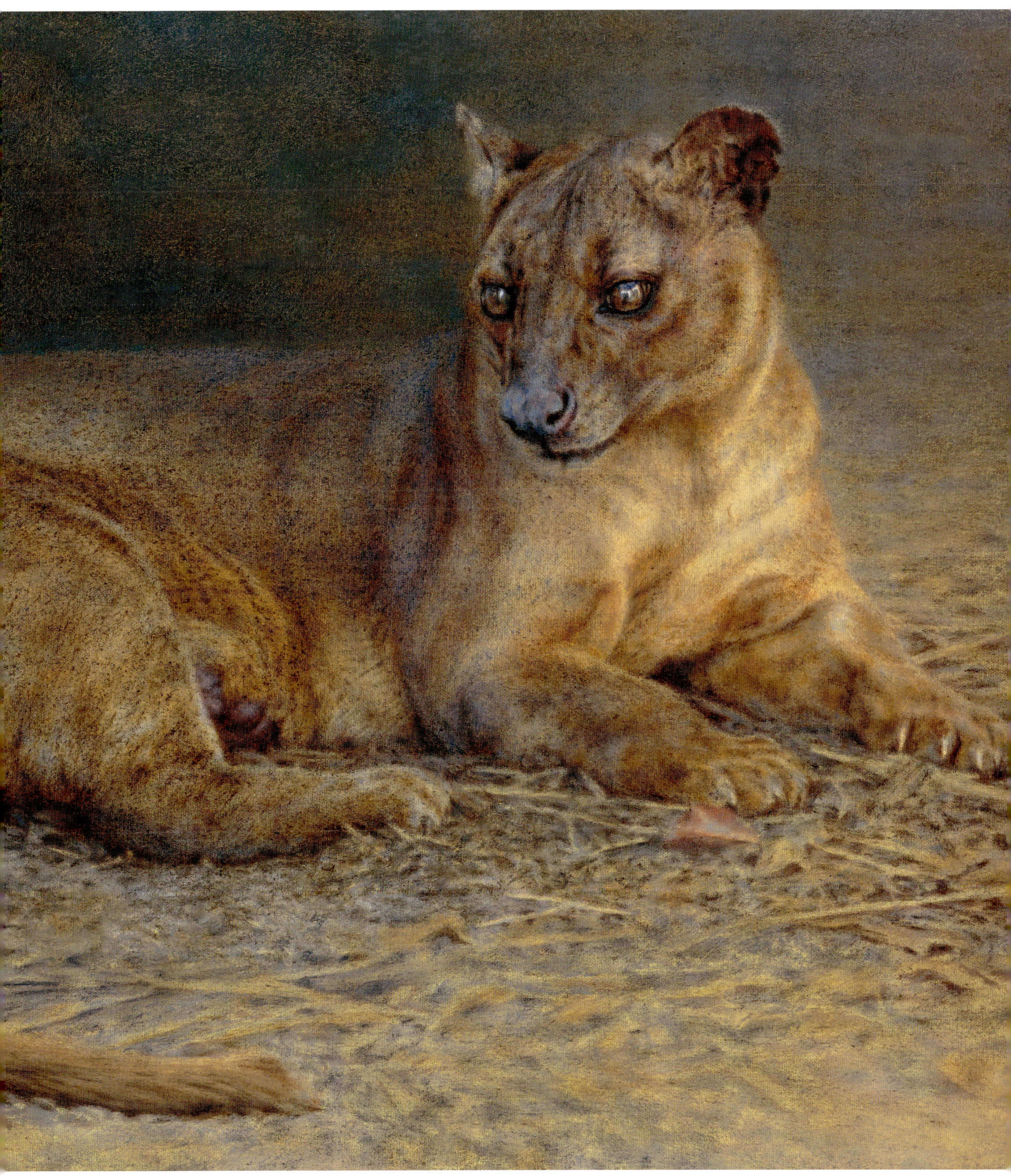

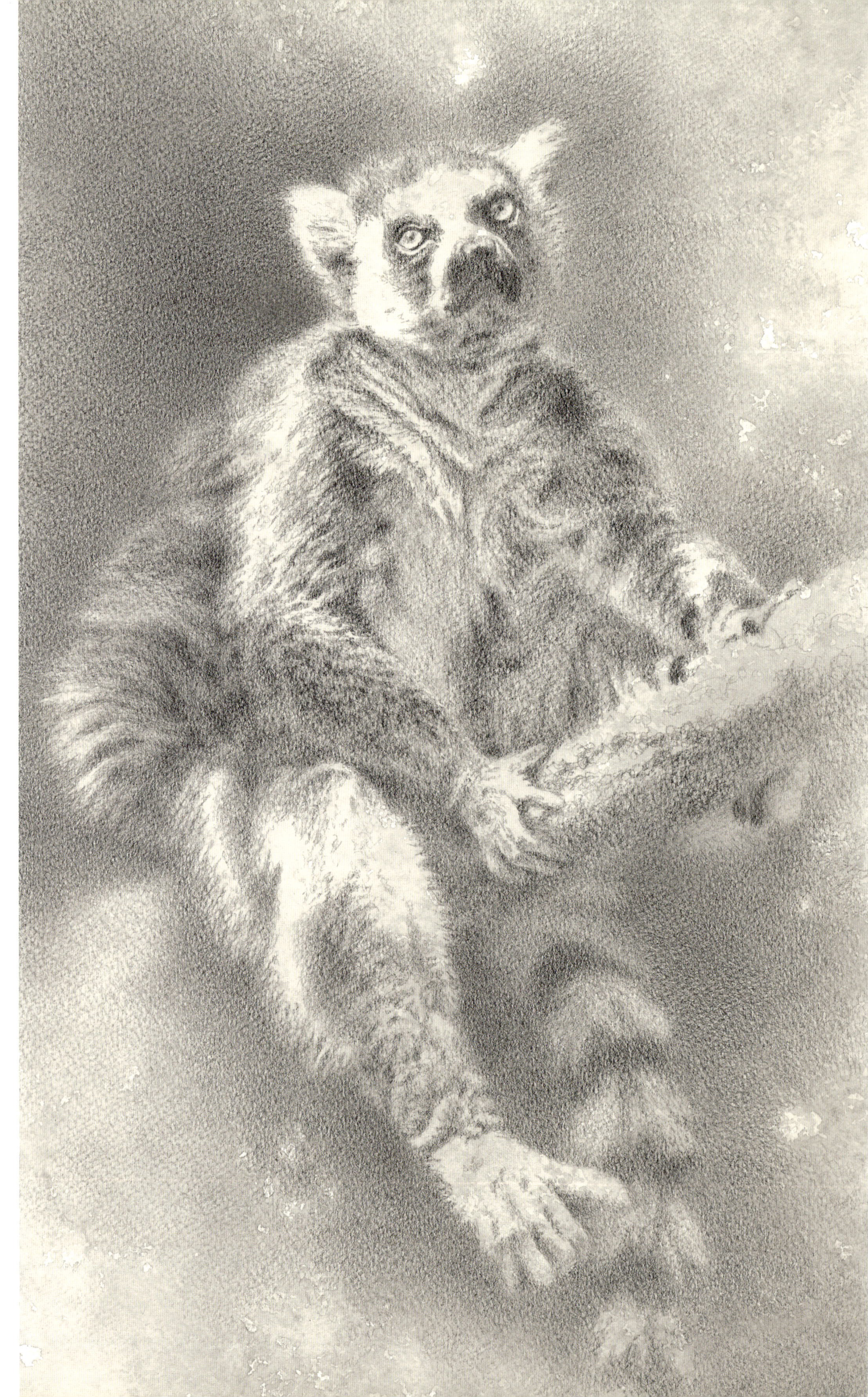

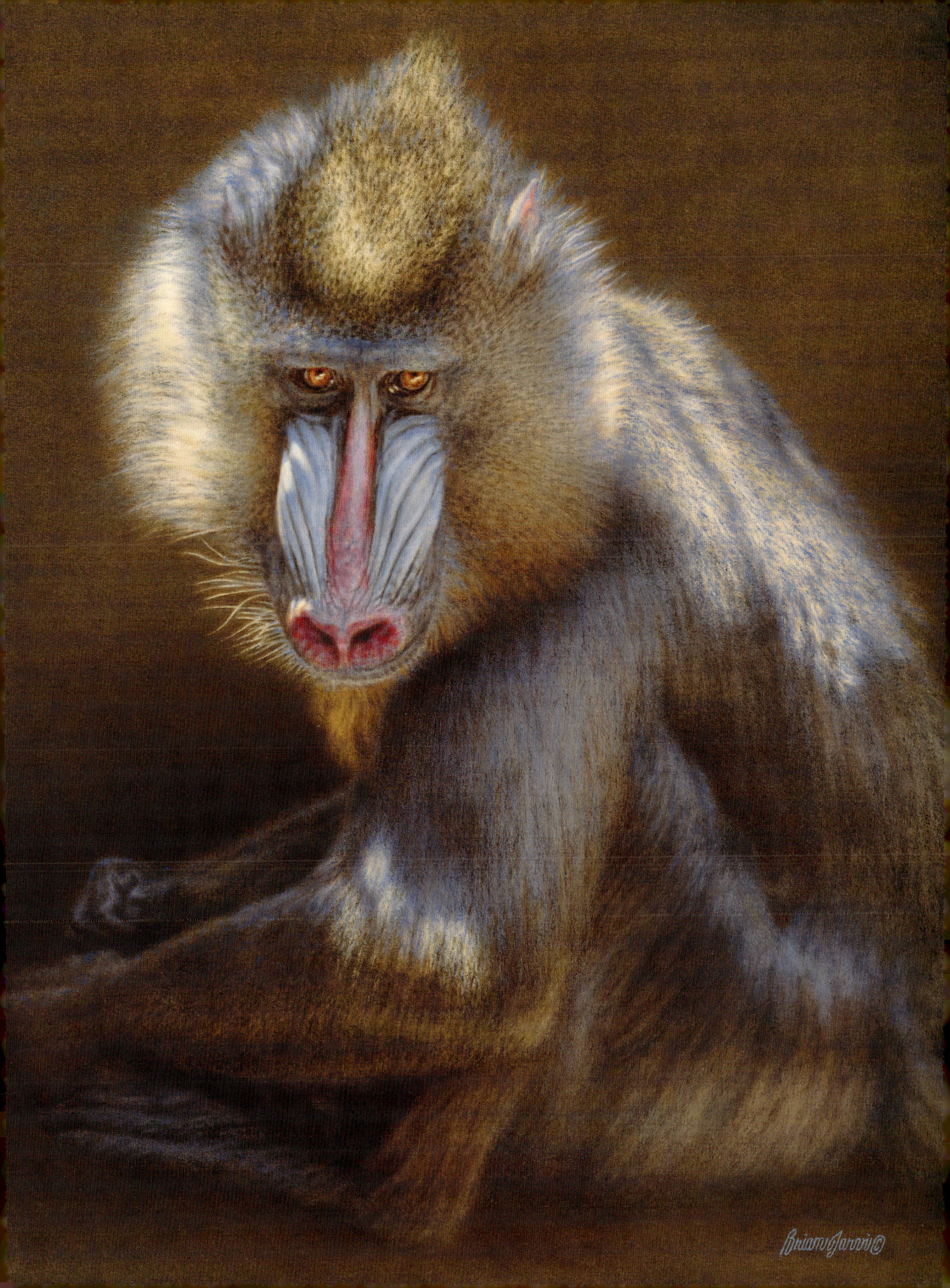
Brian Jarvi©

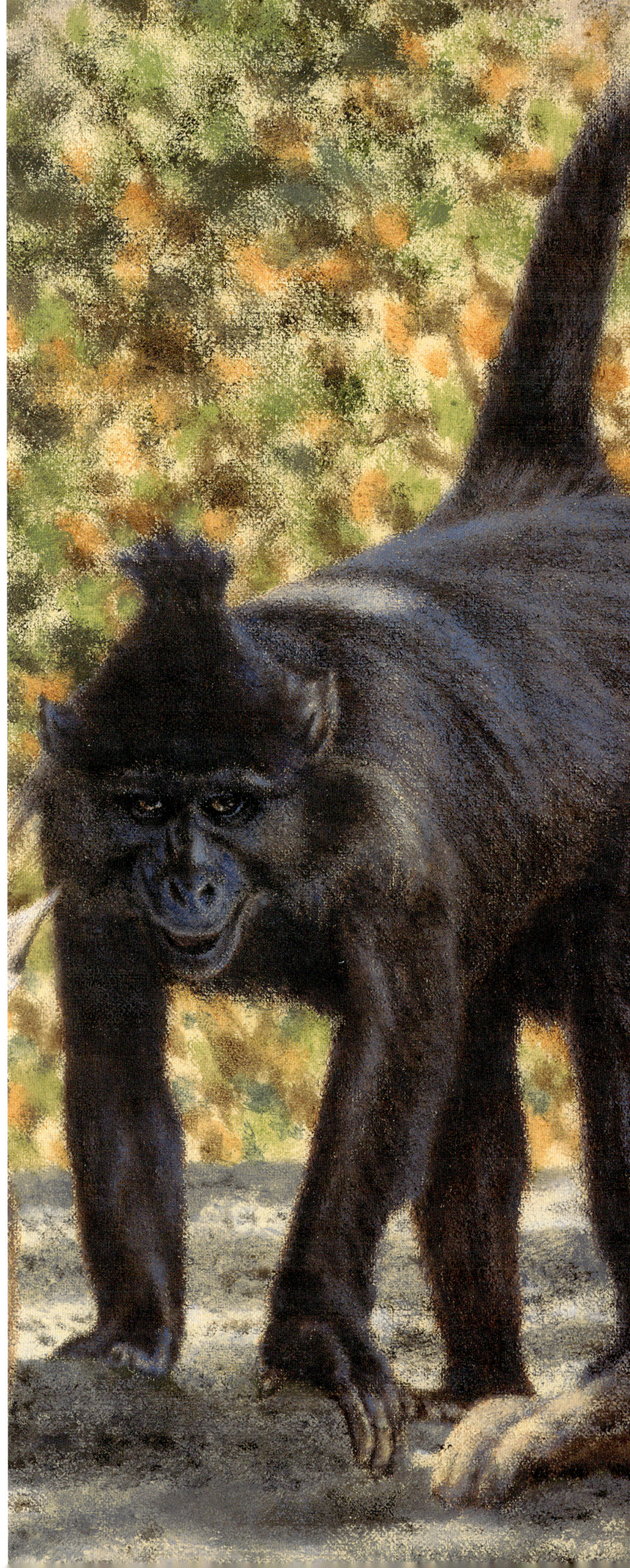

African Menagerie, Panel 5
Detail (Top Right)
Mandrill, Black-Crested Mangabey

When Jarvi began researching the black-crested mangabey, there was something oddly familiar about it. He began to realize the animal reminded him of the Wicked Witch of the West's flying primates in *The Wizard of Oz*. From then on, Jarvi referred to him as the Oz monkey.

OVERLEAF, LEFT
African Menagerie, Panel 5
Detail (Upper Center)
Patas Monkey

OVERLEAF, RIGHT
African Menagerie, Panel 5
Detail (Middle Right)
Common Eland

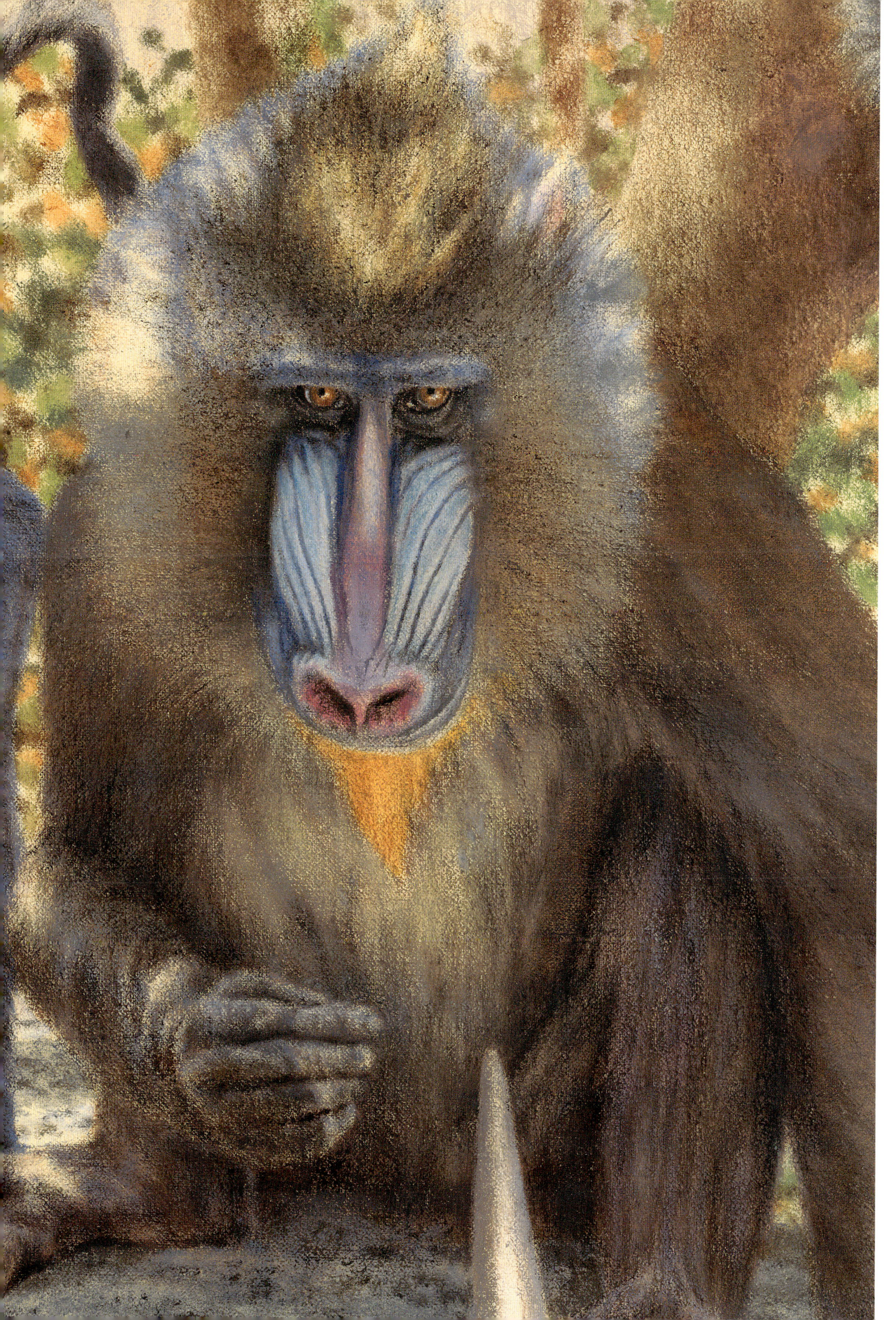

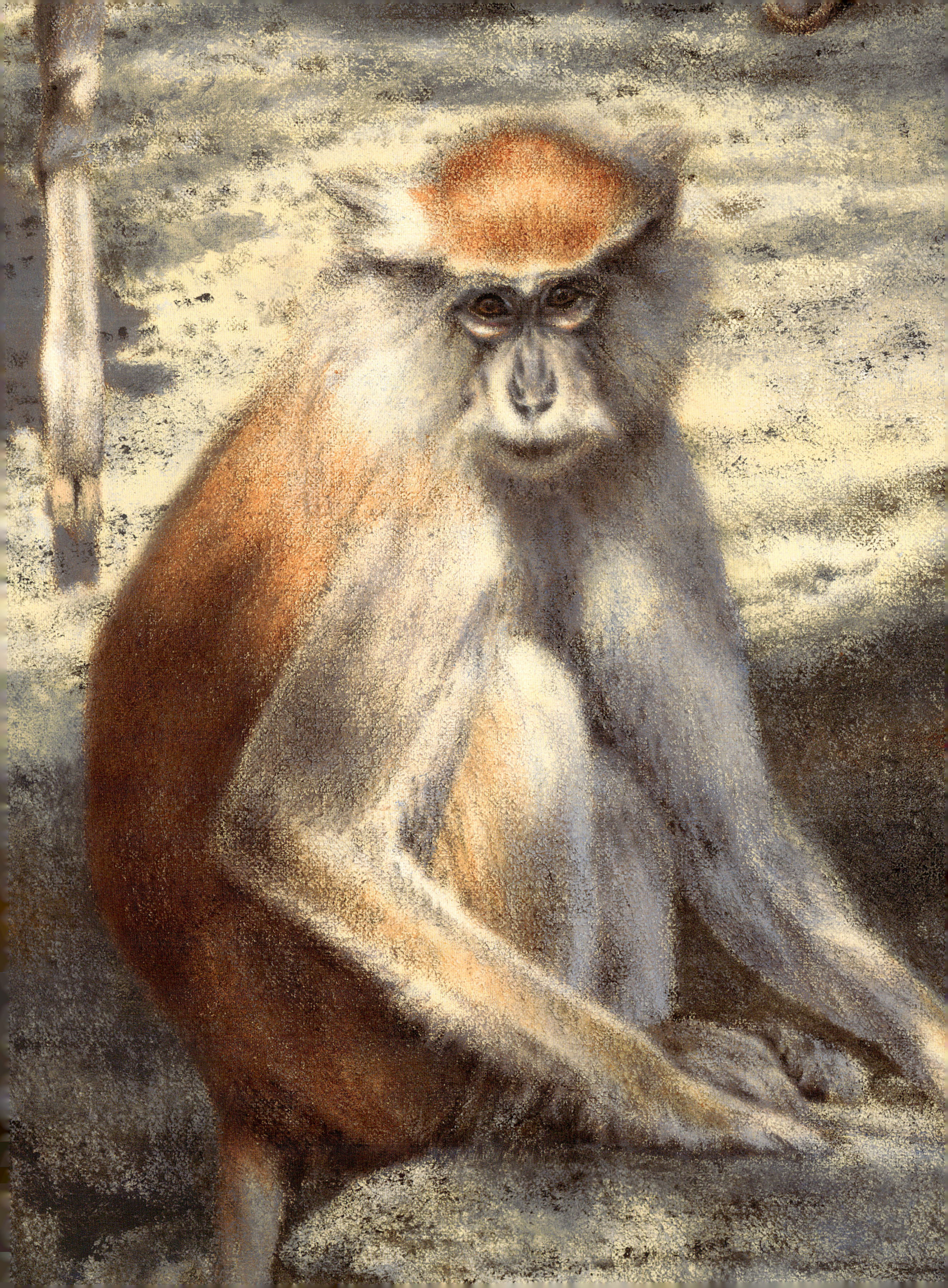

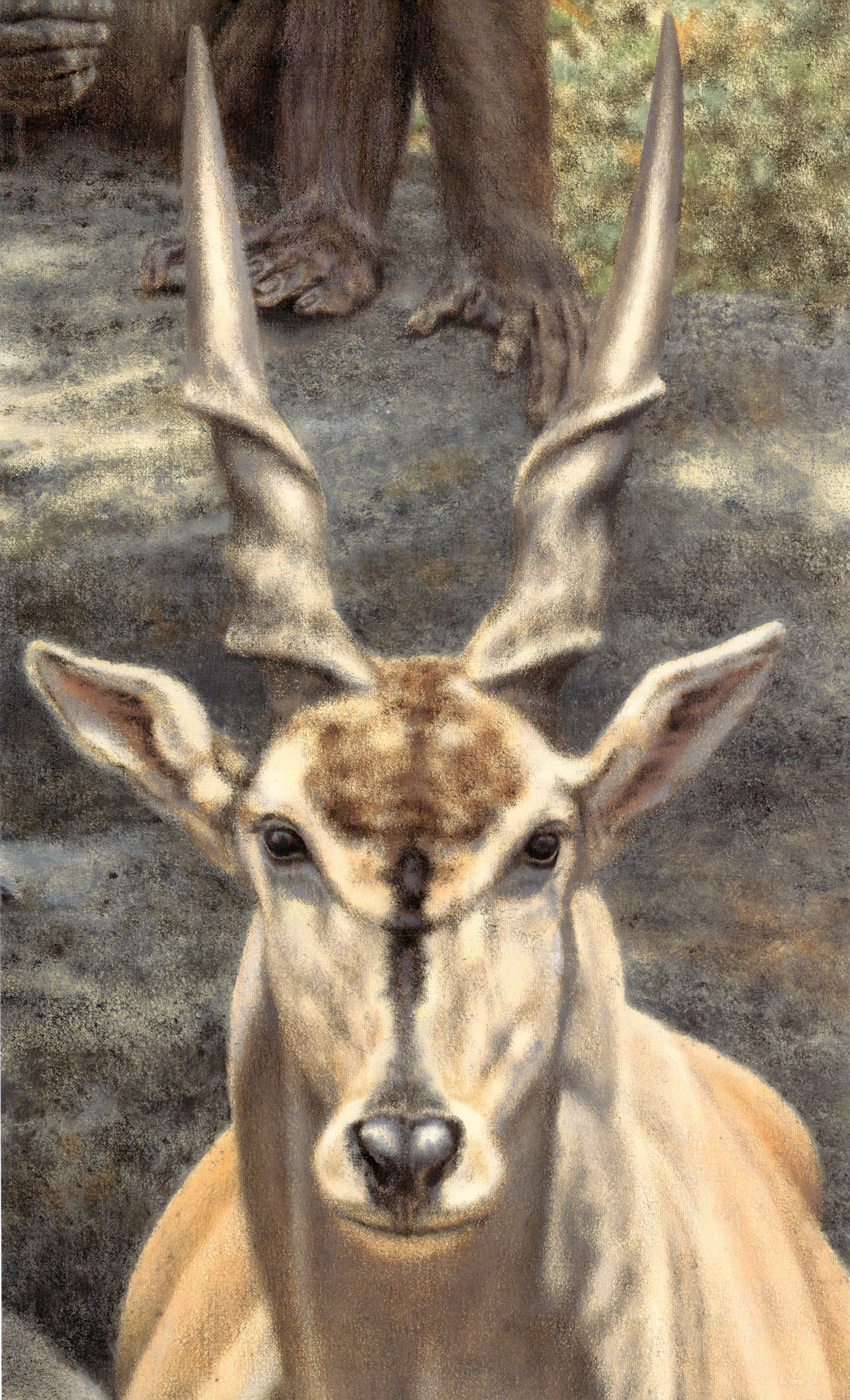

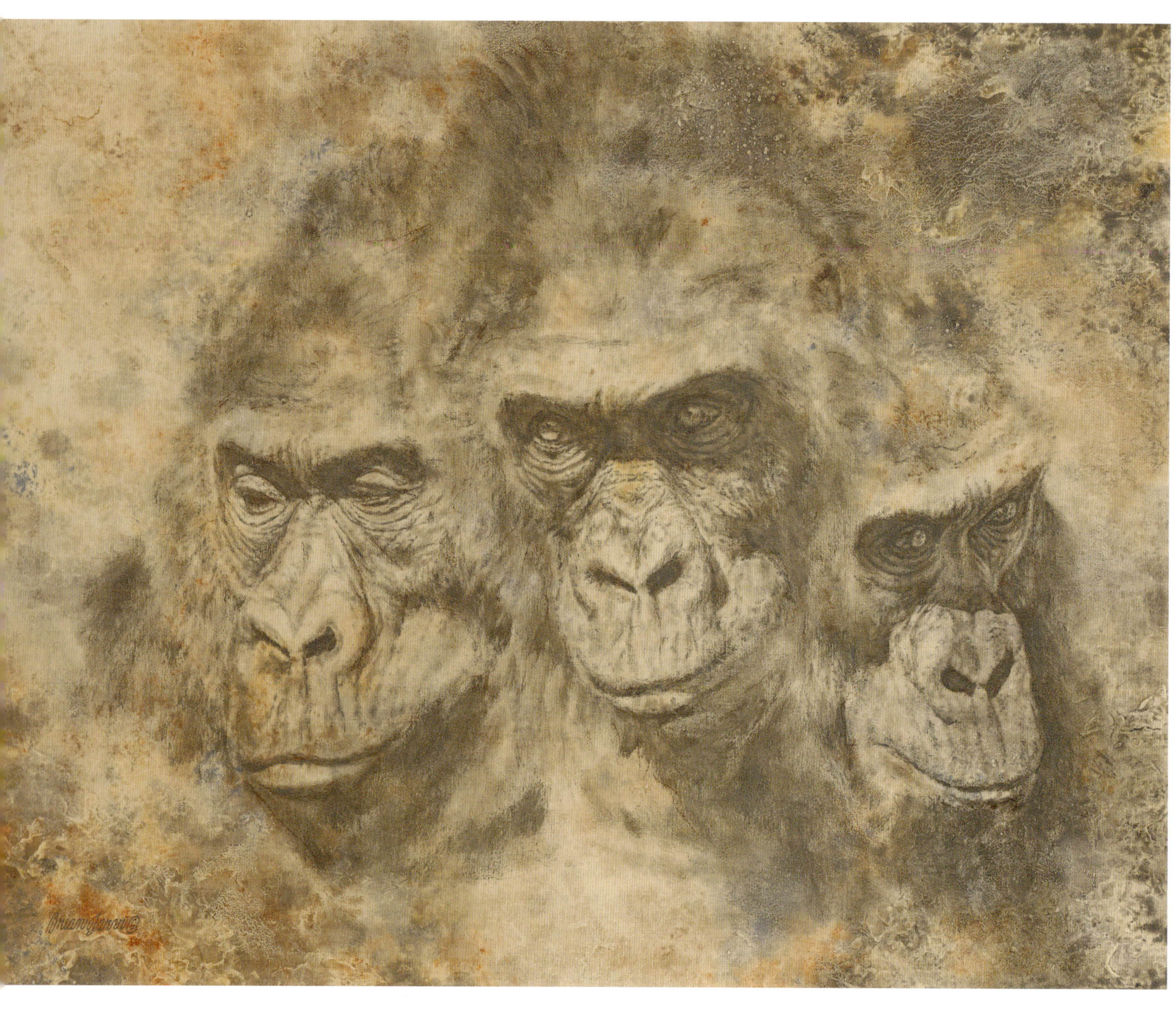

ABOVE

Primeval Study, 2014
Western Lowland Gorilla
Oil on Belgian Linen
25 x 30"
Collection of Annette Bishop

In 2014, this piece was juried into the Society of Animal Artists' *Art and the Animal* exhibition.

OPPOSITE

African Menagerie, Panel 5
Detail (Middle Right)
Western Lowland Gorilla

Auguste Rodin was known for his gestures and philosophical approaches to narrative forms, such as in his classic *The Thinker*. Jarvi dubbed this detail of the western lowland gorilla, a species that could vanish from the wild by the middle of this century, *The Skeptic*. The gorilla is demanding that we prove his sentiments wrong.

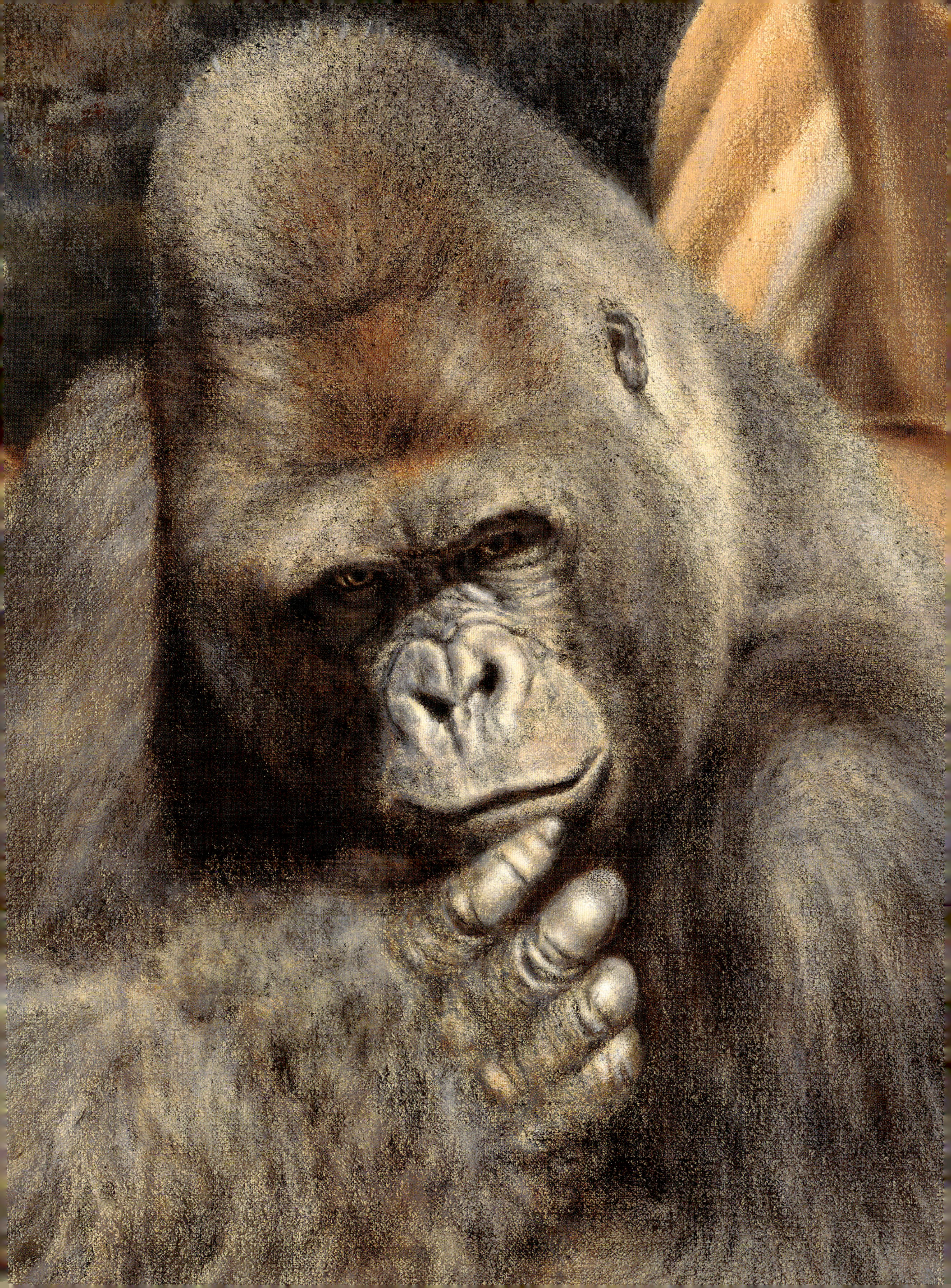

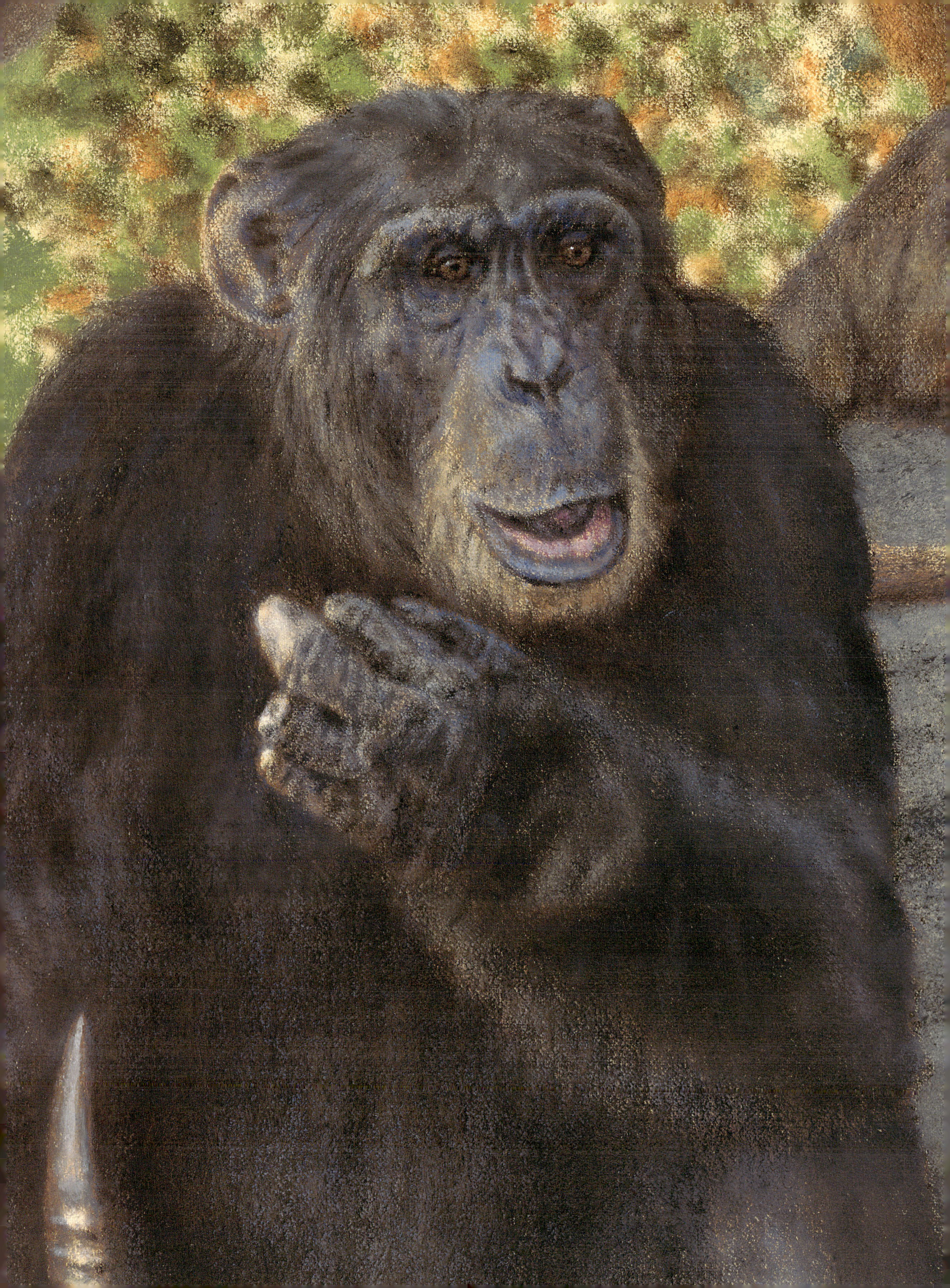

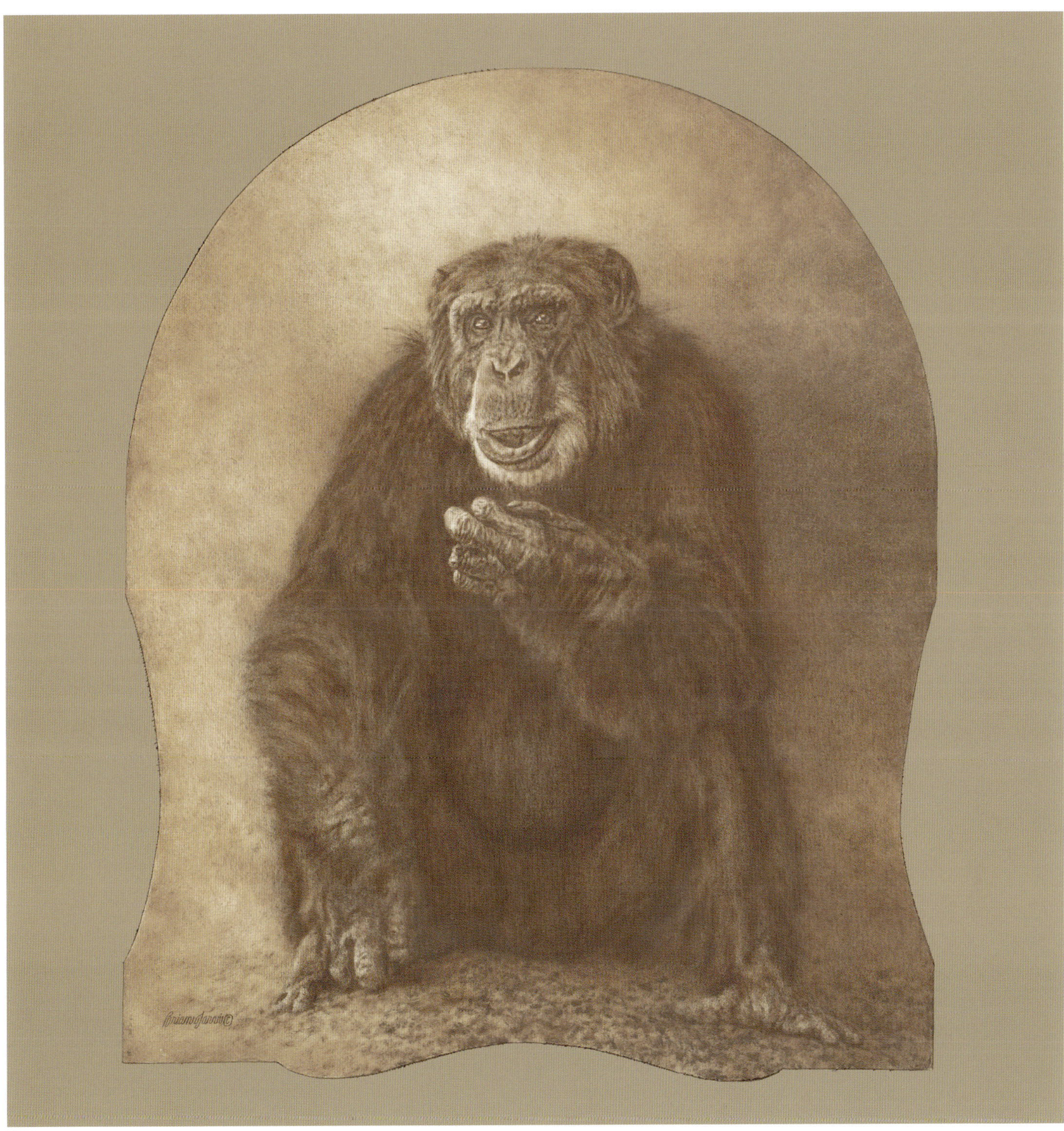

OPPOSITE
African Menagerie, Panel 5
Detail (Top Left)
Chimpanzee

Attempting to adequately express the inner pleading emotions of fellow primates is a common theme in *African Menagerie*. According to Jarvi, primates invite us to ponder the sentience and emotional range of other nonhuman beings, including giving voice to the survival of our closest relatives.

ABOVE
The Disenchanted, 2017
Chimpanzee
Oil on Belgian Linen
26 x 20"
Collection of Jan M. Gunderson and David W. Cress

The chimpanzee is another primate that offers a mirror reflection of humanity. With mischief, Jarvi's depiction suggests the chimp is delivering an irreverent message.

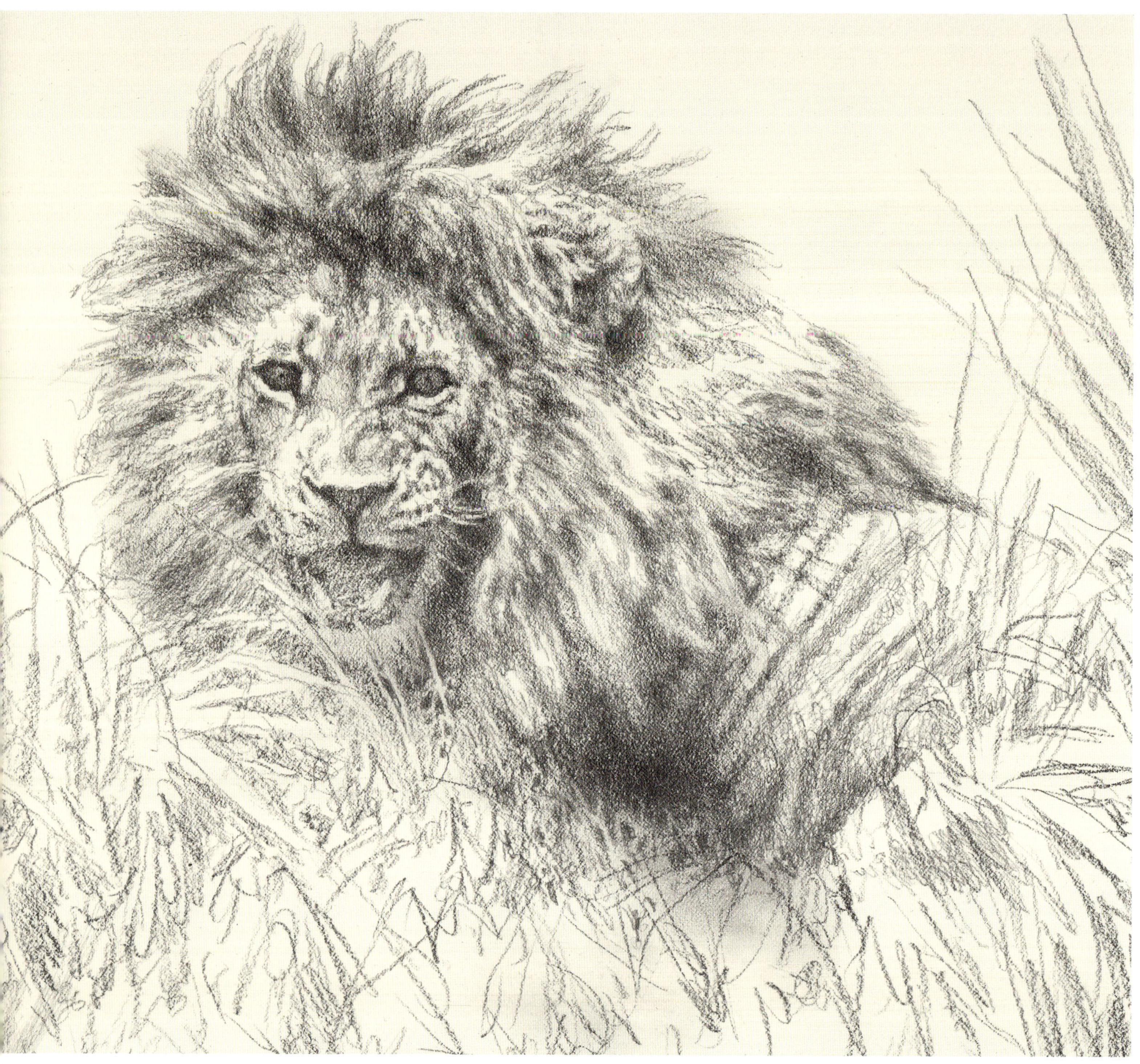

ABOVE
Aggression Study, 2016
Lion
Pencil
14 x 19½"

OPPOSITE
King's Fall, 2018
Lion
Oil on Belgian Linen
45 x 30"
Collection of Todd and Shari Nelson

In 2018, this piece was juried into the Society of Animal Artists' *Art and the Animal* exhibition.

OVERLEAF
Lion Repose, 2016
Lion
Pencil
9 x 23"
Collection of Marc and Carrie Fowler

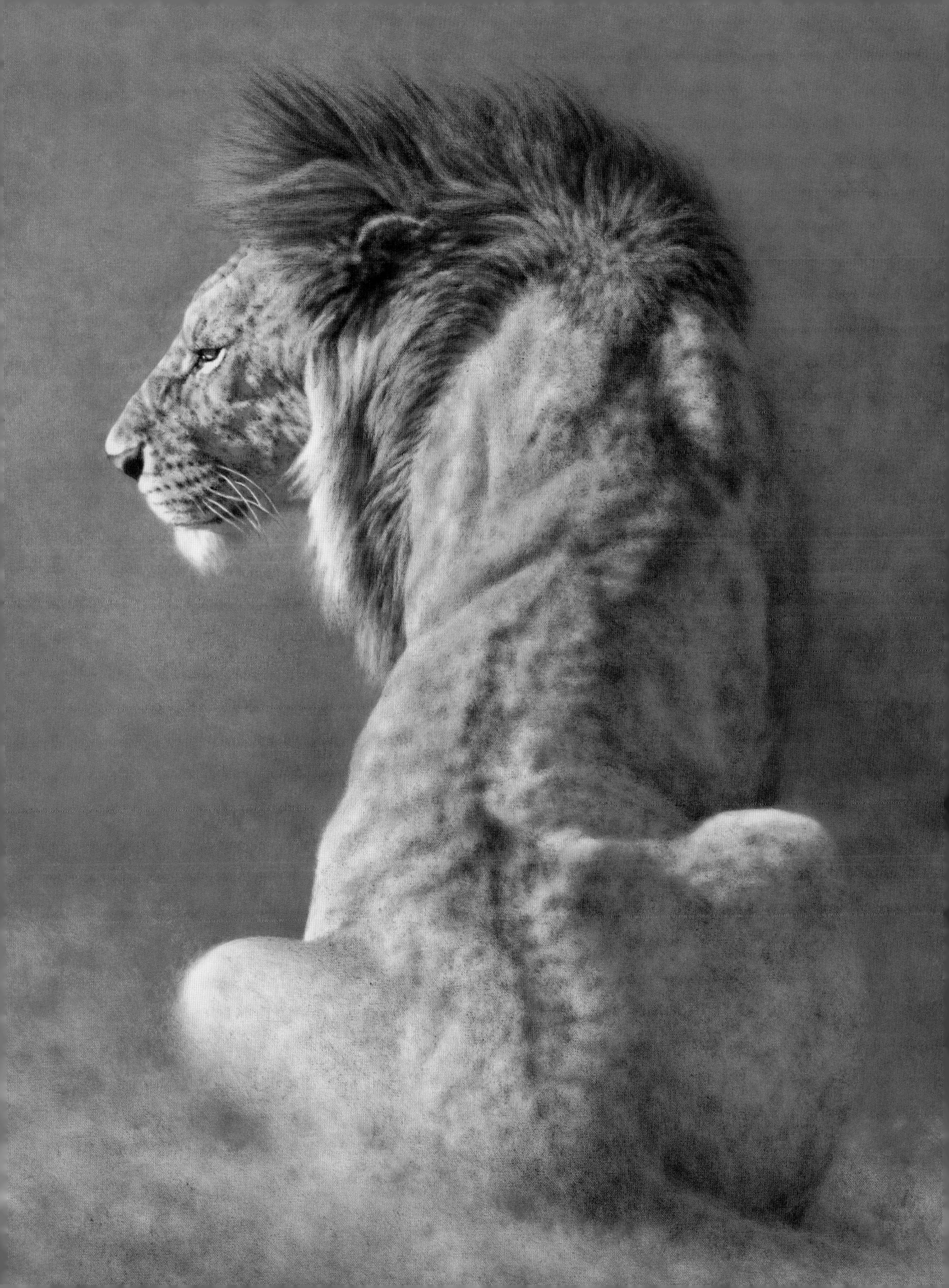

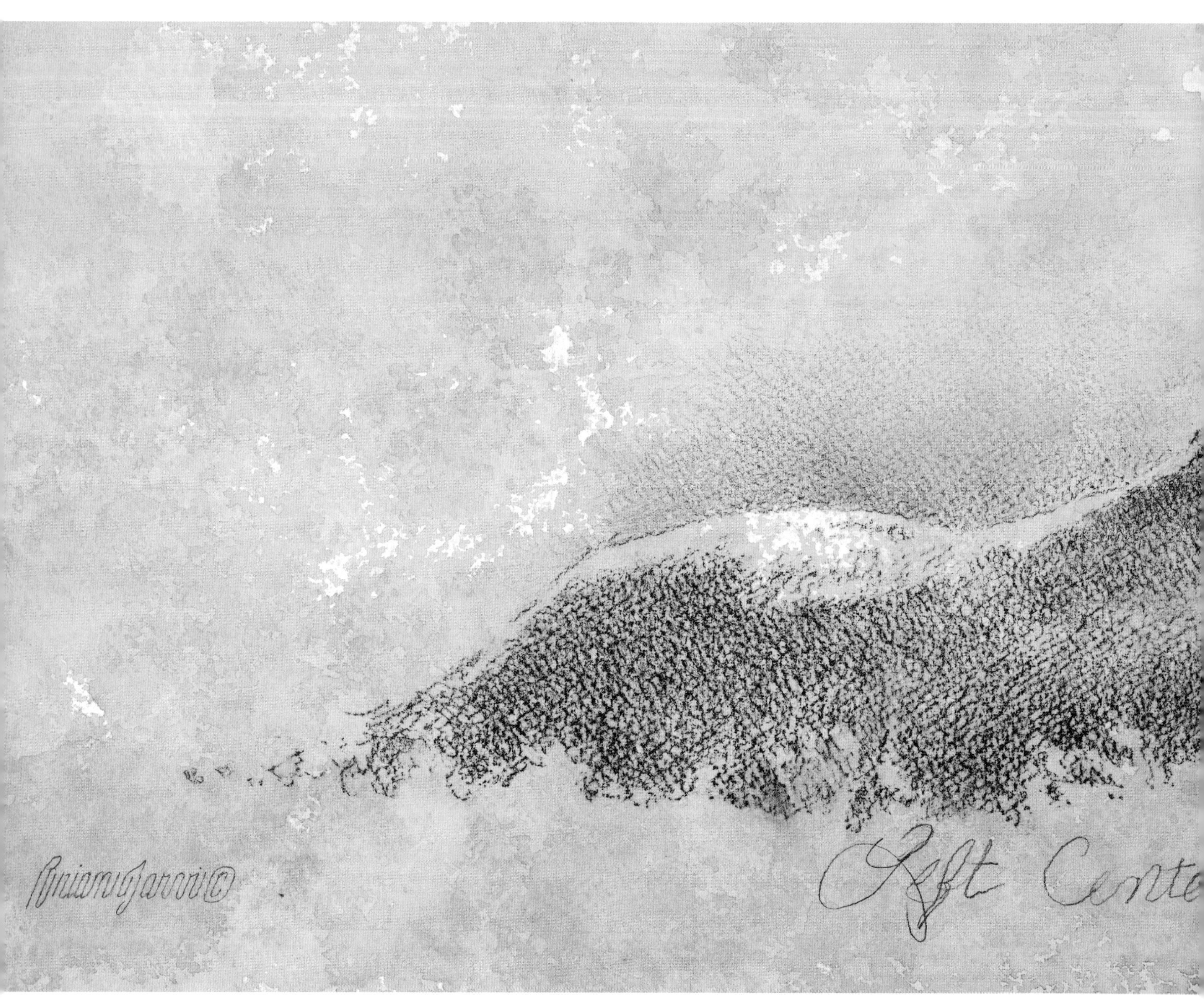

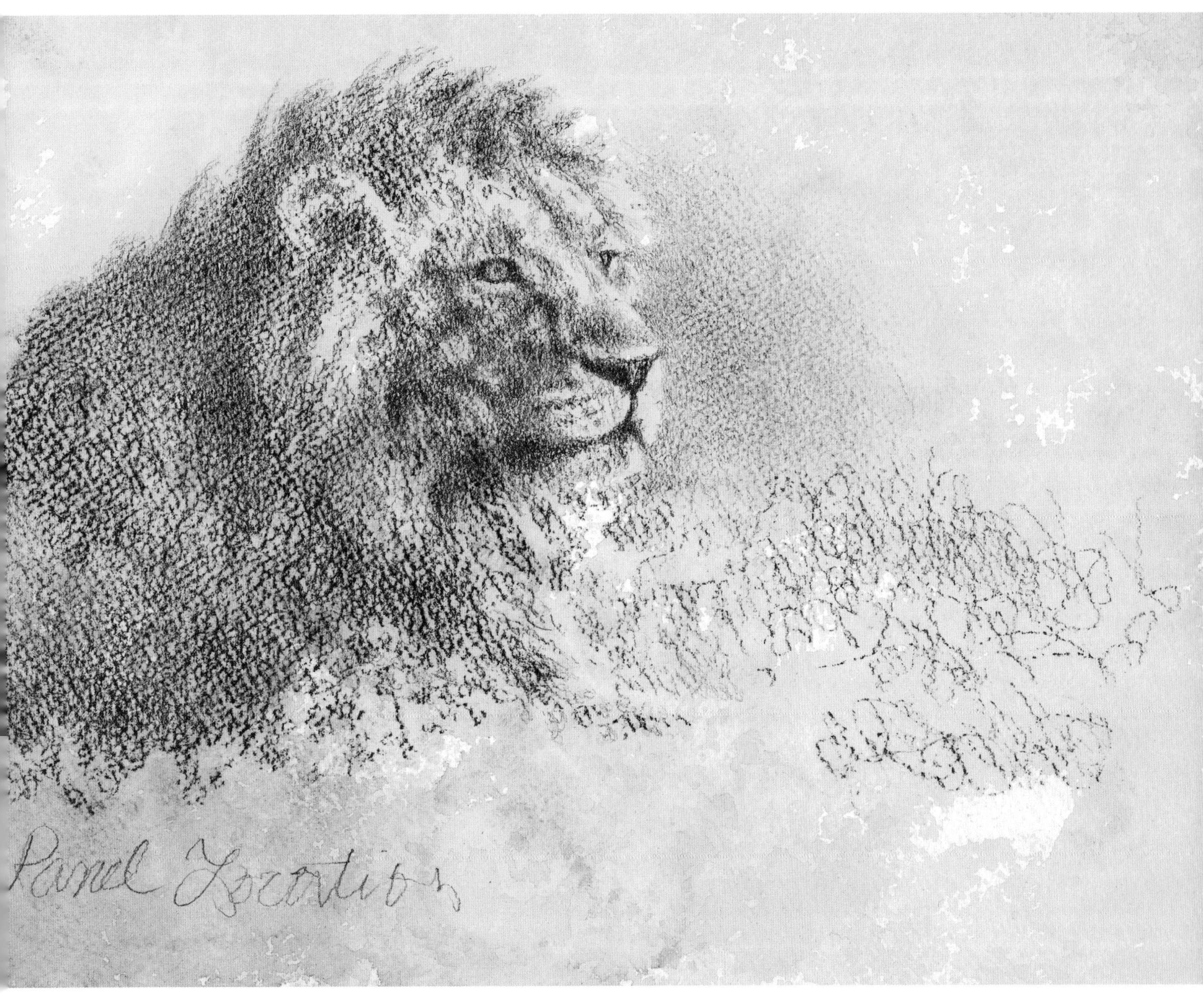
Panel Location

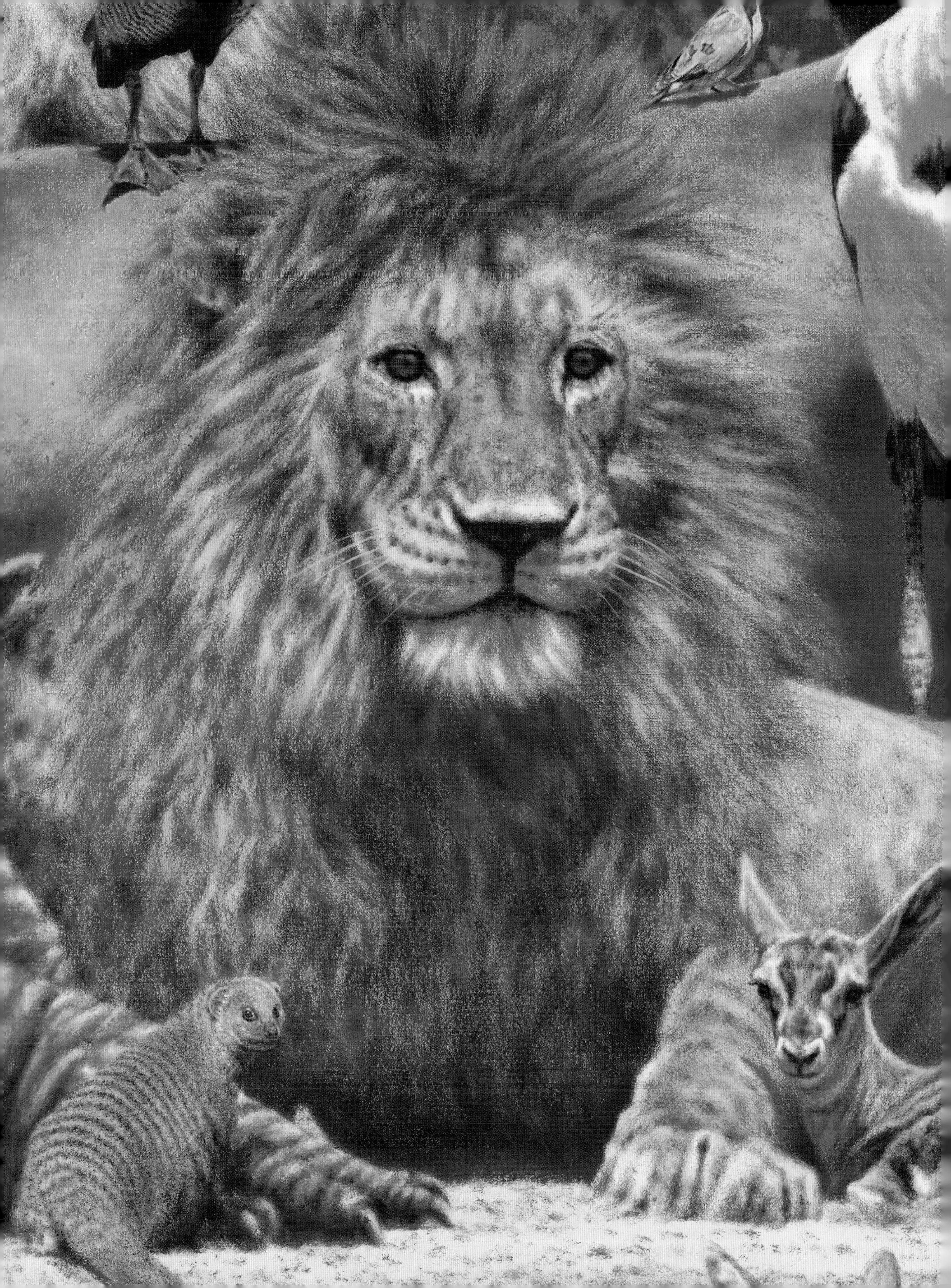

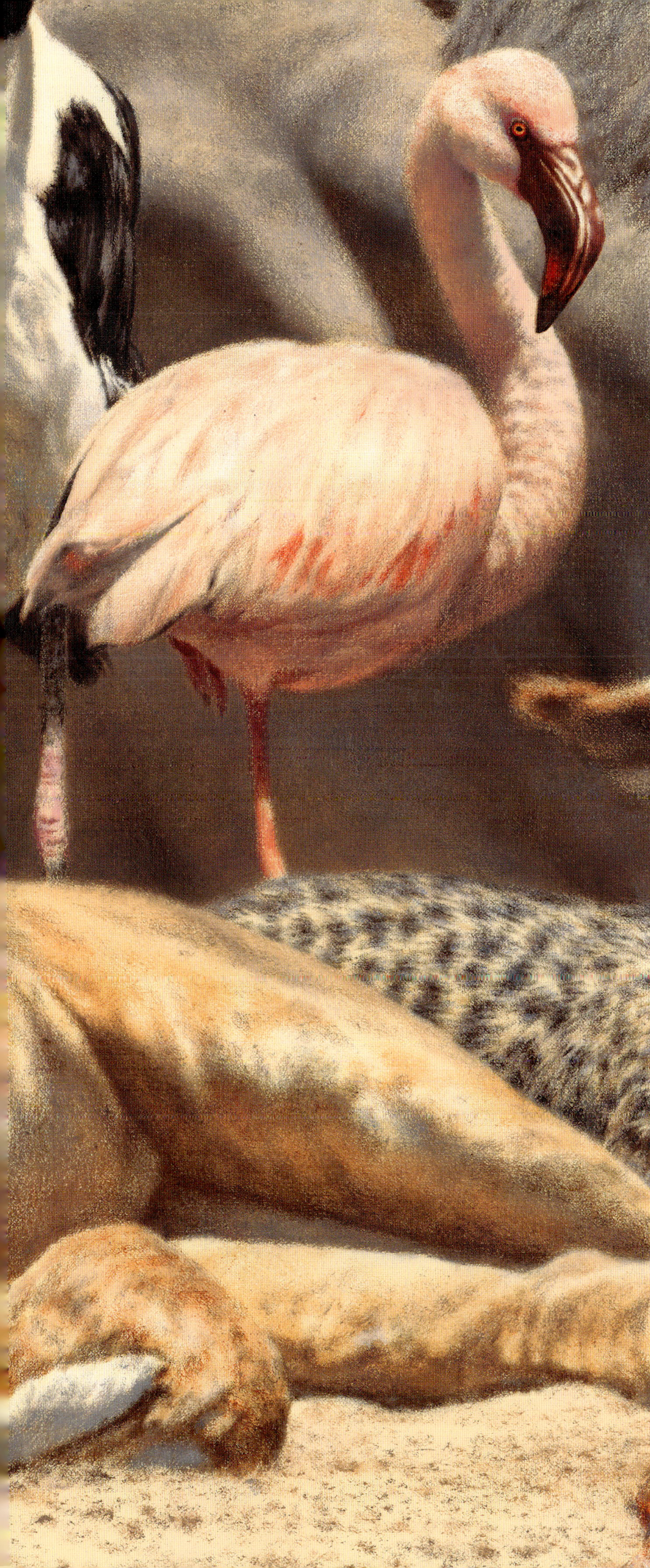

African Menagerie, Panel 5
Detail (Middle Left)
Lesser Flamingo, Lion, Thomson's Gazelle (Fawn), Banded Mongoose

There are many holy books and metaphors that hold up over the ages. Jarvi wanted to invoke the biblical theme of a lion laying down with a lamb as a meditation on peace. Human wars across Africa have exacted a deadly toll on nature. Here, scripture can be a guide, reminding us to not only reassess our adversarial relationship toward each other, but also reflect on how we treat creation.

OVERLEAF
Haley's Flamingos, 2015
Lesser Flamingo
Oil on Belgian Linen
18 x 24"
Collection of Steve and Cindy Gilbertson

Early in this study Jarvi had settled on the title *The Gathering*, but fate intervened. Soon afterward, Jarvi's daughter Haley went to live in Africa and was excited when she found a wild flamingo feather on the Skeleton Coast in Namibia. What father doesn't feel a connection to his daughters? Jarvi changed the title to *Haley's Flamingos* to honor her. While painting the more than one hundred different bird species depicted in *African Menagerie*, he often thought of his daughters—each a *rara avis* in her own right.

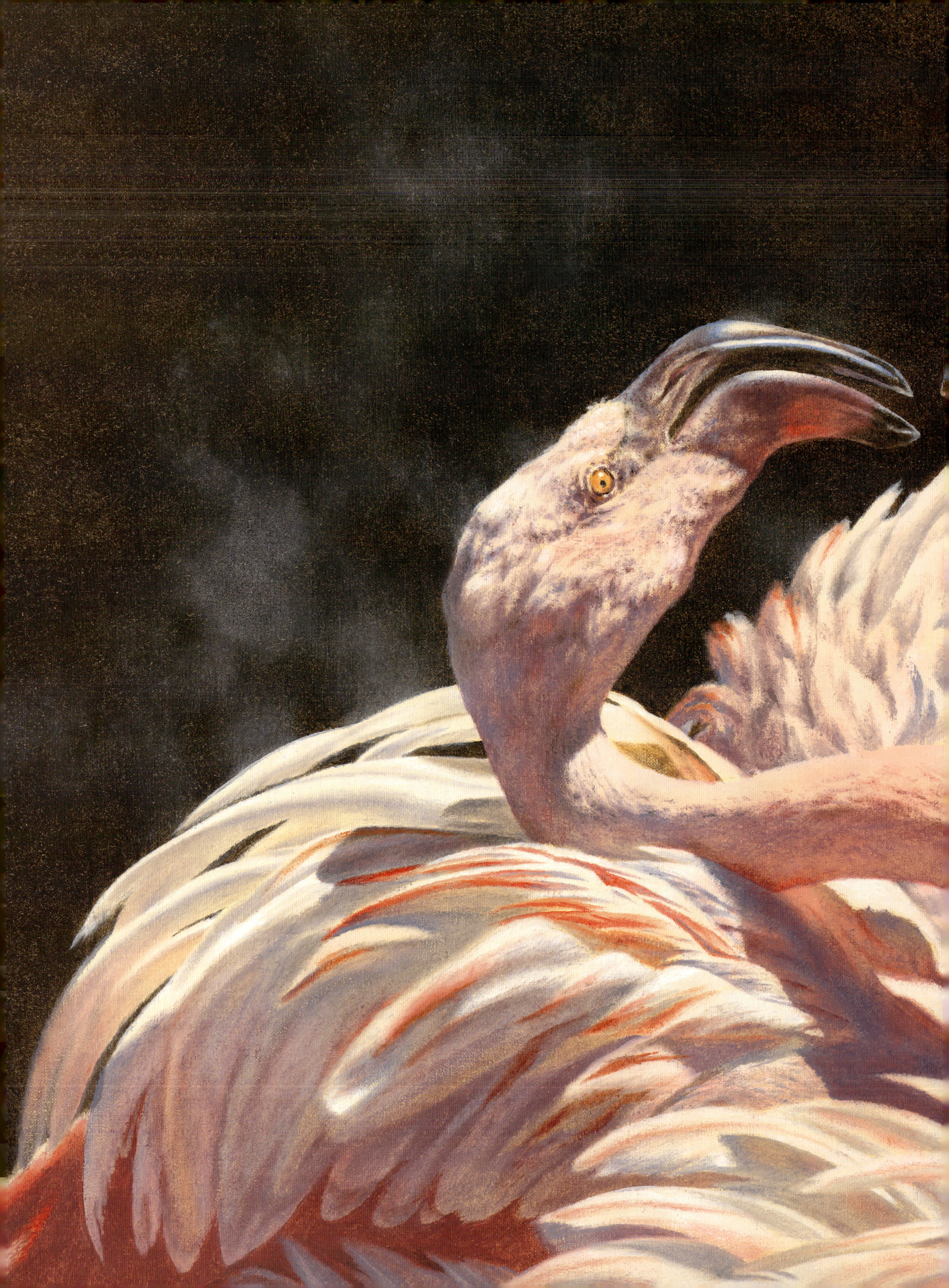

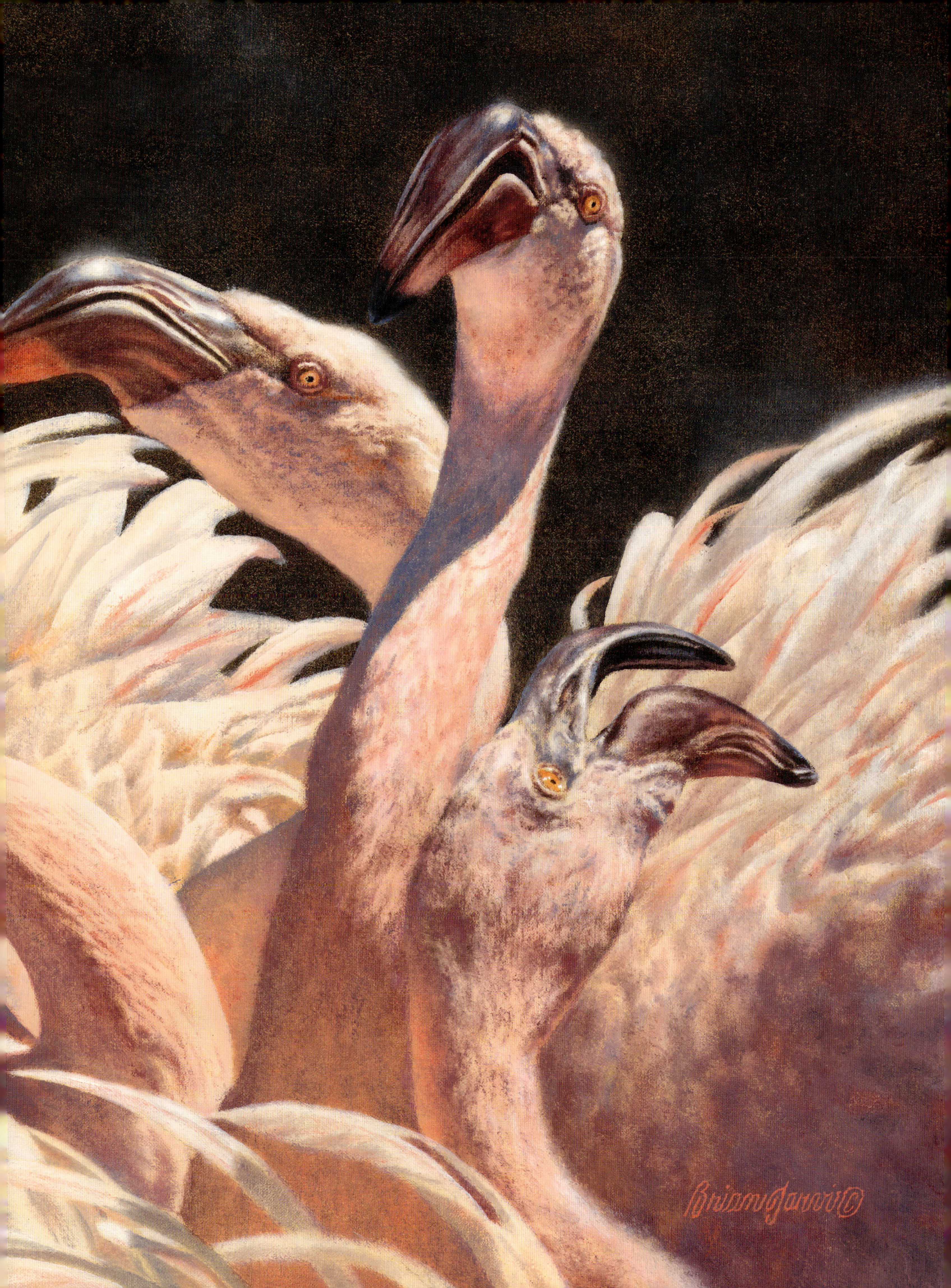
Brian Jarvi ©

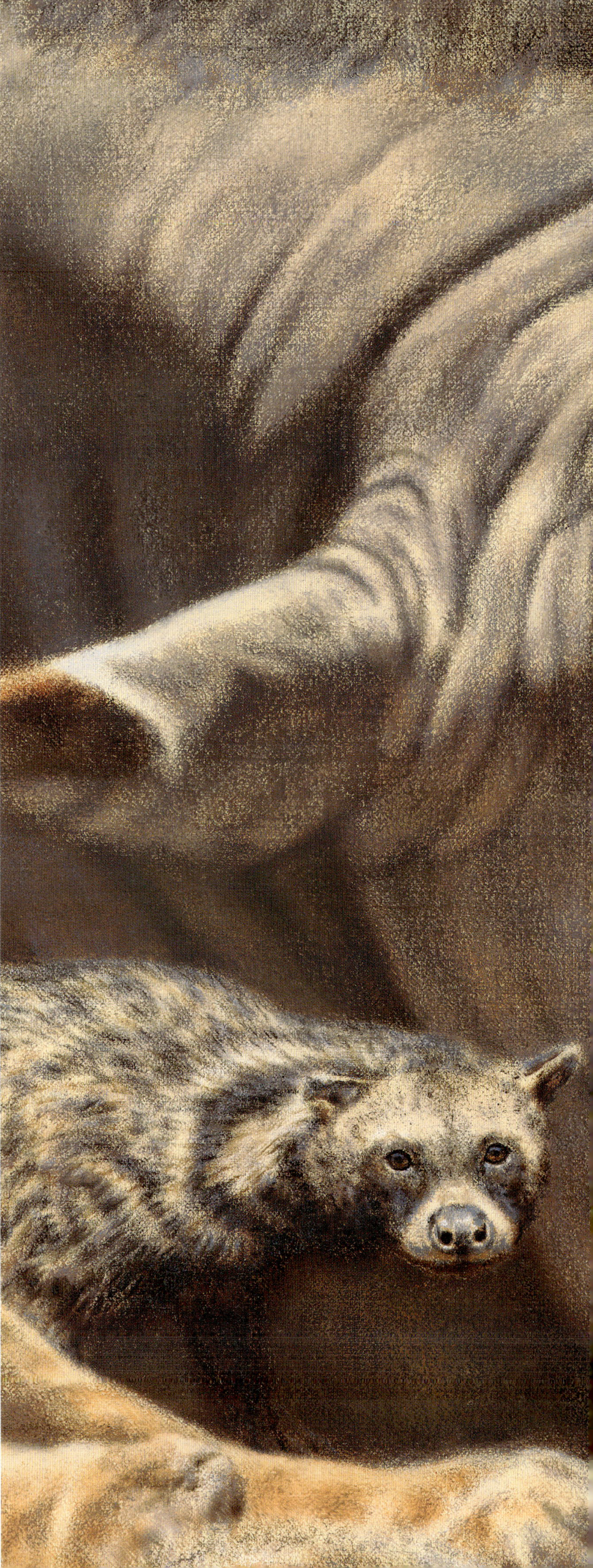

African Menagerie, Panel 5
Detail (Middle Right)
Black Rhinoceros, Caracal, African Civet, Fulvous Whistling Duck

Critically endangered, the black rhinoceros continues to confront ferocious pressure from the illegal trade of rhino horns. Rhinos, like living dinosaurs, seize our attention. Perhaps ironically, the hunting community—by channeling cash into local communities—has shown that by putting an economic value on these wondrous behemoths, protecting them at the population level makes them worth more alive than dead. Through his art, Jarvi is proud to assist conservation organizations that are making a difference on the ground.

OVERLEAF, LEFT
African Menagerie, Panel 5
Detail (Bottom Right)
Dik-Dik, White-Backed Vulture

OVERLEAF, RIGHT
Bateleur Study, 2011
Bateleur
Oil on Belgian Linen
20 x 16"

In 2011, this piece was juried into the Leigh Yawkey Woodson Art Museum's *Birds in Art* exhibition.

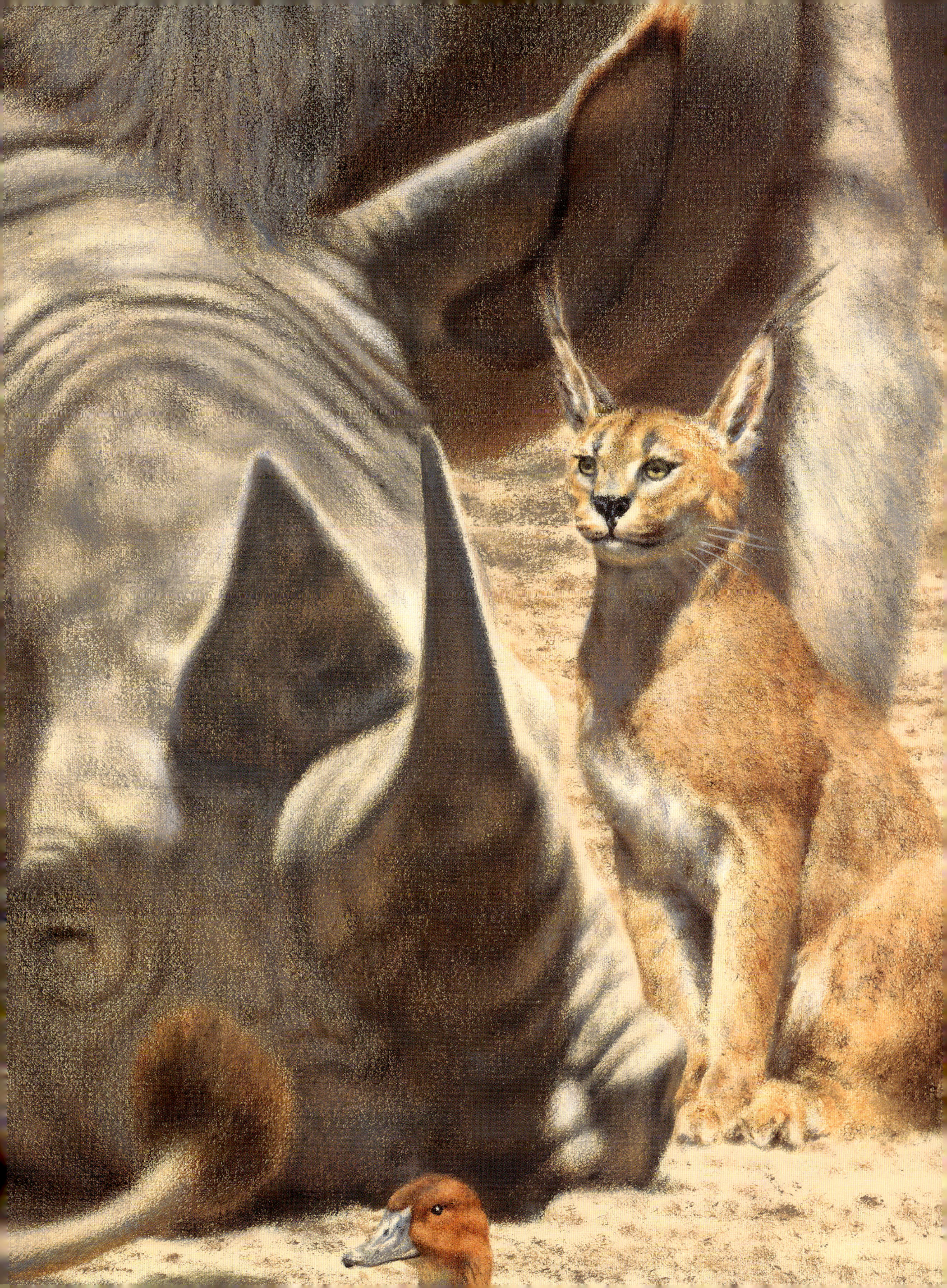

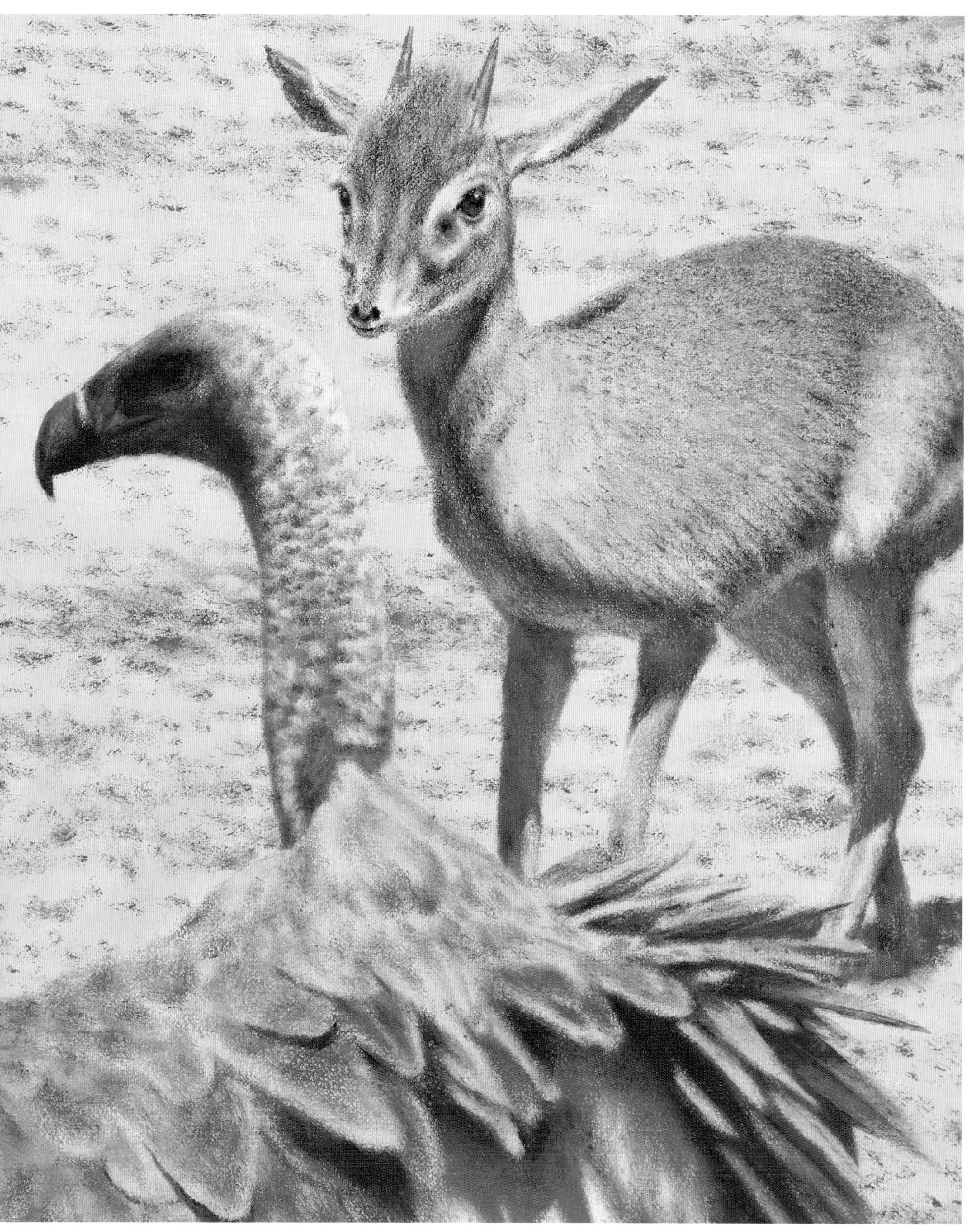

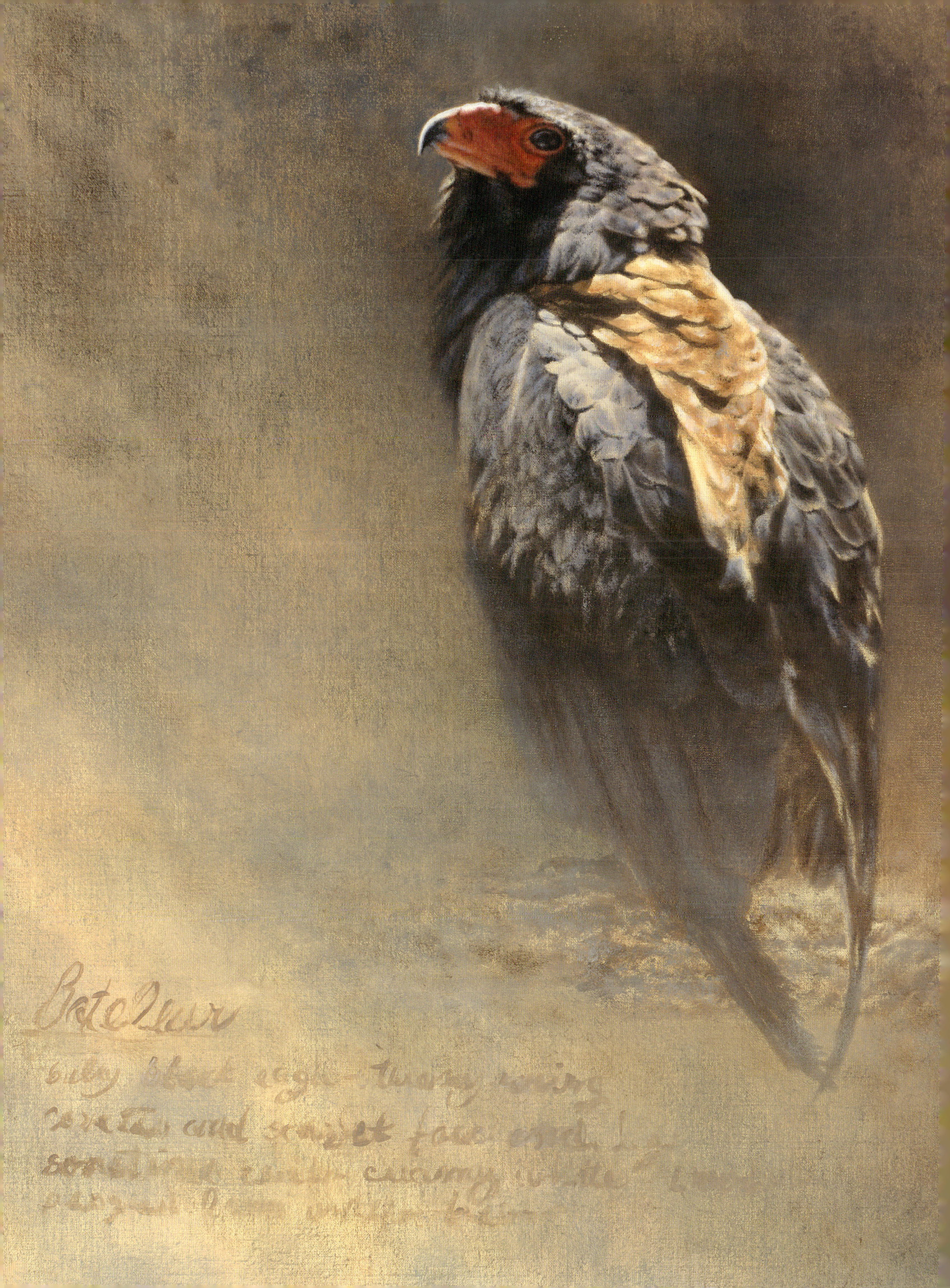
Bateleur

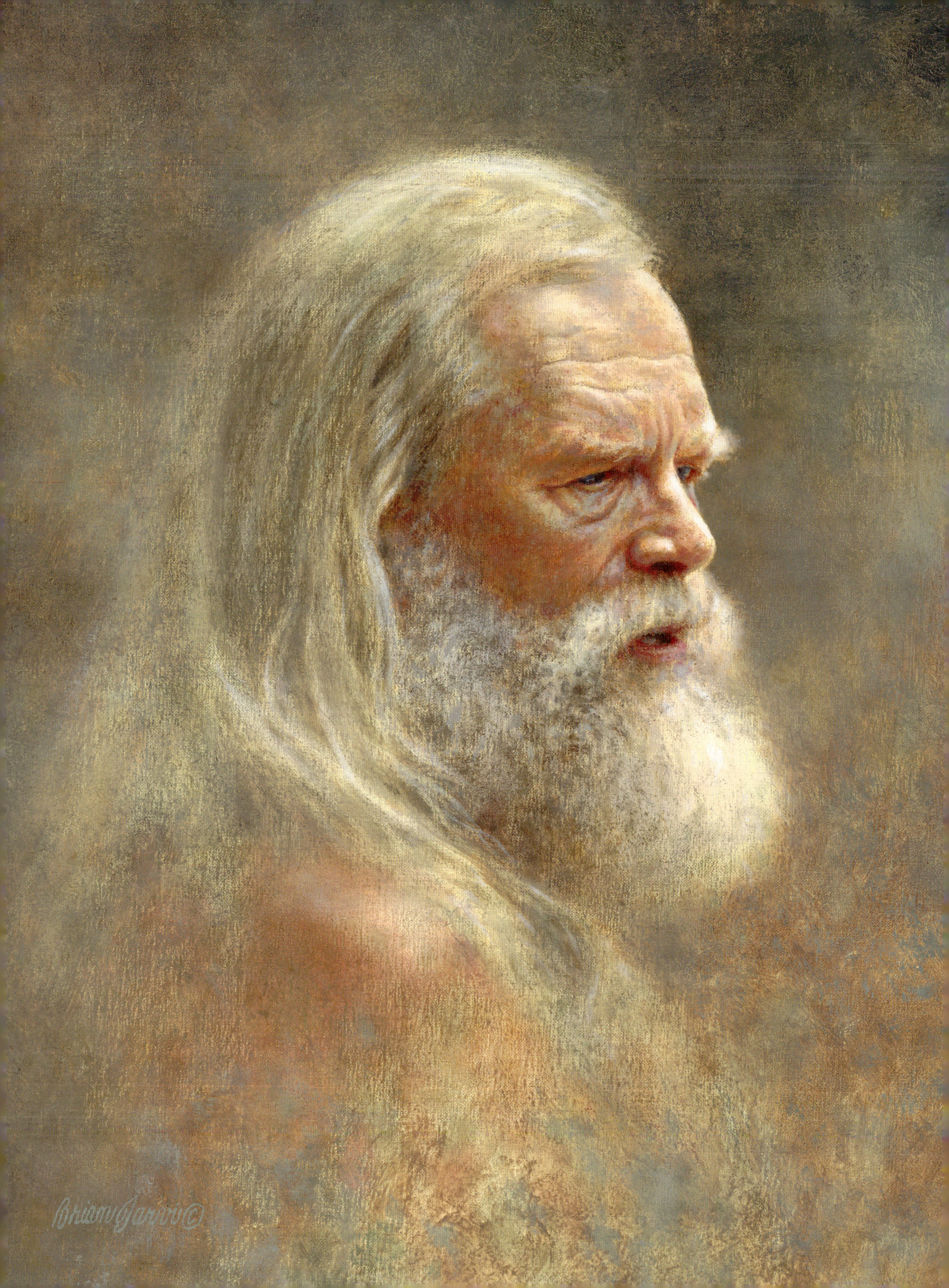

OPPOSITE
Omega Man Study I, 2015
Homo Sapiens
Oil on Belgian Linen
16 x 12"
Collection of Bruce and Karen Ogle

RIGHT
Meerkat Study, 2015
Meerkat
Pencil
12 x 8"
Collection of Lois Lang

OVERLEAF
African Menagerie, Panel 5
Detail (Bottom)
Homo Sapiens, Black-Backed Jackal, Bateleur, Meerkat, Lappet-Faced Vulture, Zebra Duiker, Hooded Vulture, Rüppell's Vulture

According to Jarvi, humans are inseparable from the animal kingdom. We are part of the same web of life, interwoven into the same life-support system. Omega Man is each of us, a symbol of our common heritage. In *African Menagerie*, he is portrayed as listening to the animals, heeding their message, and becoming awakened to their plight and our need to act. The natural bounty of Africa is a bellwether for wildlife in North America. Omega Man is a reminder that future generations will either praise us or judge us for having this incredible thing called biodiversity in the palms of our hands. Do we let it slip away, or clutch it as the treasure it is, safekeeping it for them? The choice is ours.

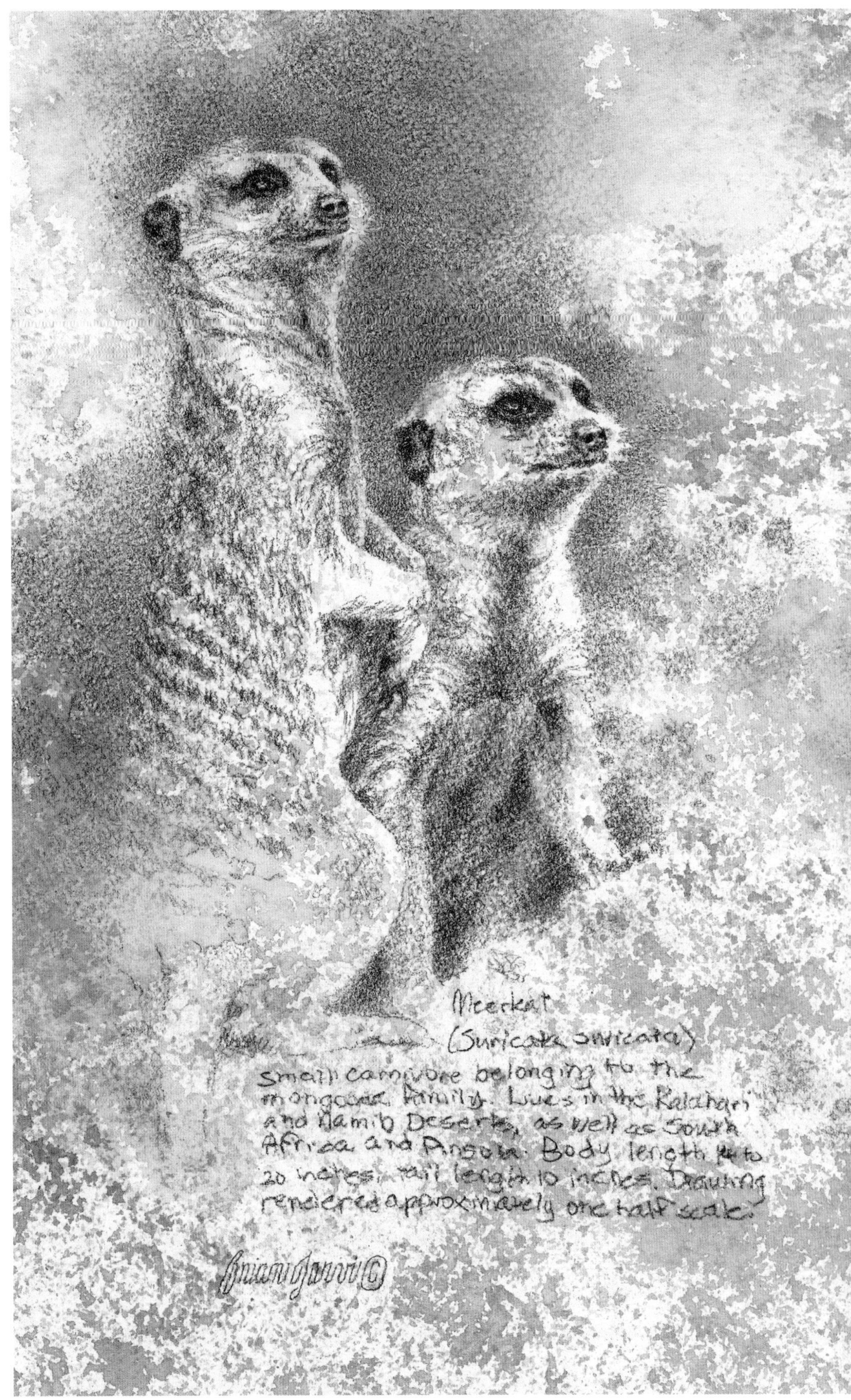

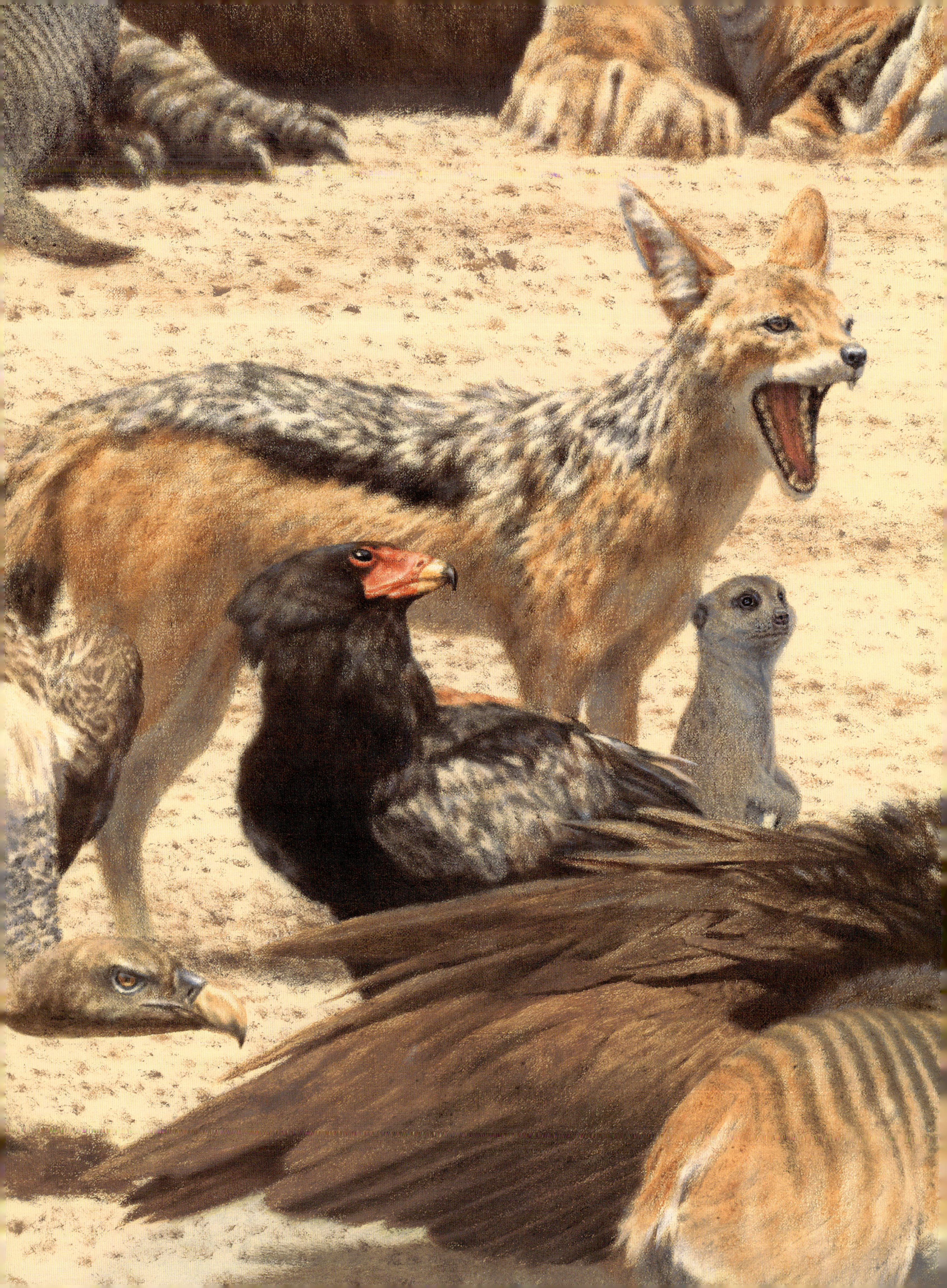

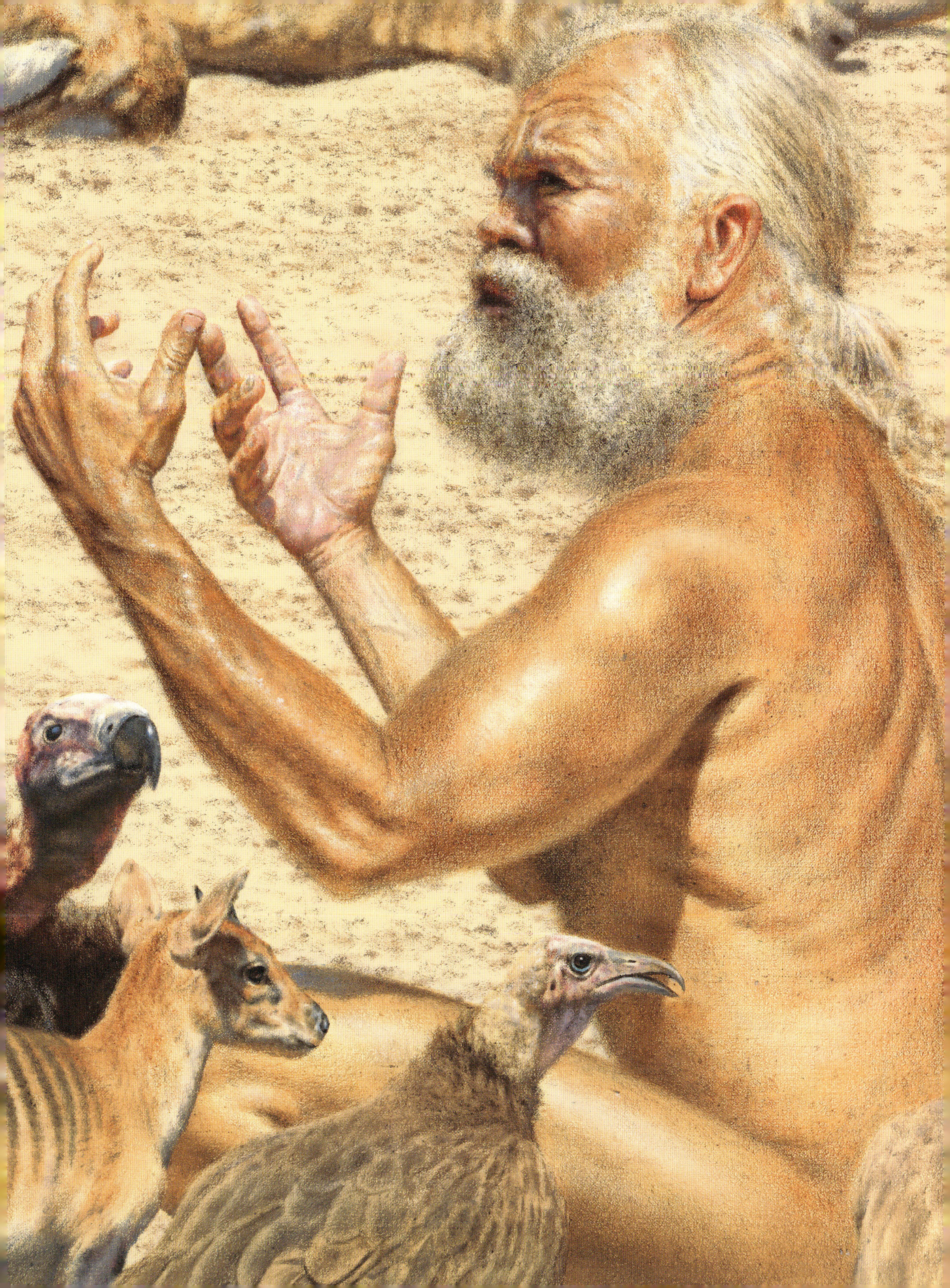

African Menagerie, Panel 5
Detail (Middle Left)
Waterbuck, Sable Antelope, Oryx, Saddle-Billed Stork, White-Faced Whistling Duck, Emerald-Spotted Dove

Jarvi tried to approach *African Menagerie* from several different dimensions of thinking beyond the visual two dimensions. We love what we come to know, and we are willing to protect what we love. Jarvi has seen it happen to his hunter and non-hunter friends alike.

OVERLEAF
African Menagerie, Panel 5
Detail (Top Left)
Giant Eagle Owl, Gymnogene, Madagascar Crested Ibis, Von der Decken's Hornbill, Heuglin's Robin

The assemblage in Africa of raptors, passerines, wading birds, and ground runners is mind-blowing. Jarvi's love for birds goes back to his earliest childhood memories along the St. Louis River.

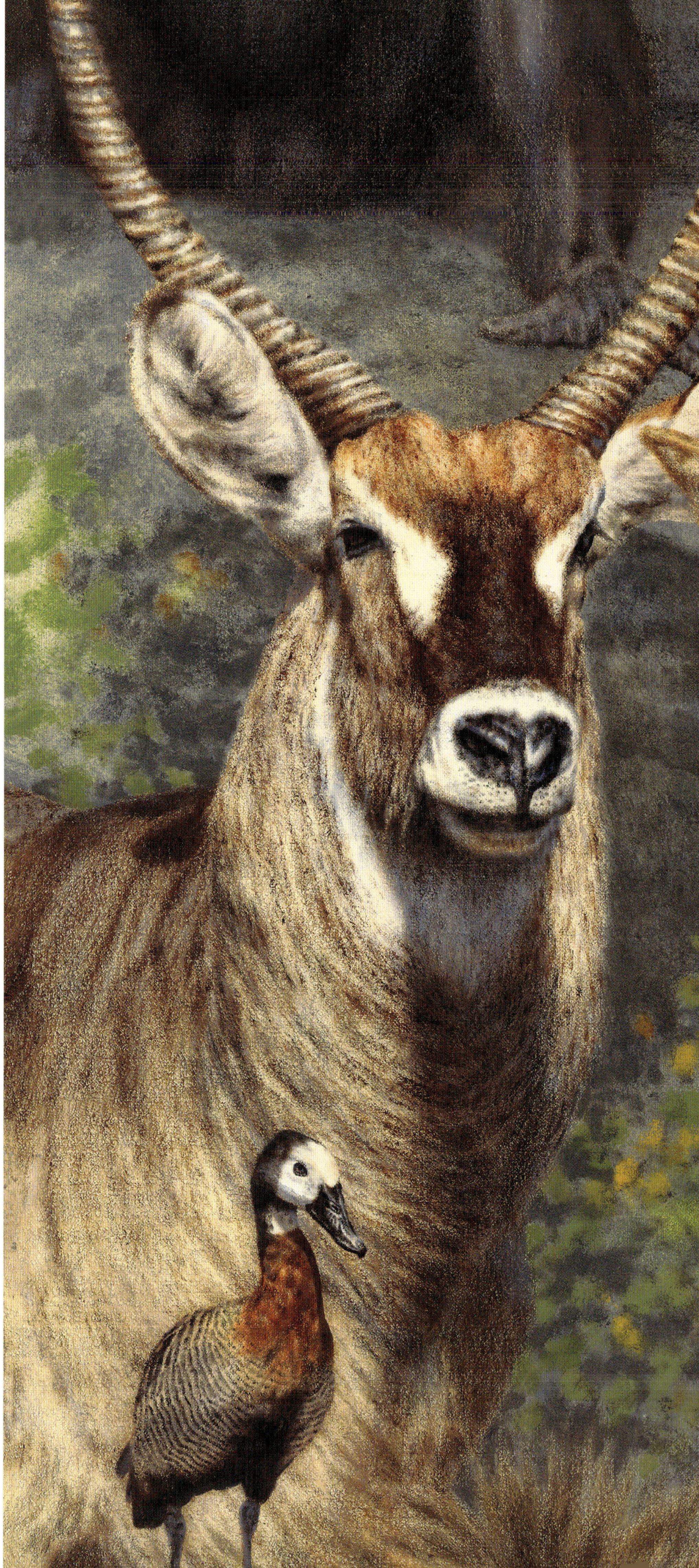

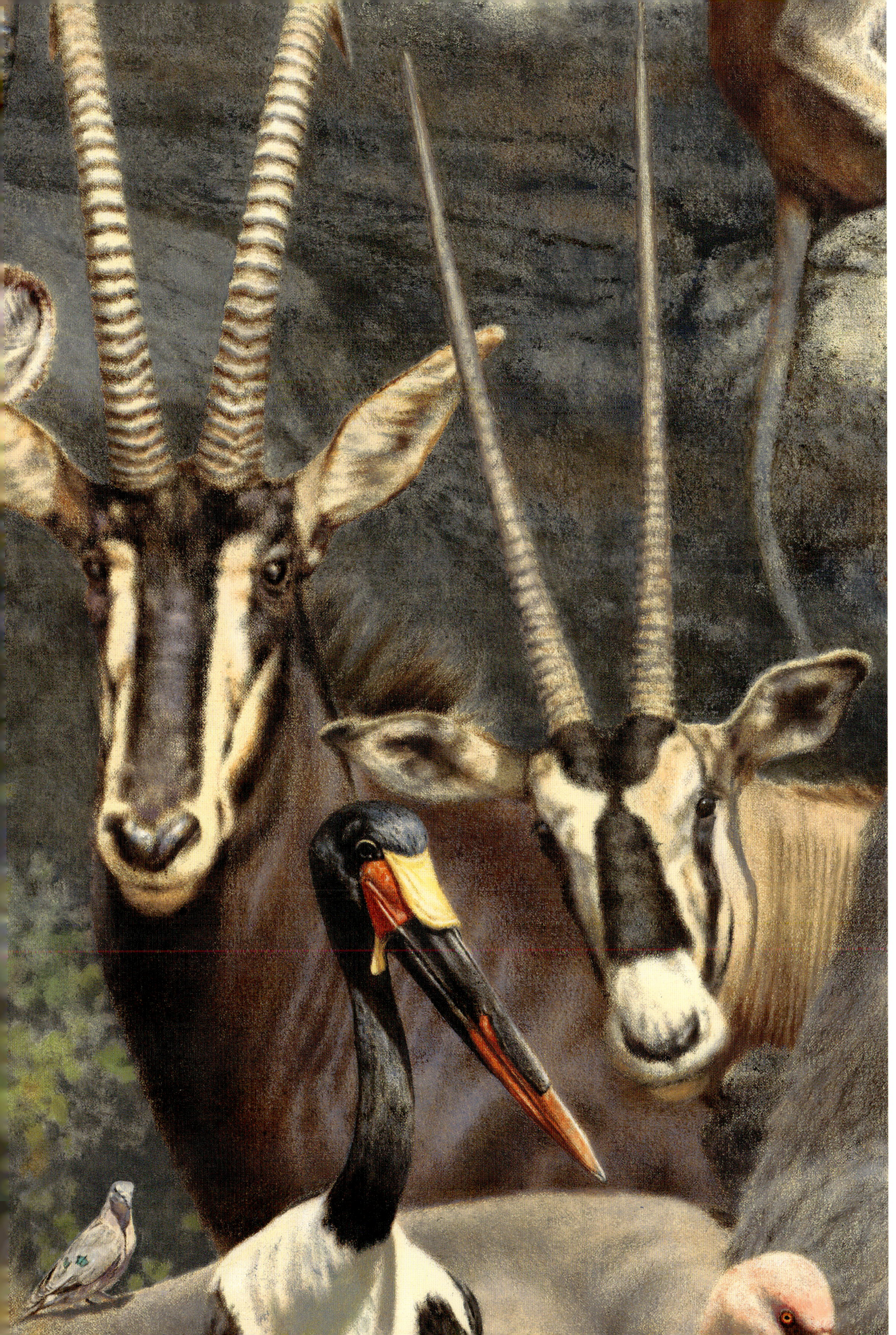

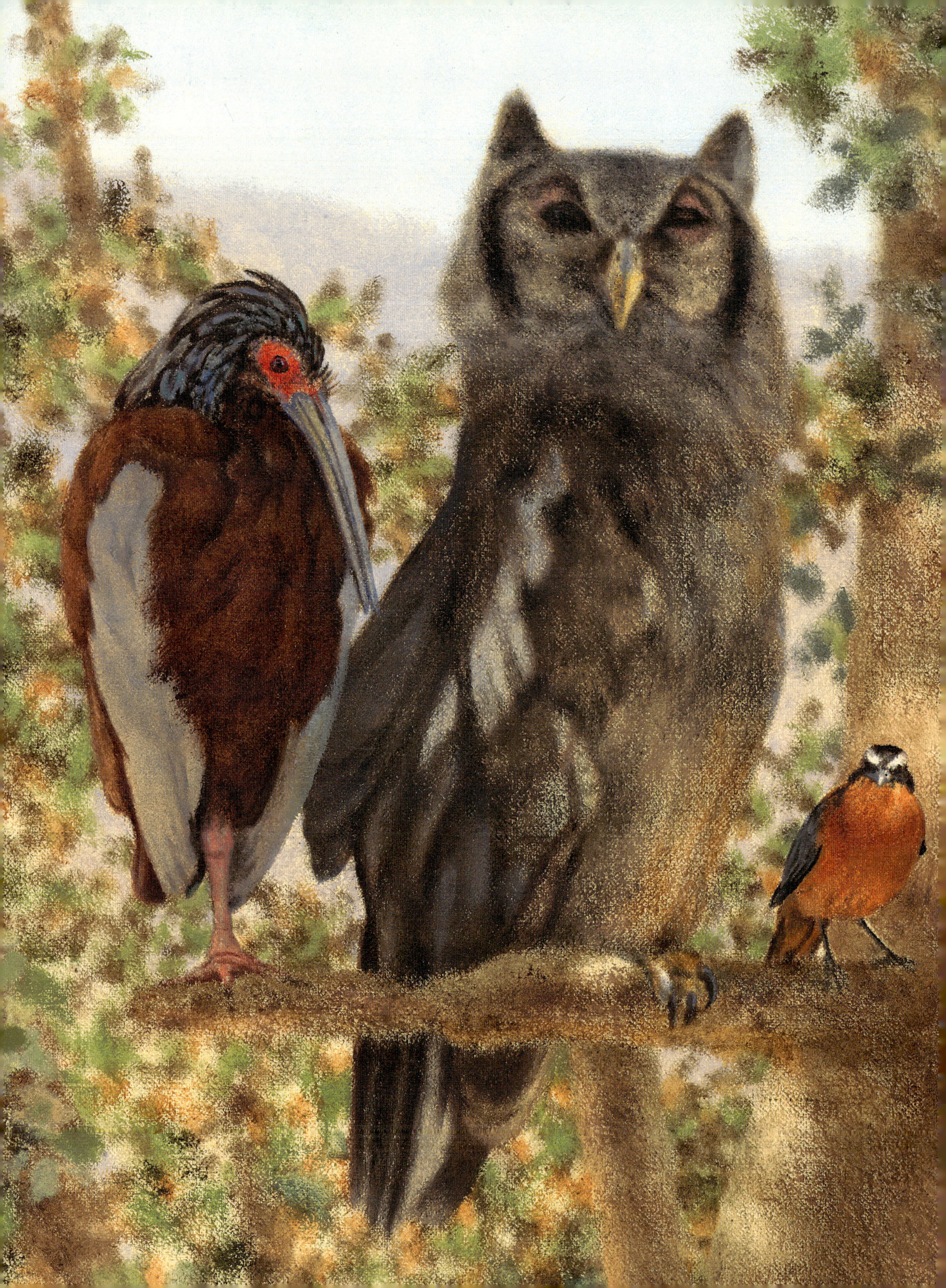

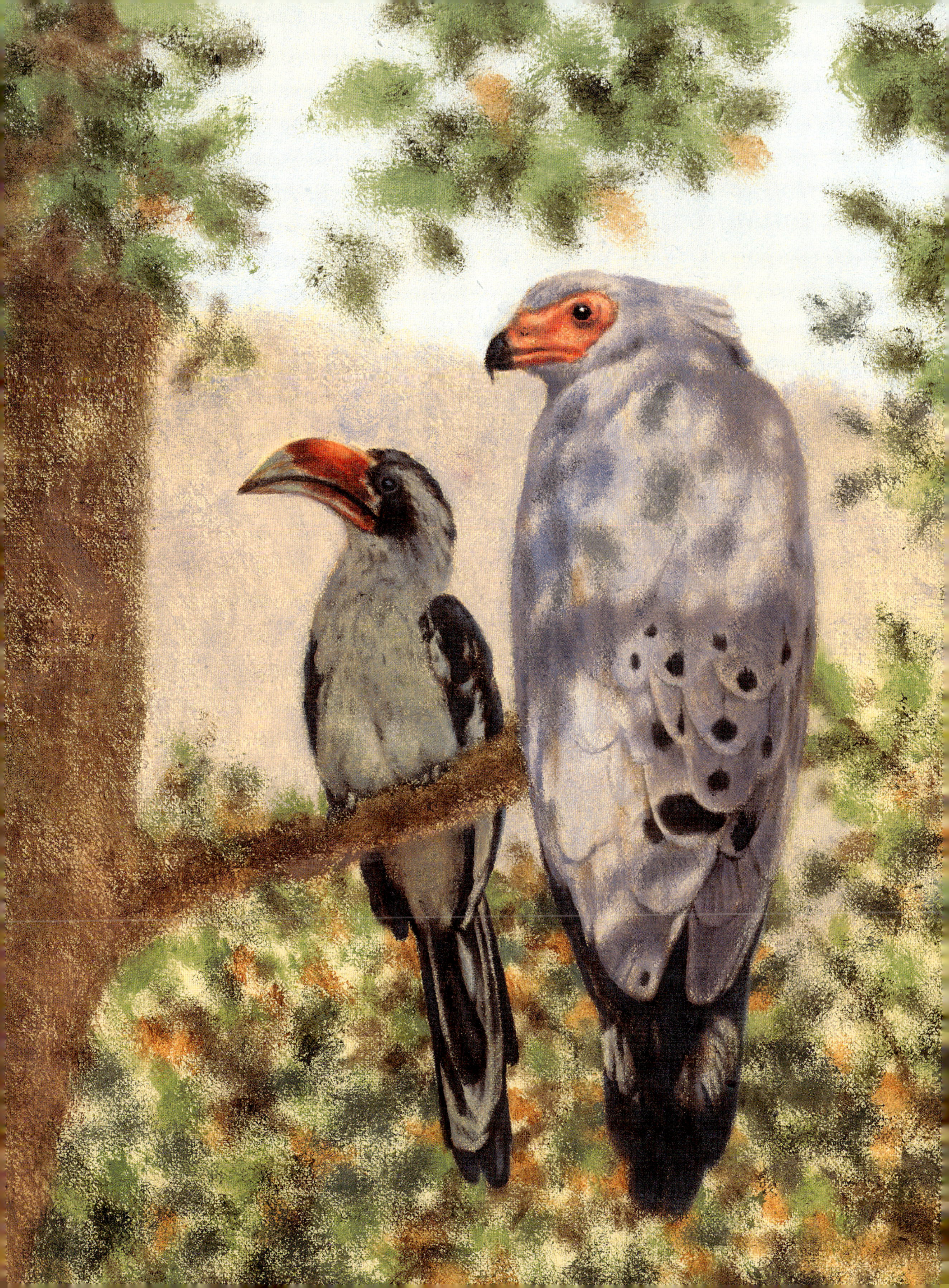

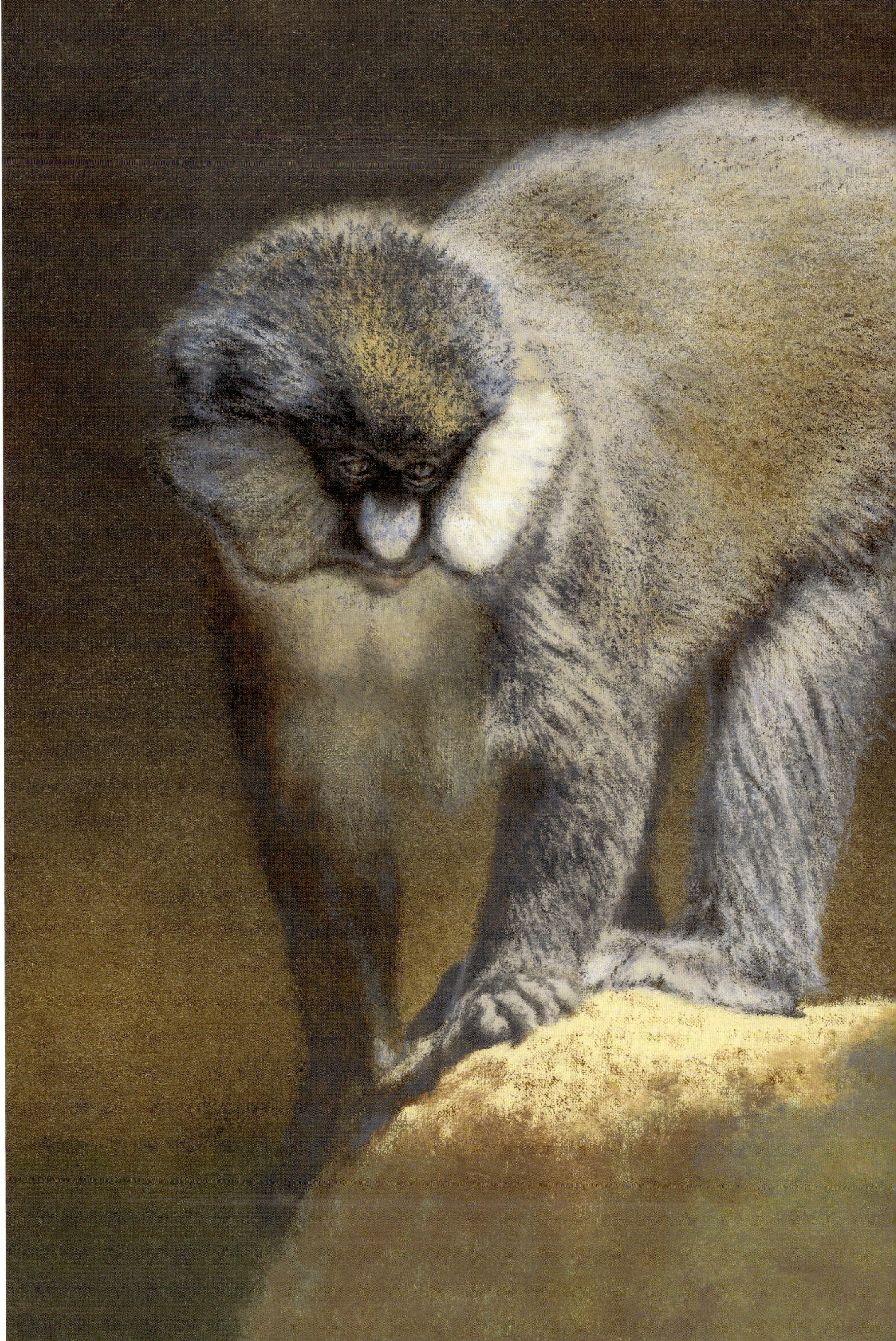

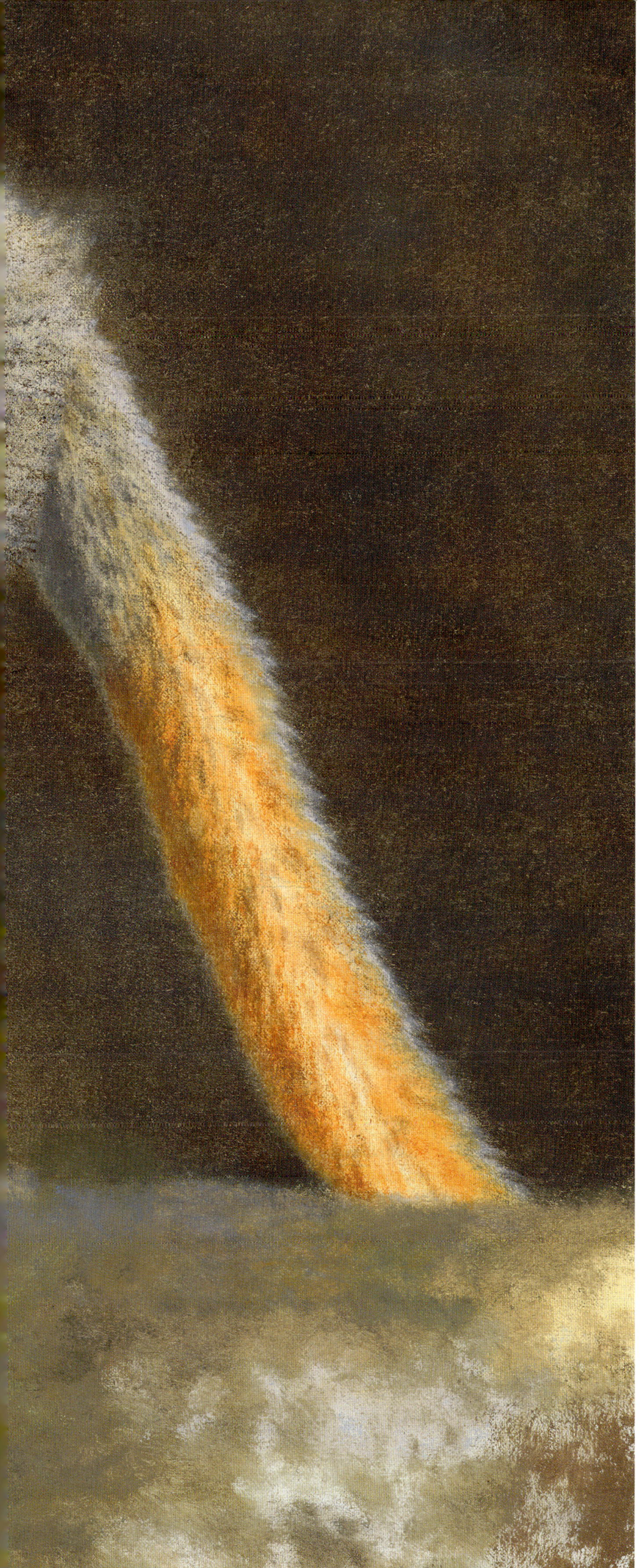

Guenon Pre-Study, 2014
Schmidt's Spot-Nosed Guenon
Oil on Belgian Linen
16 x 20"

This piece was juried into the 2014 Artists for Conservation international exhibition.

OVERLEAF, LEFT
African Menagerie, Panel 6
Detail (Middle Left)
Crowned Crane, Schmidt's
Spot-Nosed Guenon

OVERLEAF, RIGHT
Crowned Crane Study, 2016
Crowned Crane
Oil on Belgian Linen
17 x 12"

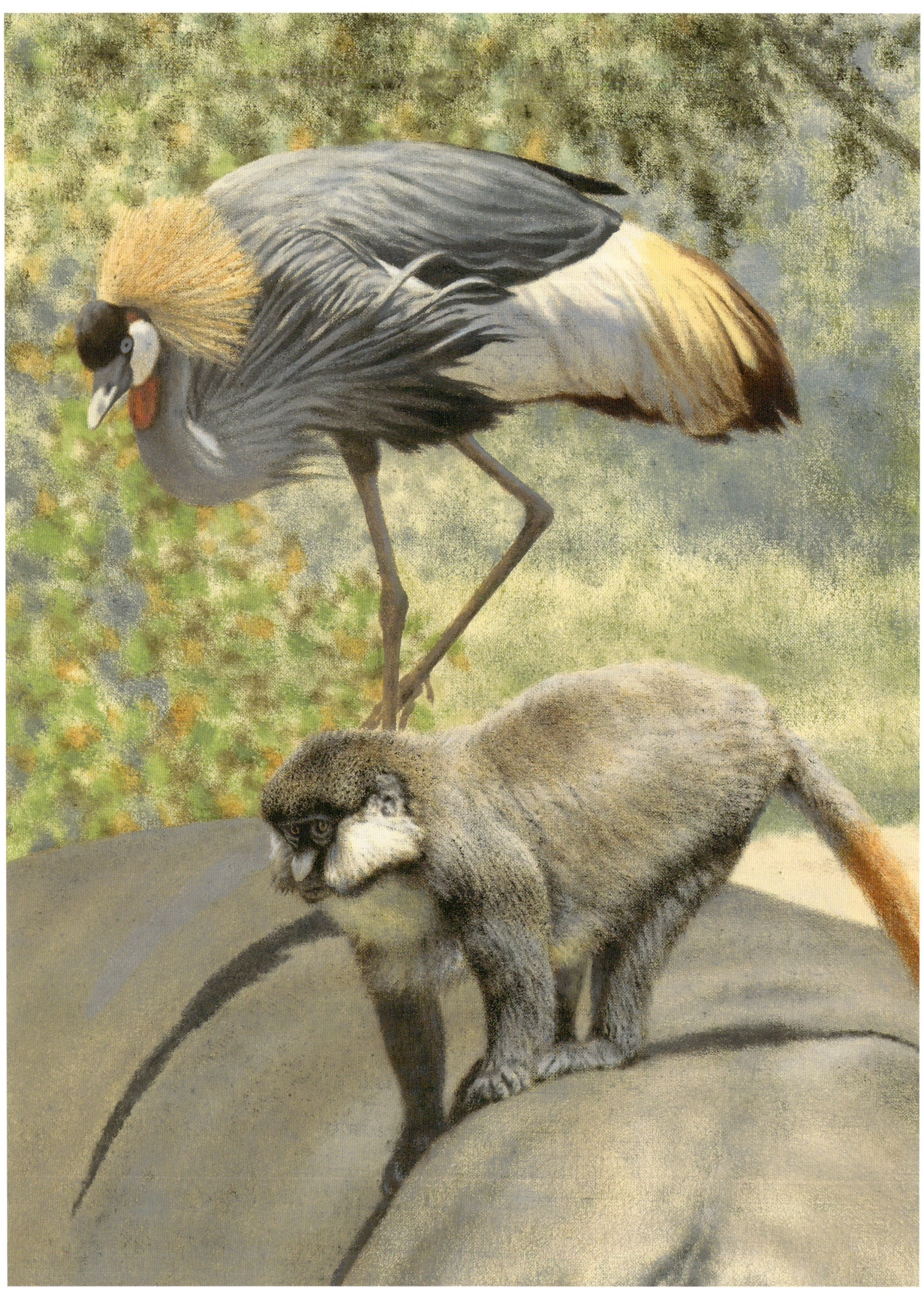

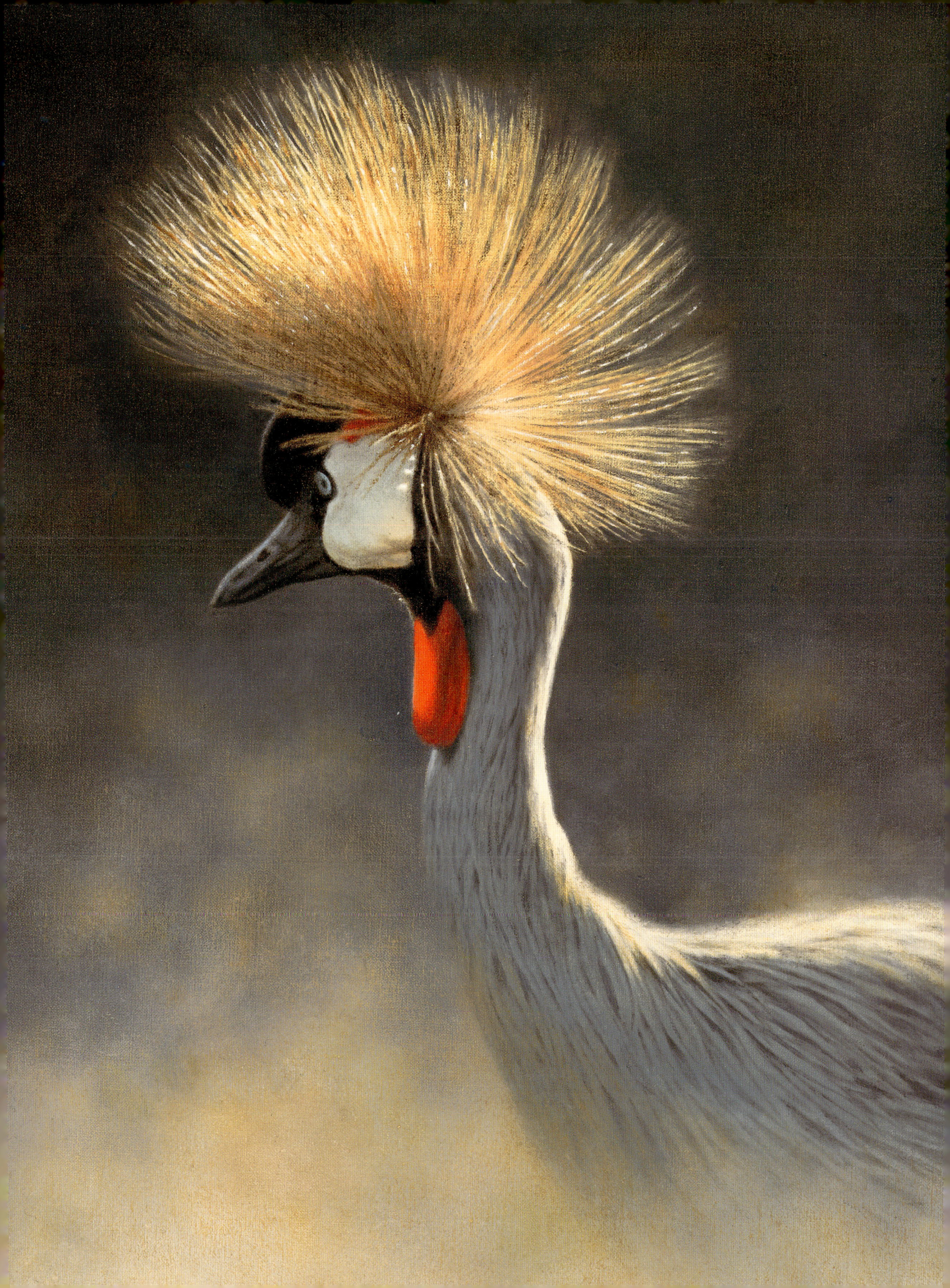

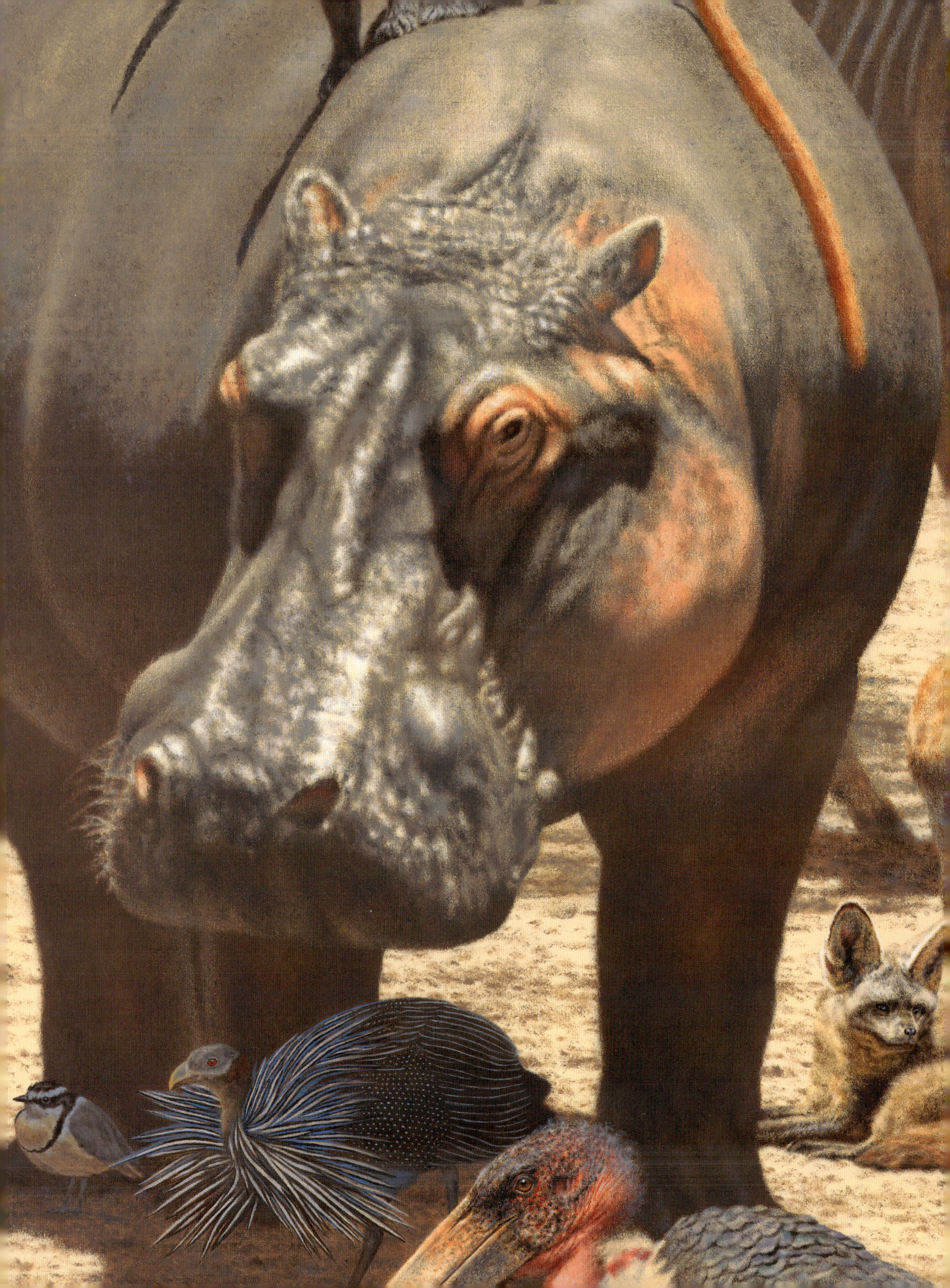

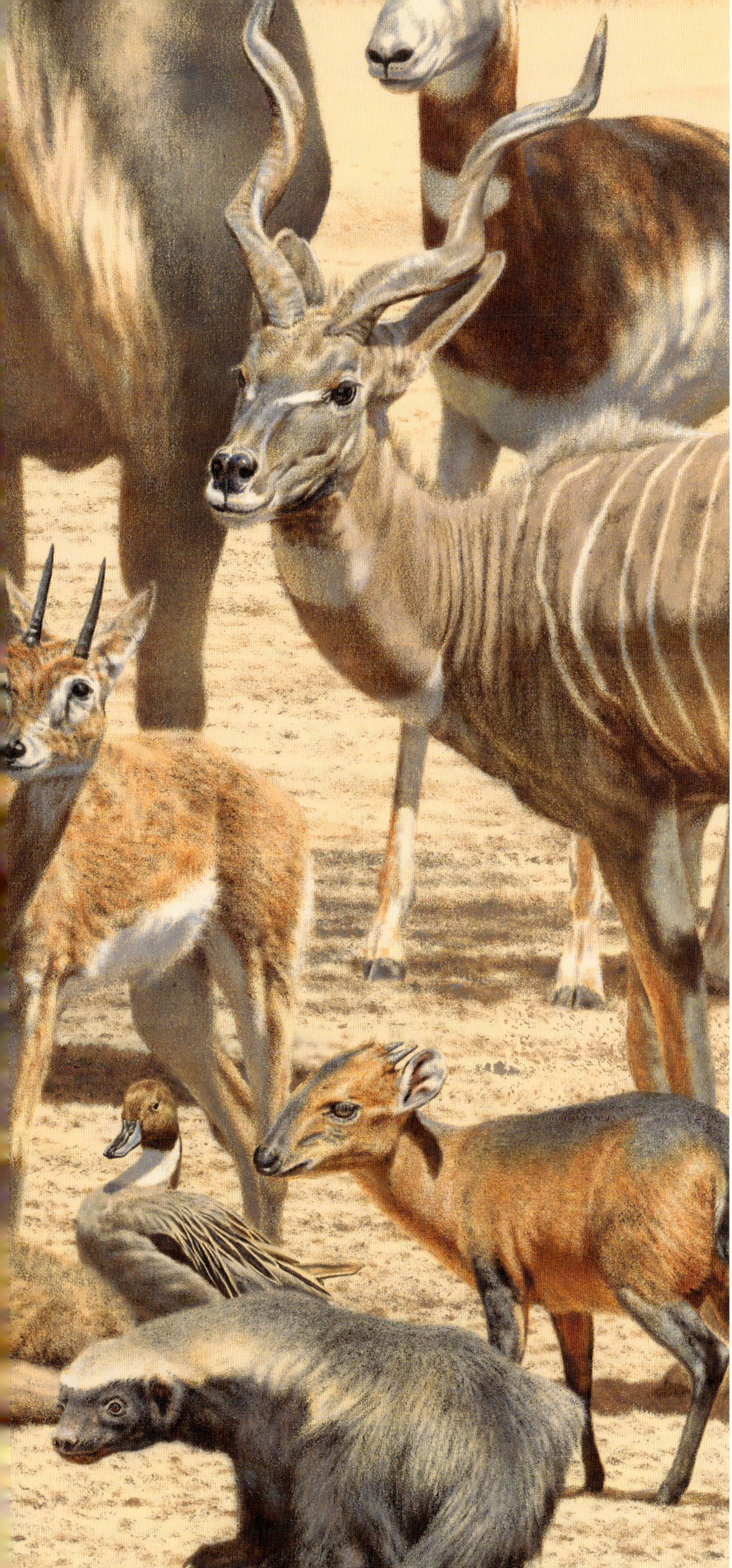

African Menagerie, Panel 6
Detail (Middle)
Lesser Kudu, Hippopotamus, Oribi, Red-Flanked Duiker, Pintail, Bat-Eared Fox, Vulturine Guineafowl, Egyptian Plover, Honey Badger, Marabou Stork

OVERLEAF
African Menagerie, Panel 6
Detail (Top)
Tawny Eagle, Wahlberg's Eagle, Peregrine Falcon, Pied Crow

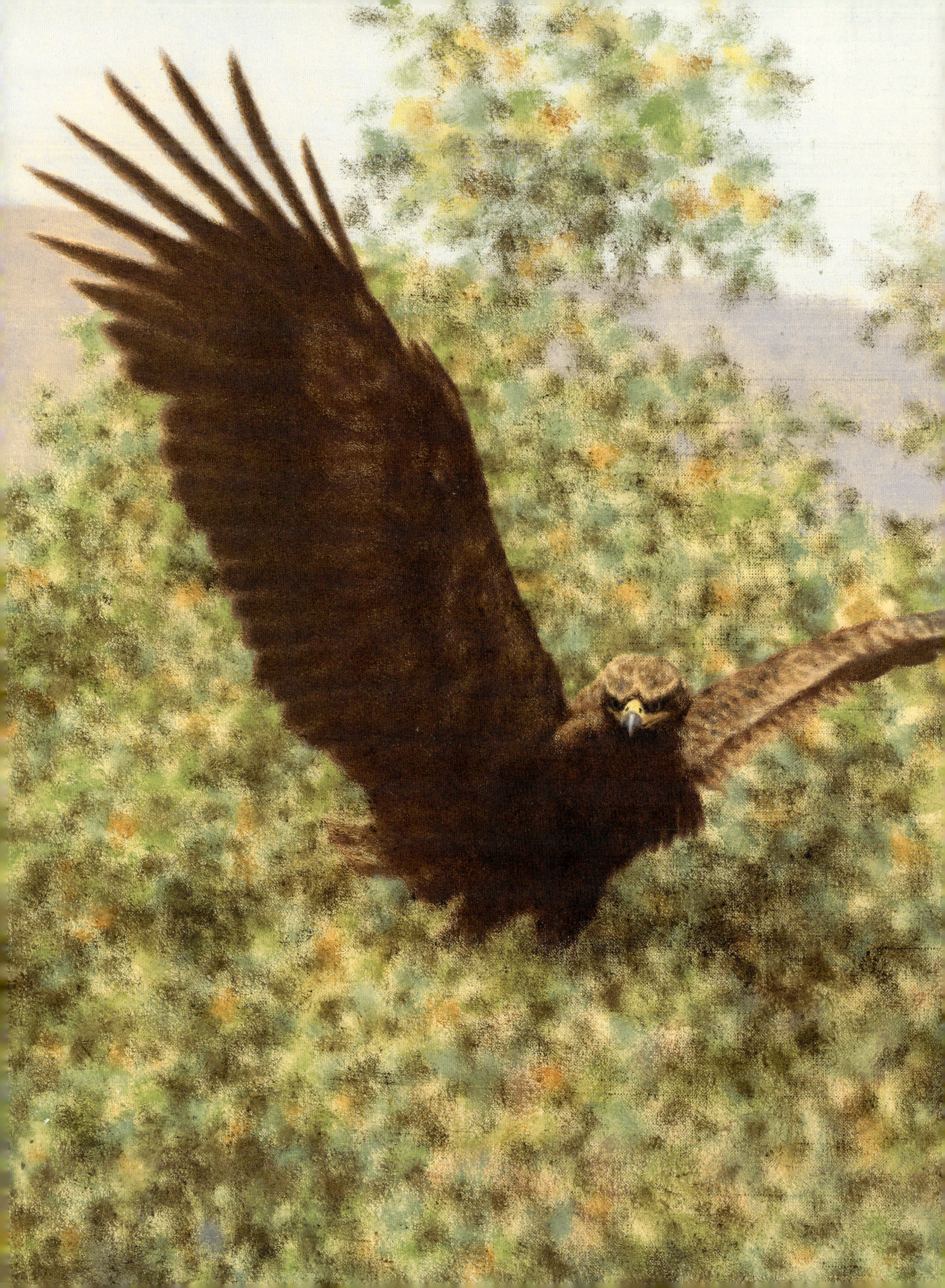